Luminous Stars in Nearby Galaxies

Luminous Stars in Nearby Galaxies

Special Issue Editor
Roberta M. Humphreys

MDPI • Basel • Beijing • Wuhan • Barcelona • Belgrade

Special Issue Editor
Roberta M. Humphreys
School of Physics
and Astronomy,
University of Minnesota
USA

Editorial Office
MDPI
St. Alban-Anlage 66
4052 Basel, Switzerland

This is a reprint of articles from the Special Issue published online in the open access journal *Galaxies* (ISSN 2075-4434) from 2019 to 2020 (available at: https://www.mdpi.com/journal/galaxies/special_issues/Luminous).

For citation purposes, cite each article independently as indicated on the article page online and as indicated below:

LastName, A.A.; LastName, B.B.; LastName, C.C. Article Title. *Journal Name* **Year**, *Article Number*, Page Range.

ISBN 978-3-03936-280-6 (Pbk)
ISBN 978-3-03936-281-3 (PDF)

Cover image courtesy of John C. Martin, Barber Observatory, University of Illinois - Springfield.

© 2020 by the authors. Articles in this book are Open Access and distributed under the Creative Commons Attribution (CC BY) license, which allows users to download, copy and build upon published articles, as long as the author and publisher are properly credited, which ensures maximum dissemination and a wider impact of our publications.
The book as a whole is distributed by MDPI under the terms and conditions of the Creative Commons license CC BY-NC-ND.

Contents

About the Special Issue Editor vii

Preface to "Luminous Stars in Nearby Galaxies" ix

Roberta M. Humphreys
The Complex Upper HR Diagram
Reprinted from: *Galaxies* **2019**, *7*, 75, doi:10.3390/galaxies7030075 1

Paul A Crowther
Massive Stars in the Tarantula Nebula: A Rosetta Stone for Extragalactic Supergiant HIIRegions
Reprinted from: *Galaxies* **2019**, *7*, 88, doi:10.3390/galaxies7040088 15

Michael S. Gordon and Roberta M. Humphreys
Red Supergiants, Yellow Hypergiants, and Post-RSG Evolution
Reprinted from: *Galaxies* **2019**, *7*, 92, doi:10.3390/galaxies7040092 38

Michaela Kraus
A Census of B[e] Supergiants
Reprinted from: *Galaxies* **2019**, *7*, 83, doi:10.3390/galaxies7040083 60

Kerstin Weis and Dominik J. Bomans
Luminous Blue Variables
Reprinted from: *Galaxies* **2020**, *8*, 20, doi:10.3390/galaxies8010020 96

Olga Maryeva, Roberto F. Viotti, Gloria Koenigsberger, Massimo Calabresi, Corinne Rossi and Roberto Gualandi
The History Goes On: Century Long Study of Romano's Star [†]
Reprinted from: *Galaxies* **2019**, *7*, 79, doi:10.3390/galaxies7030079 123

Kris Davidson
Radiation-Driven Stellar Eruptions
Reprinted from: *Galaxies* **2020**, *8*, 10, doi:10.3390/galaxies8010010 144

Kathryn Neugent and Phil Massey
The Wolf–Rayet Content of the Galaxies of the Local Group and Beyond
Reprinted from: *Galaxies* **2019**, *7*, 74, doi:10.3390/galaxies7030074 173

About the Special Issue Editor

Roberta M. Humphreys is Professor Emerita at the University of Minnesota. Her research interests are in observational stellar evolution and Galactic structure. She is best known for her research on massive stars in the Milky Way and in nearby resolved galaxies. In 1979, she and Kris Davidson identified an empirical upper luminosity boundary or upper limit in the luminosity vs temperature diagrams, i.e. the Hertzsprung-Russell (HR) Diagram. This empirical boundary, often referred to in the astronomical literature as the Humphreys-Davidson Limit, was not predicted by theory or the stellar structure models and evolutionary tracks. The lack of evolved stars above a certain luminosity implies an upper limit to the masses of stars, that can evolve to become red supergiants thus altering the previously expected evolution of the most massive stars across the HR Diagram. Her later research has been focused on the final stages of massive stars evolution often dominated by high mass loss events as observed in eta Carinae, and the warm and cool hypergiants. She is an Honorary Fellow of the Royal Astronomical Society, a recipient of the Humboldt Senior Scientist Award, and a Fellow of the American Association for the Advancement of Science.

Preface to "Luminous Stars in Nearby Galaxies"

Perhaps the greatest uncertainty in all of astrophysics, and especially in stellar structure and evolution, is distance. This is especially true for the most massive, most luminous stars that may be located at incredibly vast distances in our own galaxy. Studies on stellar populations in nearby galaxies thus have the advantage that all the stars of interest are at approximately the same distance, a distance that is relatively well known, especially in comparison with the uncertain distances of individual stars in our own galaxy. Surveys and the subsequent spectroscopy of massive stars in different stages of stellar evolution in relatively nearby resolved galaxies have revealed a complex distribution in the luminosity–temperature plane, that is (the HR diagram). The fundamentals of massive star evolution are mostly understood, but the roles of mass loss, episodic mass loss, rotation, and binarity are still in question. Moreover, the final stages of these stars of different masses and their possible relation to each other are not understood. The purpose of this volume is to provide a current review of the different populations of evolved massive stars. The emphasis is on massive stars in the Local Group, the Magellanic Clouds, and the nearby spirals M31 and M33.

Roberta M. Humphreys
Special Issue Editor

Review

The Complex Upper HR Diagram

Roberta M. Humphreys

Minnesota Institute for Astrophysics, School of Physics and Astronomy, 116 Church St. SE, University of Minnesota, Minneapolis, MN 55455, USA; roberta@umn.edu

Received: 24 July 2019; Accepted: 14 August 2019; Published: 23 August 2019

Abstract: Several decades of observations of the most massive and most luminous stars have revealed a complex upper HR Diagram, shaped by mass loss, and inhabited by a variety of evolved stars exhibiting the consequences of their mass loss histories. This introductory review presents a brief historical overview of the HR Diagram for massive stars, highlighting some of the primary discoveries and results from their observation in nearby galaxies. The sections in this volume include reviews of our current understanding of different groups of evolved massive stars, all losing mass and in different stages of their evolution: the Luminous Blue Variables (LBVs), B[e] supergiants, the warm hypergiants, Wolf–Rayet stars, and the population of OB stars and supergiants in the Magellanic Clouds.

Keywords: massive stars; local group; supergiants; Magellanic Clouds; M31; M33

1. Introduction

The reviews and papers in this Special Issue focus on the properties of the most luminous stars in nearby galaxies, those galaxies in which the brightest individual stars are resolved and can be observed. The most luminous stars are also the most massive and because of their intrinsic brightness and relatively short lifetimes they provide our first probes of the progress of stellar evolution in different environments.

The study of massive stars in other galaxies offers many advantages. Foremost of course is distance. Studies of stellar populations in nearby galaxies have the advantage that all the stars are at approximately the same distance, a distance that is relatively well known, especially in comparison with the uncertain distances of individual stars in our own galaxy. In the Milky Way, our observations are also limited to a relatively small volume by interstellar extinction, which can be high and uncertain at increasing distances. In external galaxies, extinction by dust is still a problem but the foreground extinction is well determined from maps of the interstellar "cirrus" or dust along the line of sight. Internal extinction within the galaxy can be variable and must still be corrected.

In this introductory review, I present a brief historical overview with emphasis on some of the main developments in the study of massive stars in nearby galaxies. Some of the first work on stars in other galaxies was driven by the identification of Cepheids for the extragalactic distance scale, see, for example, the comprehensive survey of NGC 2403 by Tammann & Sandage [1]. Other types of variables were also recognized including the first discussion of a class of luminous variables by Hubble & Sandage [2] in the Local Group spirals M31 and M33 that we now call Luminous Blue Variables (LBVs). This work was done on photographic plates and the magnitudes of individual stars were often measured by hand.

The study of individual stars and their placement on the HR Diagram requires several types of data: accurate multicolor photometry for spectral energy distributions (SEDs) and the measurement of interstellar extinction; infrared observations of circumstellar dust and mass loss; and, most important, spectroscopy, for classification, temperatures, luminosity indicators, emission lines, and evidence for mass loss from line profiles. When observing stars in other galaxies, we also have to be concerned about foreground contamination: stars in the Milky Way seen projected against the distant galaxy. Photometry is not sufficient to remove these Galactic stars, especially those of intermediate temperature and late spectral type; spectra are required.

The primary galaxies discussed in this article are the Local Group spirals M31 and M33, and the Large and Small Magellanic Clouds due to their relative proximity and the number of surveys of their luminous stellar populations. Other Local Group galaxies are also included as well as results for stars in a few nearby spiral galaxies.

2. Early Work—The Magellanic Clouds

The early objective prism surveys by the Harvard College Observatory and the resulting Henry Draper Catalog and Extension (HD and HDE) provided the first survey with spectral types of the brightest stars in the Magellanic Clouds [3,4]. The HDE however did not cover the Small Cloud. Feast, Thackeray & Wesselink [5] published the first detailed list of the brightest stars in the Magellanic Clouds with spectral classifications based on slit spectra of individual stars. Their paper included magnitudes, colors, positions, radial velocities and notes on the individual stars relative to emission lines and other features for 50 stars in the SMC and 105 in the LMC. The HDE catalog was the primary source for the LMC stars, plus stars selected as blue based on their colors in the 30 Dor region. The SMC list relied on "B" type stars from the HD plus emission line stars from the survey by Henize [6]. Their work did not include the red or M-type supergiants. The reddest stars in their study were the very luminous F- and G-type supergiants. Their HR Diagram, reproduced here in Figure 1, while not complete, shows the visually brightest stars with M_v approaching -10 mag.

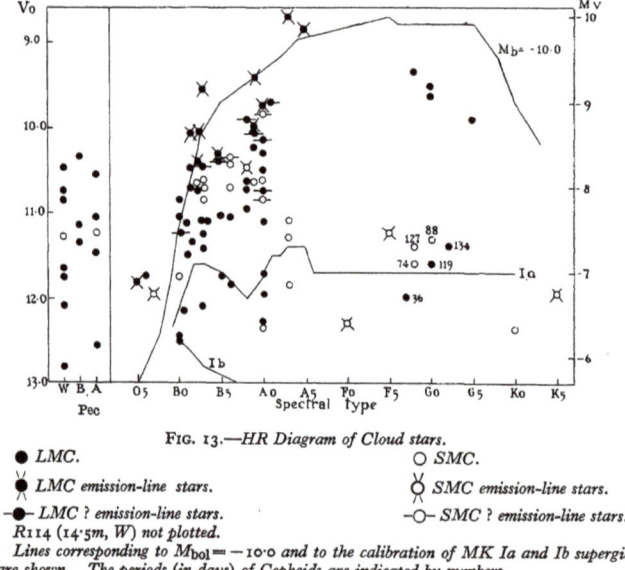

Figure 1. The HR Diagram for the brightest stars in the LMC and SMC from [5]. LMC stars are closed symbols and those in the SMC are shown as open symbols.

Additional work on the fainter stellar populations of hot or OB-type stars and red M-type stars was provided by the lower resolution objective prism surveys throughout the 1960s and 1970s. These included surveys for OB type stars in the LMC and SMC by Sanduleak [7,8] and Fehrenbach & Duflot [9] and for the M-type stars by Westerlund [10,11] and Blanco et al. [12,13]. These necessary photographic surveys provided extensive finding lists for further spectral classification leading to the first comprehensive HR Diagrams for these two galaxies and the comparison of their massive and luminous stellar populations with the Milky Way.

3. The HR Diagram

Fundamental data for the most luminous stars have application to numerous astrophysical questions specifically with respect to stellar evolution, the final stages of the most massive stars as the progenitors of supernovae, and the dependence of their basic parameters on the host galaxy. In the 1970s, there was also considerable interest in the luminosity calibration of the brightest stars and their potential as extragalactic distance indicators [14].

The massive star population in the Milky Way, although restricted to a relatively small volume, within ≈3 kpc of the Sun, is critical as a reference population and as representative of our local region of the Galaxy, despite uncertain distances and possible incompleteness. Humphreys [15] published an HR Diagram for the Galactic supergiants and O stars with spectral types and photometry in stellar associations and clusters with known distances to derive their luminosities. For a more complete population in the upper HR Diagram, the less luminous early B-type main sequence and giant stars were also included. Compared to previous work, numerous surveys, especially of the Southern sky, had added greatly to the number of confirmed supergiants and O stars [16–18] and red supergiants [19,20]. For comparison with evolutionary tracks, the derived absolute visual magnitudes and spectral types were transformed to absolute bolometric luminosities and effective temperatures based on the available calibrations.

Humphreys [21] subsequently published HR Diagrams for the LMC. The basic data for the confirmed supergiants and early-type stars came from the extensive catalogs of spectral types and photometry by Ardeberg et al. [22] and Brunet et al. [23], from Feast, Thackeray & Wesselink [5], and from Walborn [24] for the early O-type stars. The data for the M-type supergiants came from the spectroscopic survey in the same paper.

An empirical comparison of these HR Diagrams, for the massive stars in our region of the Milky Way and for the LMC, revealed comparable populations of massive stars based on the distribution of their spectral types and luminosities across the HR Diagrams. The most important result was the recognition by Humphreys & Davidson [25] of an empirical upper luminosity boundary or upper limit in the luminosity vs. temperature diagrams.

4. The Humphreys–Davidson Limit

The original HR Diagrams from the Humphreys and Davidson 1979 paper are reproduced here in Figures 2 and 3, with the upper boundary shown as a solid line. The original eyeball fit was first drawn to approximate the upper boundary of the supergiant luminosities in the Milky Way. The same line was then transferred to the LMC Diagram, which also matched the observed upper envelope to the LMC luminosities. The upper boundary for both galaxies is an envelope of declining luminosity and decreasing temperature for the hottest stars and a relatively tight upper limit to the luminosities of the cooler stars (less than ≈10,000 K) near $M_{Bol} \approx -9.5$ Mag.

FIG. 3.—The "theoretical" H-R diagram, M_{bol} versus log T_e, for the galactic supergiants. The position of the ZAMS and evolutionary tracks with mass loss are shown. The solid line defines the approximate upper boundary of the supergiant luminosities. The positions of two peculiar stars, η Car and P Cyg, and three supergiant infrared sources are also indicated.

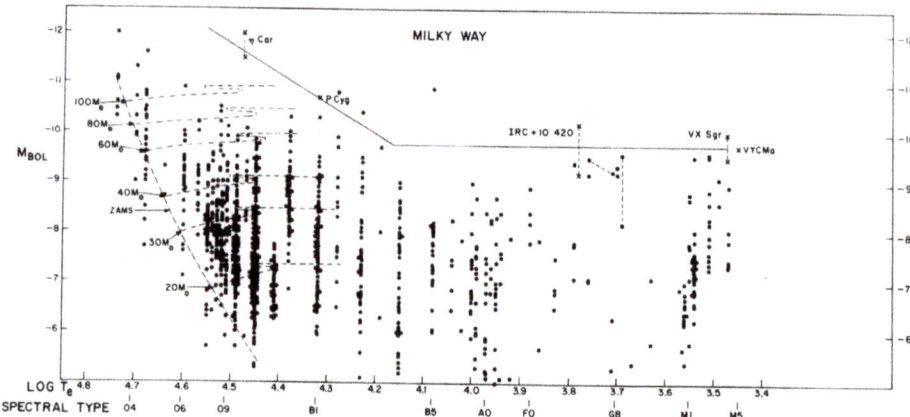

Figure 2. The HR Diagram, M_{Bol} vs. log T, for the luminous stars in the Milky Way from [25].

FIG. 4.—Same as Fig. 3 for the LMC supergiants. The positions of S Dor and two other peculiar stars are indicated.

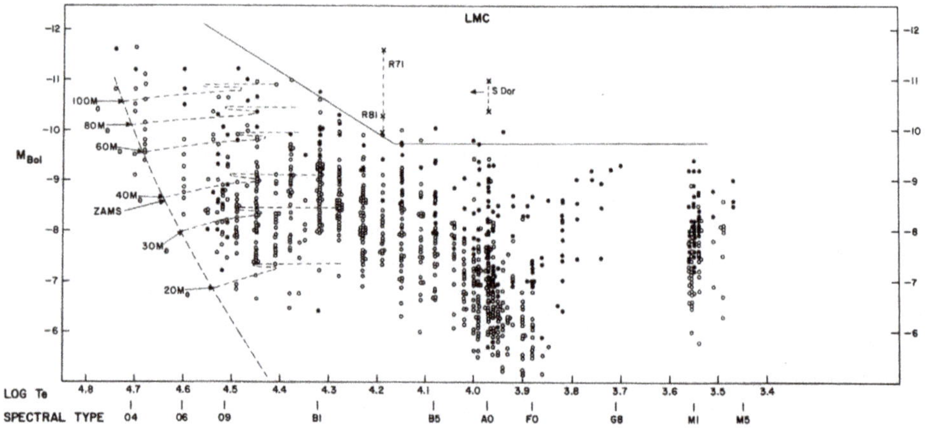

Figure 3. The HR Diagram, M_{Bol} vs. log T, for the luminous stars in the LMC from [25].

This empirical boundary is often referred to in the literature as the "Humphreys–Davidson" Limit. It was not predicted by theory or the stellar structure models and evolutionary tracks at that time. The lack of evolved stars, post main sequence stars, above a certain luminosity implies an upper limit to the masses of stars that can evolve to become red supergiants thus altering the previously expected evolution of the most massive stars across the HR Diagram. In the original study, this limit corresponded to an initial mass near 60 $M_\odot$. Improved models with mass loss and rotation suggest a mass more like 40–50 $M_\odot$ today.

Spectroscopy of luminous star candidates in other Local Group galaxies confirmed the upper luminosity boundary in galaxies of different types, but, in most cases, surveys and population studies were minimal: M31 [26], IC 1613 and NGC 6822 [27]. Humphreys & Sandage [28] completed a major survey for the brightest blue and red stars in M33, that provided the basis for spectroscopy and identification of the most luminous supergiants in that nearby spiral [29], and in subsequent studies. Similar HR Diagrams for the massive stars in the SMC were published a few years later by Humphreys [30] based on a combination of previous work and new observations.

At the time the Humphreys–Davidson paper was published, it was apparent that significant mass loss occurred in both the blue and red supergiants and there was increasing interest in the role that mass loss may have on their evolution. The lack of evolved cooler counterparts to the most massive evolved hot stars in both galaxies suggested that their post main sequence evolution was of special interest. A few high luminosity stars were known for their instabilities, variability and evidence for high mass loss such as eta Car and P Cyg in the Milky Way and S Dor in the LMC. Several luminous blue variables, spectroscopically similar to eta Car and P Cyg, were now recognized in other galaxies [31,32]. As a group, they were known as S Doradus Variables or as Hubble–Sandage Variables in M31 and M33. Several of these stars were included on the HR Diagrams. They were increasing evidence that the phenomena of high mass loss and instabilities, as observed in eta Car and P Cyg, were more common. Humphreys and Davidson thus suggested that the most massive hot stars could not evolve to cooler temperatures because of their instabilities resulting in high mass loss. The temperature dependence of the luminosity limit for the hottest stars was evidence that the instability was mass dependent. This mass loss could be unsteady and much greater at times resulting in high mass loss events. The relatively tight upper luminosity boundary for the cooler stars represented the upper limit to the initial masses of stars that could evolve across the HR Diagram in a stable way to become red supergiants (RSGs).

Some of the first evolutionary tracks with mass loss were also being published at that time [33–36]. It was shown that high mass loss (higher than observed) would cause the tracks to reverse and the stars evolve back to warmer temperatures.

5. Surveys and More Surveys

We emphasize the importance of spectroscopy for the identification and analysis of the most luminous stars, but surveys are essential for identifying candidates. Beginning in the 1980s, astronomers began to add multi-wavelength surveys, primarily in the near-infrared, to complement the traditional optical photometry. Elias, Frogel, & Humphreys [37] obtained near IR photometry for known and candidate red supergiants in the LMC and SMC for a comparison with the Galactic population. The all-sky near-infrared 2MASS survey from 1.2 to 2.2 µm [38] reached the brightest stars in M31 and M33 as well as the Clouds. The infrared observations allowed astronomers to look for free-free emission from the stellar winds and the presence of circumstellar dust, another indicator of mass loss, and to correct the luminosites for possible additional extinction due to circumstellar dust.

Space-based telescopes such as UIT and later GALEX added FUV and NUV imaging and photometry for more complete SEDS at the shorter wavelengths and more accurate estimates of the total luminosities for the hottest supergiants. The mid-infrared surveys with Spitzer/IRAC of the Magellanic Clouds [39,40] and M31 [41] and M33 [42] added fluxes from 3 to 8 µm and even longer wavelengths to search for colder dust, thus allowing us to investigate their mass loss histories.

With the advent of the wide-field CCD mosaic cameras, ground-based, multi-wavelength optical surveys, such as the Local Group (LGGS) by Massey et al. [43], added to the fundamental data. These ground-based surveys however are seeing-limited and lack spatial resolution, thus the images are often multiple. Here, again, spectroscopy or higher resolution imaging with Adaptive Optics on large telescopes can identify and even separate the stars. These ground-based surveys are enhanced by observations with the Hubble Space Telescope such as the PHAT surveys in M31 and M33 [44]. Numerous imaging programs of other galaxies intended for other purposes with HST have provided lists of resolved stars for further observation in, for example, M101 [45] and NGC 2403 and M81 [46,47].

6. Stellar Population Comparisons

One of the outstanding stellar evolution questions is how stellar populations may depend on their environment, namely the properties of the host galaxy, its mass and luminosity, the fraction of interstellar gas and dust, and especially on the chemical composition or metallicity. As the most luminous and visually brightest stars, massive stars provide the first indicator or measurement of how star formation, evolution, and the terminal state, depend on these factors. Mass loss is known to alter stellar evolution, and in the standard picture of line-driven winds from hot stars, mass loss is also expected to be metallicity dependent and significantly less in lower metallicity systems.

The first evidence for significant differences in the properties of the massive star populations was the well-established absolute magnitude dependence of the visually "brightest blue star" on the galaxy type or luminosity in the surveys for extragalactic distance indicators [14]. The smaller, less massive galaxies had fewer of the most massive stars, thus their evolved counterparts were statistically less likely to become the most luminous observed blue stars. These results suggest that the star formation rate for these most massive stars was less in the smaller galaxies. In contrast, the luminosities of the brightest red stars showed little or no dependence on the host galaxy. The stars of somewhat lower initial mass thus existed in sufficient numbers in the smaller galaxies to produce evolved descendants in the red supergiant region, and the luminosities of their brightest members reflected the upper luminosity boundary.

The metallicity dependence of the overall characteristics of the massive star populations in different galaxies has been less apparent. Comparison of the HR Diagrams of the massive stars in our region of the Milky Way with the LMC revealed very similar populations [25]. The LMC oxygen abundance is lower than Solar but by no more than a factor of two. Consequently, a comparison with the outer regions of the Milky Way, also with reduced metallicity, not surprisingly, showed little variation. Among our Local Group galaxies with comprehensive stellar surveys for luminous stars, the SMC has the lowest metallicity, about 1/10th Solar, and a reduced oxygen abundance by about a factor of five.

Early tell-tale evidence for an observable metallicity affect was the distribution of the spectral types of the M supergiants [48]. A preliminary survey of red supergiants in the LMC and SMC compared to known Galactic RSGs, revealed a dramatic shift in their spectral types in the SMC to much earlier spectral types, compared to the LMC and Milky Way stars. Except for one M2-type star in the SMC, all of the others were type M0 or earlier. This apparent shift, attributed to weaker TiO bands, was due to the lower SMC metallicity resulting in lower opacities in the atmospheres. The spectra thus arise in warmer layers. This result was confirmed in later surveys that extended to fainter magnitudes [37,49].

Stellar wind theory predicts a measurable dependence of the mass loss rate on metallicity [50], decreasing with declining heavy element abundances, but the measured rates [51] are somewhat higher than expected. Clumping in the stellar winds is another complication which when included in the mass loss models reduces the mass loss rates [52–54]. Measurement of the stellar wind properties and mass loss rates in the luminous, hot OB-type stars in nearby galaxies, especially in the Magellanic Clouds, has progressed with the advent of very large telescopes equipped with high resolution spectrographs. Paul Crowther's article in this issue on the FLAMES survey of the luminous, hot stars in the Clouds discusses their winds and mass loss rates.

7. The Most Luminous Stars of Different Types

Since the early work of the 1970s and 1980s to identify the most luminous and brightest stars in nearby galaxies, numerous surveys and studies of the massive stars, primarily in Local Group galaxies, have greatly expanded the completeness of the population samples. These include, in the LMC and SMC, surveys for the yellow and red supergiants [55,56], and, in M31 and M33, studies of the luminous star population [57–61] and surveys for the yellow and red supergiants [62–66]. Surveys and follow-up spectroscopy of the luminous blue and red stars in the Local Group irregulars NGC 6822 and IC 1613 are less complete [27,67–69], but they provide an additional sample of the massive stars in two smaller galaxies and also with reduced metallicity.

Table 1 presents a summary of the most luminous stars of different spectral types or temperature ranges in six Local Group Galaxies with their morphological types and integrated visual luminosities. Initial samples of the massive stars in the well-studied nearby spirals, M101, M81 and NGC 2403, outside our Local Group, have also been observed [45–47,70–72] and are included here. The bolometric luminosities of the stars are listed for the three highest in each spectral type group for each galaxy. The adopted distance moduli (Table 2) were used to determine the luminosities. The O-type stars are not included because many are eventually recognized as binary or are in multiple systems. Likewise, those stars in extremely crowded fields, including many OB-type stars, are not listed since they may be blended. Consequently, this table does not include those stars that will be the intrinsically most luminous, most massive members of their home galaxies. A few stars of special interest are identified by name in the table. Some of them are discussed in other articles in this issue. Note that the Galactic stars are not included. The upper HR Diagram for the Milky Way needs to re-examined when the Gaia survey is complete with improved distances and very likely with a larger volume sample.

Table 1. The Most Luminous Stars (M_{Bol}) of Different Spectral Types.

Galaxy Type M_v	M31 Sb I-II −21.5	M33 Sc II −18.9	LMC Im −18.5	SMC Im −16.8	NGC 6822 Im −15.2	IC 1613 Ir −14.8	NGC 2403 Sc III −19.3	M81 Sb I-II −20.9	M101 Sc I −21.0
Spectral Type	M_{Bol}	M_{Bol}	M_{Bol}	M_{Bol}	M_{Bol}	M_{Bol}	M_{Bol}	M_{Bol}	M_{Bol}
O9.5–B5	−10.6(2)	−10.1	−10.9	−10.3(2)	−10.3	−9.5	−10.0(2)	−10.6	...
	−10.2	−10.0	−10.7	−10.2	−9.9	−8.8	−9.5	−10.0	...
	−10.0(2)	−9.8	−10.0	−9.8(2)	−9.5	−8.7	−9.4	−9.7	...
B8–A8	−9.3	−10.1(B324)	−9.6(HD33579)	−9.6(HD7583)	−8.1	−8.4	−9.8	−9.3	−10.4(2)
	−9.1	−9.5(2)	−9.0	−9.2	−7.7	−6.9	−9.5	−9.1(2)	−10.3
	−8.9(2)	−9.4	−8.9	−9.0	−7.5	−6.7	−9.4	−8.9	−9.9
FGK	−9.8	−9.6(2)	−9.3	−9.2	...	...	−9.3	−9.7	−9.8
	−9.7	−9.5(Var A)	−9.2	−9.1	...	...	−8.8	−9.6	−9.3
	−9.5	−9.2(3)	−8.9(3)	−8.1	...	...	−8.5	−9.4	−9.2
RSGs	−9.4(2)	−9.6	−9.6(MOH-G64)	−9.2	−9.2	−9.4	−8.5	−9.5:	−9.9:
	−9.3	−9.5(2)	−9.1	-9.1	−9.1	−8.8	−8.4(2)	−9.2(2)	−9.5
	−9.1(2)	−9.4(2)	−9.0	−8.9	−8.7	−8.7	...	...	−9.2

Some comments with respect to the data in this table are helpful. It is clear that data for some of the spectral type groups are lacking or incomplete such as for the YSGs (FGK) in NGC 6822 and IC 1613. Those stars are undoubtedly present, but have just not been identified in the published surveys, and they may also be of somewhat lower luminosity in those two galaxies. A survey to identify and classify the hotter supergiants in M101 has not been completed, and, although a survey identifying RSGs in NGC 2403 exists [72], confirming spectroscopy and photometry is lacking. In general, the numbers for the hot supergiant group (O9.5–B5) may not include the most luminous members because many of these stars are in crowded regions.

The six Local Group galaxies in Table 1 present a diverse group of galaxy types with a wide range of luminosities. The well known dependence of the most luminous "blue" stars on the parent galaxy is especially notable for the evolved, post-main sequence A-type supergiants in the two lowest luminosity and lowest mass galaxies, NGC 6822 and IC 1613. Together with the SMC, these are also the galaxies with the lowest metallicity, but the heavy element abundance in NGC 6822, however, is intermediate between the SMC and LMC. Reduced metallicity was also expected to reduce the opacity in the stellar atmospheres and increase their absolute visual magnitudes but the data for these galaxies show the opposite effect.

Thus, although metallicity undoubtedly plays a role in stellar evolution, the luminosity and mass of the parent galaxy is the primary determinant, to the first order, for the luminosities of the most luminous stars. This is a size of sample effect. Assuming a similar slope for the mass function, the most massive galaxies will have a larger progenitor population of massive stars, and consequently, at any given time, we will therefore be more likely to observe their most luminous, evolved counterparts.

We also see another effect with this group of galaxies related to the galaxy type. M31 (Sb I–II) and M81 (Sb I -II), both high luminosity spirals, have lower luminosity evolved A-type blue stars compared with the Sc-type spirals and Magellanic irregulars. This second effect very likely reflects a dependence on the lower star formation rate, not mass, in the Sb spirals. Both of these spirals have above solar metallicites, although M81 has a metallicity gradient in its disk similar to the Milky Way and the Sc spirals M33 and M101, and solar-type abundances in the outer parts.

This degeneracy, wherein the luminosity of the most luminous blue stars are dependent on the host galaxy, complicated their use as direct distance indicators. The most luminous red supergiants, however, exhibit a nearly constant upper luminosity, related to the Humphreys–Davidson limit, over a wide range of galaxy types and luminosities. The presence of extensive circumstellar dust and uncertain reddening in the most luminous RSGs though limits their usefulness as distance indicators in the visual, but their luminosities in the infrared need to be further studied. Kudritzki [73] introduced a new distance determination method for B- and A-type supergiants, the Flux-weighted Gravity-Luminosity Relation (FGLR), that depends on quantitative analysis of the Balmer lines measured in low resolution spectra with good S/N ratio. It has been successfully applied to several nearby galaxies [47,74,75], demonstrating that the visually brightest stars, the A-type supergiants, have potential as distance indicators at very large distances.

8. The Complex Upper HR Diagram

The upper luminosity boundary to the HR Diagram, or Humphreys–Davidson Limit, complicates our understanding of massive star evolution. Above some initial mass, $\approx$40–50 M$_\odot$, the stars do not evolve across the HR Diagram to become RSGs. Their evolution to cooler temperatures is most likely halted by proximity to the modified Eddington Limit and the accompanying high mass loss episodes. The opacity "Modified Eddington Limit", well-known since the mid-1980's [76], see the papers and discussion in [77,78], describes an instability that arises when L/M approaches the classical L/M_{Edd} value [79]. Models based on this and other instabilities have been proposed to reproduce the upper limit in the HR Diagram [80–82]. See Humphreys & Davidson [83] for a review.

The stars then evolve back to warmer temperatures. Having shed a lot of mass, they are now increasingly subject to atmospheric instabilities before their terminal state. They thus have a different evolutionary path and different mass loss histories than their somewhat lower mass counterparts in the same galaxy or even the same cluster. These lower mass supergiants, from $\approx$9 or 10 up to 30–40 M$_\odot$ or so, evolve across the HR Diagram, becoming red supergiants which alters their interior structure and with enhanced mass loss. Some may also evolve back to a warmer state, perhaps as warm hypergiants, to become LBVs, or B[e] supergiants, where their final fate is most likely as core-collapse supernovae.

Consequently, the upper HR Diagram is populated by a diversity of evolved, luminous and variable stars of different types that challenge our understanding of their physics, evolution and eventual fate. Many of them are distinguished by their emission line spectra, and evidence for stellar winds and mass loss. In addition, some of them exhibit periods of enhanced mass loss, such as the LBVs/S Dor variables and the warm and cool hypergiants with their resolved ejecta. In addition, Wolf–Rayet stars of various types, Oe and Of stars, the B[e] supergiants, and the Fe II emission line stars occupy the same parts of the HR Diagram. They may or may not be related. They could be stars of similar initial mass but in different stages of their evolution or have experienced different mass loss histories, and some may be binaries. This diversity is one of the challenges to understanding massive stars, their evolution, and eventual fate.

The HR Diagrams below (Figure 4) for M31 and M33 show the distribution in the luminosity-temperature plane of three of these mass losing classes of stars, the LBVs, B[e] supergiants and the warm hypergiants.

One of the outstanding questions is the final fate of the most massive stars. It used to be simply assumed that all stars much above initial masses of 10 M$_\odot$ or so, would end their brief lives as some kind of supernova, but their final stage as core-collapse SNe is now in question. Smartt [84,85] suggested an upper mass limit of $\approx$18 M$_\odot$ for the red supergiant progenitors of the Type II-P SNe, while Jennings et al. [86] found a lack of massive supernova progenitors in M31 and M33 and suggested an upper mass of 35–45 M$_\odot$. Do the most massive stars collapse directly to black holes instead, long suspected for extreme, very massive stars such as eta Carinae?

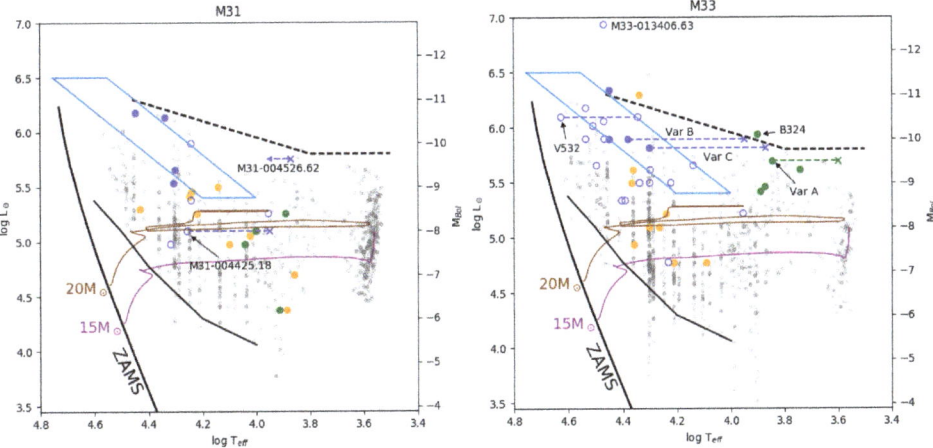

Figure 4. The schematic HR Diagrams for M31 and M33 showing the positions of the confirmed LBVs and candidate LBVs shown respectively, as filled and open blue circles, the warm hypergiants as green circles and the B[e] supergiants as orange circles. The LBV transits during their high mass loss state are shown as dashed blue lines. The LBV/S Dor instability strip is outlined in blue. The 15 and 20 M$_\odot$ tracks are from [87] with rotation are shown to provide a reference for the lower mass B[e]sgs, which are probable rotators. (Higher mass tracks are not shown due to crowding.) The supergiant population is shown in the background in light gray. Reproduced from [61].

In addition, the supernova surveys have identified numerous non-terminal giant eruptions, in which the object greatly increases its total luminosity possibly expelling several solar masses and the star survives. Some of these events are confused with true SNe and thus have been called "supernova impostors". This is a diverse group of objects with a range of luminosities and possible progenitors. A few impostors appear to be normal LBV/S Dor variables in their eruptive or maximum light state. Most are giant eruptions, possibly similar to eta Car [83] from evolved massive stars, while some are red transients (ILRTs [88]) from a lower mass population. The origin of the instability in these giant eruptions is unknown, but proximity to the Eddington Limit is crucial [89]. It is not known what role these high mass loss events may play in the final stages of massive star evolution or their relation to the evolved massive star population and to other stars with instabilities such as the LBVs and the warm hypergiants.

Thus, we observe a complex upper HR Diagram with different evolutionary paths dependent on initial mass, and several types of evolved stars not only experiencing continuous mass loss, but also high mass loss events. The study of luminous stars in the nearer galaxies provides us with an improved census of these evolved stars, their relative numbers, physical properties and behavior, and clues to their evolutionary state and possible relationship to each other on the HR Diagram. In this Special Issue on luminous stars in nearby galaxies, the reviews focus on different examples of evolved massive stars; the luminous O and B-type stars in the Magellanic Clouds, LBVs, B[e] supergiants, the warm hypergiants, and the Wolf–Rayet stars.

Although the articles are about different types of evolved massive stars, many are in the same galaxies. We have therefore adopted the following distance moduli based on the Cepheid scale for these nearby galaxies for consistency and for ease of cross-referencing and comparison.

Table 2. Adopted Distance Moduli.

Galaxy	Distance Modulus (mag)	Comment
LMC	18.5	Cepheids
SMC	18.9	Cepheids
M31	24.4	Cepheids [90]
M33	24.5	Cepheids [91]
NGC 6822	23.4	Cepheids [92]
IC 1613	24.3	Cepheids [93]
NGC 2403	27.5	Cepheids [94]
M81	27.8	Cepheids [94]
M101	29.1	Cepheids [94]

Funding: This research received no external funding.

Conflicts of Interest: The authors declare no conflict of interest.

References

1. Tammann, G.A.; Sandage, A. The stellar content and distance of the galaxy NGC 2403 in the M81 group. *Astrophys. J.* **1968**, *151*, 825–860. [CrossRef]
2. Hubble, E.; Sandage, A. The brightest variable stars in extragalactic nebulae. I. M31 and M33. *Astrophys. J.* **1953**, *118*, 353–361. [CrossRef]
3. Cannon, A.J. The Henry Draper extension. *Ann. Astron. Obs. Harvard Coll.* **1936**, *100*, 367.
4. Cannon, A.J.; Pickering, E.C. The Henry Draper catalog. *Ann. Astron. Obs. Harvard Coll.* **1924**, 91–99.
5. Feast, M.W.; Thackeray, A.D.; Wesselink, A.J. The brightest stars in the Magellanic Clouds. *Mon. Not. R. Astron. Soc.* **1960**, *121*, 337–385. [CrossRef]
6. Henize, K. Catalogues of Hα-emission Stars and Nebulae in the Magellanic Clouds. *Astrophys. J.* **1956**, *2*, 315–344. [CrossRef]
7. Sanduleak, N. A finding list of proven or probable Small Magellanic Clouds members. *Astron. J.* **1968**, *73*, 246–250. [CrossRef]
8. Sandulaeak, N. *A Deep Objective-Prism Survey for Large Magellanic Cloud Members*; The Cerro Tololo Inter-American Observatory: Coquimbo Region, Chile, 1970.
9. Fehrenbach, C.; Duflot, M. Large Magellanic Cloud. List of LMC members and lis t of galactic stars. Charts for recognizing these stars on the maps published by the Smithsonian Institution Astrophysical Observatory. *Astron. Astrophys. Suppl.* **1970**, *10*, 231.
10. Westerlund, B. An infrared survey of the Magellanic Clouds. I. Four regions in the Large Cloud. *Uppsala Astron Obs. Ann.* **1960**, *4*, 7.
11. Westerlund, B. Population I in the Large Magellanic Cloud. *Astron. J.* **1961**, *5*, 1–28.

12. Blanco, V.M.; McCarthy, M.F.; Blanco, B. Carbon and late M-type stars in the Magellanic Clouds. *Astrophys. J.* **1980**, *242*, 938–964. [CrossRef]
13. Blanco, V.M.; McCarthy, M.F. The distribution of carbon and M-type giants in the Magellanic Clouds. *Astron. J.* **1983**, *88*, 1442–1457. [CrossRef]
14. Sandage, A.; Tammann, G.A. Steps toward the Hubble constant. II. The brightest stars in late-type spiral galaxies. *Astrophys. J.* **1974**, *191*, 603–621. [CrossRef]
15. Humphreys, R.M. Studies of luminous stars in nearby galaxies. I. Supergiants and O stars in the Milky Way. *Astrophys. J.* **1978**, *38*, 309–350. [CrossRef]
16. Humphreys, R.M. Spectroscopic and photometric observations of luminous stars in Carina-Centaurus (l = 282d–305d). *Astron. Astrophys.* **1973**, *9*, 85–96.
17. Humphreys, R.M. Spectroscopic and photometric observations of luminous stars in the Centaurus-Norma (l = 305–340) section of the Milky Way. *Astron. Astrophys. Suppl.* **1975**, *19*, 243–247.
18. Walborn, N.R. Spectral classification of OB stars in both hemispheres and the absolute-magnitude calibration. *Astron. J.* **1972**, *77*, 312–318. [CrossRef]
19. Humphreys, R.M.; Strecker, D.W.; Ney, E.P. Spectroscopic and Photometric Observations of M Supergiants in Carina. *Astrophys. J.* **1972**, *172*, 75–88. [CrossRef]
20. Humphreys, R.M.; Ney, E.P. Visual and infrared observations of late-type supergiants in the southern sky. *Astrophys. J.* **1974**, *194*, 623–628. [CrossRef]
21. Humphreys, R.M. Studies of luminous stars in nearby galaxies. II–M supergiants in the Large Magellanic Cloud. *Astrophys. J.* **1979**, *39*, 389–403. [CrossRef]
22. Ardeberg, A.; Brunet, J.P.; Maurice, E.; Prevot, L. Spectrographic and photometric observations of supergiants and foreground stars in the direction of the Large Magellanic Cloud. *Astron. Astrophys. Suppl.* **1972**, *6*, 249–309.
23. Brunet, J.-P.; Imbert, M.; Martin, N.; Mianes, P.; Prévot, L.; Rebeirot, E.; Rousseau, J. Studies of the LMC stellar content. I. A catalogue of 272 new O-B2 stars. *Astron. Astrophys. Suppl.* **1975**, *21*, 109–136.
24. Walborn, N.R. Spectral classification of O and B0 supergiants in the Magellanic Clouds. *Astrophys. J.* **1977**, *215*, 53–57. [CrossRef]
25. Humphreys, R.M.; Davidson, K. Studies of luminous stars in nearby galaxies. III—Comments on the evolution of the most massive stars in the Milky Way and the Large Magellanic Cloud. *Astrophys. J.* **1979**, *232*, 409–420. [CrossRef]
26. Humphreys, R.M. Studies of luminous stars in nearby galaxies. IV—Baade's field IV in M31. *Astrophys. J.* **1979**, *234*, 854–860. [CrossRef]
27. Humphreys, R.M. Studies of luminous stars in nearby galaxies. V—The local group irregulars NGC 6822 and IC 1613. *Astrophys. J.* **1980**, *238*, 65–78. 086/157958. [CrossRef]
28. Humphreys, R.M.; Sandage, A. On the stellar content and structure of the spiral Galaxy M33. *Astrophys. J.* **1980**, *44*, 319–381. [CrossRef]
29. Humphreys, R.M. Studies of luminous stars in nearby galaxies. VI–The brightest supergiants and the distance to M33. *Astrophys. J.* **1980**, *241*, 587–597. [CrossRef]
30. Humphreys, R.M. Studies of luminous stars in nearby galaxies. VIII–The Small Magellanic Cloud. *Astrophys. J.* **1983**, *265*, 176–193. [CrossRef]
31. Humphreys, R.M. The spectra of AE Andromedae and the Hubble-Sandage variables in M31 and M33. *Astrophys. J.* **1975**, *200*, 426–429. [CrossRef]
32. Humphreys, R.M. Luminous variable stars in M31 and M33. *Astrophys. J.* **1978**, *219*, 445–451. [CrossRef]
33. Chiosi, C.; Nasi, E.; Sreenivasan, S.R. Massive stars evolution with mass-loss. *Astron. Astrophys.* **1978**, *63*, 103–124.
34. De Loore, C.; De Greve, J.P.; Lamers, H.J.G.L.M. Evolution of massive stars with mass loss by stellar wind. *Astron. Astrophys.* **1977**, *61*, 251–259.
35. Maeder, A. The most massive stars in the Galaxy and the LMC–Quasi-homogeneous evolution, time-averaged mass loss rates and mass limits. *Astron. Astrophys.* **1980**, *92*, 101–110.
36. Maeder, A. Grids of evolutionary models for the upper part of the HR diagram. Mass loss and the turning of some red supergiants into WR stars. *Astron. Astrophys.* **1981**, *102*, 401–410.

37. Elias, J.H.; Frogel, J.A.; Humphreys, R.M. M supergiants in the Milky Way and the Magellanic Clouds: Colors, spectral types, and luminosities. *Astrophys. J.* **1985**, *57*, 57–131. [CrossRef]
38. Skrutskie, M.F.; Cutri, R.M.; Stiening, R.; Weinberg, M.D.; Schneider, S.; Carpenter, J.M.; Beichman, C.; Capps, R.; Chester, T.; Elias, J.; et al. The Two Micron All Sky Survey (2MASS). *Astron. J.* **2006**, *131*, 1163–1183. [CrossRef]
39. Bonanos, A.; Massa, D.L.; Sewilo, M.; Lennon, D.J.; Panagia, N.; Smith, L.J.; Meixner, M.; Babler, B.L.; Bracker, S.; Meade, M.R.; et al. Spitzer SAGE Infrared Photometry of Massive Stars in the Large Magellanic Cloud. *Astron. J.* **2009**, *138*, 1003–1021. -6256/138/4/1003. [CrossRef]
40. Bonanos, A.; Lennon, D.J.; Köhlinger, F.; Van Loon, J.T.; Massa, D.L.; Sewilo, M.; Evans, C.J.; Panagia, N.; Babler, B.L.; Block, M.; et al. Spitzer SAGE-SMC Infrared Photometry of Massive Stars in the Small Magellanic Cloud. *Astron. J.* **2010**, *140*, 416–429. 6256/140/2/416. [CrossRef]
41. Mould, J.; Barmby, P.; Gordon, K.; Willner, S.P.; Ashby, M.L.N.; Gehrz, R.D.; Humphreys, R.; Woodward, C.E. A Point-Source Survey of M31 with the Spitzer Space Telescope. *Astrophys. J.* **2008**, *687*, 230–241. [CrossRef]
42. McQuinn, K.B.W.; Woodward, C.E.; Willner, S.P.; Polomski, E.F.; Gehrz, R.D.; Humphreys, R.M.; van Loon, J.T.; Ashby, M.L.N.; Eicher, K.; Fazio, G.G. The M33 Variable Star Population Revealed by Spitzer. *Astrophys. J.* **2007**, *664*, 850–861. [CrossRef]
43. Massey, P.; Olsen, K.A.G.; Hodge, P.W.; Strong, S.B.; Jacoby, G.H.; Schlingman, W.; Smith, R.C. A Survey of Local Group Galaxies Currently Forming Stars. I. UBVRI Photometry of Stars in M31 and M33. *Astron. J.* **2006**, *131*, 2478–2496. [CrossRef]
44. Dalcanton, J.J.; Williams, B.F.; Lang, D.; Lauer, T.R.; Kalirai, J.S.; Seth, A.C.; Dolphin, A.; Rosenfield, P.; Weisz, D.R.; Bell, E.F.; et al. The Panchromatic Hubble Andromeda Treasury. *Astrophys. J.* **2012**, *200*, 1–37. [CrossRef]
45. Grammer, S.H.; Humphreys, R.M.; Gerke, J. The Massive Star Population in M101. III. Spectra and Photometry of the Luminous and Variable Stars. *Astron. J.* **2015**, *149*, 1–15. [CrossRef]
46. Humphreys, R.M.; Stangl, S.; Gordon, M.S.; Davidson, K.; Grammer, S.H. Luminous and Variable Stars in NGC 2403 and M81. *Astron. J.* **2019**, *157*, 1–16. [CrossRef]
47. Kudritzki, R.-P.; Urbaneja, M.A.; Gazak, Z.; Bresolin, F.; Przybilla, N.; Gieren, W.; Pietrzyński, G. Quantitative Spectroscopy of Blue Supergiant Stars in the Disk of M81: Metallicity, Metallicity Gradient, and Distance. *Astrophys. J.* **2012**, *747*, 15. [CrossRef]
48. Humphreys, R.M. M supergiants and the low metal abundances in the Small Magellanic Cloud. *Astrophys. J.* **1979**, *231*, 384–389. [CrossRef]
49. Massey, P.; Olsen, K.A.G. The Evolution of Massive Stars. I. Red Supergiants in the Magellanic Clouds. *Astron. J.* **2003**, *126*, 2867–2886. [CrossRef]
50. Vink, J.S.; de Koter, A.; Lamers, H.J.G.L.M. Mass-loss predictions for O and B stars as a function of metallicity. *Astron. Astrophys.* **2001**, *369*, 574–588.:20010127. [CrossRef]
51. Mokiem, M.R.; de Koter, A.; Vink, J.S.; Puls, J.; Evans, C.J.; Smartt, S.J.; Crowther, P.A.; Herrero, A.; Langer, N.; Lennon, D.J.; et al. The empirical metallicity dependence of the mass-loss rate of O- and early B-type stars. *Astron. Astrophys.* **2007**, *473*, 603–614.:20077545. [CrossRef]
52. Crowther, P.A.; Hillier, D.J.; Evans, C.J.; Fullerton, A.W.; De Marco, O.; Willis, A.J. Revised Stellar Temperatures for Magellanic Cloud O Supergiants from Far Ultraviolet Spectroscopic Explorer and Very Large Telescope UV-Visual Echelle Spectrograph Spectroscopy. *Astrophys. J.* **2002**, *579*, 774–799. [CrossRef]
53. Hillier, D.J.; Lanz, T.; Heap, S.R.; Hubeny, I.; Smith, L.J.; Evans, C.J.; Lennon, D.J.; Bouret, J.C. A Tale of Two Stars: The Extreme O7 Iaf+ Supergiant AV 83 and the OC7.5 III((f)) star AV 69. *Astrophys. J.* **2003**, *588*, 1039–1063. [CrossRef]
54. Fullerton, A.W.; Massa, D.L.; Prinja, R.K. The Discordance of Mass-Loss Estimates for Galactic O-Type Stars. *Astrophys. J.* **2006**, *637*, 1025–1039. [CrossRef]
55. Neugent, K.F.; Massey, P.; Skiff, B.; Drout, M.R.; Meynet, G.; Olsen, K.A. Yellow Supergiants in the Small Magellanic Cloud: Putting Current Evolutionary Theory to the Test. *Astrophys. J.* **2010**, *719*, 1784–1795. [CrossRef]
56. Neugent, K.F.; Massey, P.; Skiff, B.; Meynet, G. Yellow and Red Supergiants in the Large Magellanic Cloud. *Astrophys. J.* **2012**, *749*, 177. [CrossRef]

57. Humphreys, R.M.; Davidson, K.; Grammer, S.; Kneeland, N.; Martin, J.C.; Weis, K.; Burggraf, B. Luminous and Variable Stars in M31 and M33. I. The Warm Hypergiants and Post-red Supergiant Evolution. *Astrophys. J.* **2013**, *773*, 46. [CrossRef]
58. Humphreys, R.M.; Weis, K.; Davidson, K.; Bomans, D.J.; Burggraf, B. Luminous and Variable Stars in M31 and M33. II. Luminous Blue Variables, Candidate LBVs, Fe II Emission Line St ars, and Other Supergiants. *Astrophys. J.* **2014**, *790*, 48. [CrossRef]
59. Massey, P.; Neugent, K.F.; Smart, B.M. A Spectroscopic Survey of Massive Stars in M31 and M33. *Astron. J.* **2016**, *152*, 62. [CrossRef]
60. Humphreys, R.M.; Gordon, M.S.; Martin, J.C.; Weis, K.; Hahn, D. Luminous and Variable Stars in M31 and M33. IV. Luminous Blue Variables, Candidate LBVs, B[e] Supergiants, and the Warm Hypergiants: How to Tell Them Apart. *Astrophys. J.* **2017**, *836*, 64. [CrossRef]
61. Humphreys, R.M.; Davidson, K.; Hahn, D.; Martin, J.C.; Weis, K. Luminous and Variable Stars in M31 and M33. V. The Upper HR Diagram. *Astrophys. J.* **2017**, *844*, 40. [CrossRef]
62. Massey, P.; Silva, D.R.; Levesque, E.M.; Plez, B.; Olsen, K.A.; Clayton, G.C.; Meynet, G.; Maeder, A. Red Supergiants in the Andromeda Galaxy (M31). *Astrophys. J.* **2009**, *703*, 420–440. X/703/1/420. [CrossRef]
63. Drout, M.R.; Massey, P.; Meynet, G.; Tokarz, S.; Caldwell, N. Yellow Supergiants in the Andromeda Galaxy (M31). *Astrophys. J.* **2009**, *703*, 441–460. /703/1/441. [CrossRef]
64. Drout, M.R.; Massey, P.; Meynet, G. The Yellow and Red Supergiants of M33. *Astrophys. J.* **2012**, *750*, 97. [CrossRef]
65. Gordon, M.S.; Humphreys, R.M.; Jones, T.J. Luminous and Variable Stars in M31 and M33. III. The Yellow and Red Supergiants and Post-red Supergiant Evolution. *Astrophys. J.* **2016**, *825*, 50. [CrossRef]
66. Massey, P.; Evans, K.A. The Red Supergiant Content of M31. *Astrophys. J.* **2016**, *826*, 224. [CrossRef]
67. Massey, P.; Armandroff, T.E.; Pyke, R.; Patel, K.; Wilson, C.D. Hot, Luminous Stars in Selected Regions of NGC 6822, M31, and M33. *Astron. J.* **1995**, *110*, 2715–2745. [CrossRef]
68. Massey, P. Evolved Massive Stars in the Local Group. I. Identification of Red Supergiants in NGC 6822, M31, and M33. *Astrophys. J.* **1998**, *501*, 153–174. [CrossRef]
69. Bresolin, F.; Urbaneja, M.A.; Gieren, W.; Pietrzyński, G.; Kudritzki, R.P. VLT Spectroscopy of Blue Supergiants in IC 1613. *Astrophys. J.* **2007**, *671*, 2028–2039. [CrossRef]
70. Humphreys, R.M.; Aaronson, M.; Liebofsky, M.; McAlary, C.W.; Strom, S.E.; Capps, R.W. The luminosities of M supergiants and the distances to M 101, NGC 2403, and M 81. *Astron. J.* **1986**, *91*, 808–821. [CrossRef]
71. Humphreys, R.M.; Aaronson, M. The Visually Brightest Early-Type Supergiants in the Spiral Galaxies NGC 2403, M81, and M101. *Astron. J.* **1987**, *94*, 1156–1169. [CrossRef]
72. Zickgraf, F.-J.; Humphreys, R.M. A Stellar Content Survey of NGC 2403 and M81. *Astron. J.* **1991**, *102*, 113–133. [CrossRef]
73. Kudritzki, R.-P.; Urbaneja, M.A.; Bresolin, F.; Przybilla, N.; Gieren, W.; Pietrzyński, G. Quantitative Spectroscopy of 24 A Supergiants in the Sculptor Galaxy NGC 300: Flux-weighted Gravity-Luminosity Relationship, Metallicity, and Metallicity Gradient. *Astrophys. J.* **2008**, *681*, 269–289. [CrossRef]
74. Urbaneja, M.A. The Araucaria Project: The Local Group Galaxy WLM—Distance and Metallicity from Quantitative Spectroscopy of Blue Supergiants. *Astrophys. J.* **2008**, *684*, 118–135. [CrossRef]
75. Vivian, U.; Urbaneja, M.A.; Kudritzki, R.P.; Jacobs, B.A.; Bresolin, F.; Przybilla, N. A New Distance to M33 Using Blue Supergiants and the FGLR Method. *Astrophys. J.* **2009**, *704*, 1120–1134. [CrossRef]
76. Humphreys, R.M.; Davidson, K. The Most Luminous Stars. *Science* **1984**, *223*, 243–249. [CrossRef] [PubMed]
77. *Instabilities in Luminous Early Type Stars*; Henny, L., de Loore, C., Eds.; D. Reidel Publishing Company: Dordrecht, The Netherlands, 1987; Volume 136.
78. *Physics of Luminous Blue Variables*; Davidson, K., Moffat, A.F.J., Lamers, H.J.G.L.M., Eds.; Kluwer Academic Publishers: Dordrecht, The Netherlands; Boston, MA, USA, 1989, Volume 157.
79. Humphreys, R.M.; Weis, K.; Davidson, K.; Gordon, M.S. On the Social Traits of Luminous Blue Variables. *Astrophys. J.* **2016**, *825*, 64. [CrossRef]
80. Lamers, H.J.G.L.M.; Fitzpatrick, E.L. The Relationship between the Eddington Limit, the Observed Upper Luminosity Limit for Massive Stars, and the Luminous Blue Variables. *Astrophys. J.* **1988**, *324*, 279–287. [CrossRef]

81. Glatzel, W.; Kiriakidis, M. Stability of Massive Stars and the Humphreys/Davidson Limit. *MNRAS* **1993**, *263*, 375–384. [CrossRef]
82. Ulmer, A.; Fitzpatrick, E.L. Revisiting the Modified Eddington Limit for Massive Stars. *Astrophys. J.* **1998**, *504*, 200–206. [CrossRef]
83. Humphreys, R.M.; Davidson, K. The Luminous Blue Variables: Astrophysical Geysers. *PASP* **1994**, *106*, 1025–1051. [CrossRef]
84. Smartt, S.J.; Eldridge, J.J.; Crockett, R.M.; Maund, J.R. The death of massive stars–I. Observational constraints on the progenitors of Type II-P supernovae. *Mon. Not. R. Astron. Soc.* **2009**, *395*, 1409–1437. [CrossRef]
85. Smartt, S.J. Observational Constraints on the Progenitors of Core-Collapse Supernovae: The Case for Missing High-Mass Stars. *PASA* **2015**, *32*. [CrossRef]
86. Jennings, Z.G.; Williams, B.F.; Murphy, J.W.; Dalcanton, J.J.; Gilbert, K.M.; Dolphin, A.E.; Weisz, D.R.; Fouesneau, M. The Supernova Progenitor Mass Distributions of M31 and M33: Further Evidence for an Upper Mass Limit. *Astrophys. J.* **2014**, *795*, 170. [CrossRef]
87. Ekstrom, S.; Georgy, C.; Eggenberger, P.; Meynet, G.; Mowlavi, N.; Wyttenbach, A.; Granada, A.; Decressin, T.; Hirschi, R.; Frischknecht, U.; et al. Grids of stellar models with rotation. I. Models from 0.8 to 120 $M_\odot$ at solar metallicity (Z = 0.014). *Astron. Astrophys.* **2012**, *537*, A146. [CrossRef]
88. Bond, H.E. Hubble Space Telescope Imaging of the Outburst Site of M31 RV. II. No Blue Remnant in Quiescence. *Astrophys. J.* **2011**, *737*, 17. [CrossRef]
89. Davidson, K. Giant eruptions of very massive stars. *J. Phys. Conf. Ser.* **2016**, *728*. [CrossRef]
90. Riess, A.G.; Fliri, J.; Valls-Gabaud, D. Cepheid Period-Luminosity Relations in the Near-infrared and the Distance to M31 from the Hubble Space Telescope Wide Field Camera 3. *Astrophys. J.* **2012**, *745*, 156. . [CrossRef]
91. Scowcroft, V.; Bersier, D.; Mould, J.R.; Wood, P.R. The effect of metallicity on Cepheid magnitudes and the distance to M33. *Mon. Not. Royal Astron. Soc.* **2009**, *396* 1287–1296. [CrossRef]
92. Feast, M.W.; Whitelock, P.A.; Menzies, J.W.; Matsunaga, N. The Cepheid distance to the Local Group galaxy NGC 6822. *Mon. Not. Royal Astron. Soc.* **2012**, *421*, 2998–3003. [CrossRef]
93. Scowcroft, V.; Freedman, W.L.; Madore, B.F.; Monson, A.J.; Persson, S.E.; Seibert, M.; Rigby, J.R.; Melbourne, J. The Carnegie Hubble Program: The Infrared Leavitt Law in IC 1613. *Astrophys. J.* **2013**, *773*, 106. [CrossRef]
94. Freedman, W.L.; Madore, B.F.: Gibson, B.K.; Ferrarese, L.; Kelson, D.D.; Sakai, S.; Mould, J.R.; Kennicutt, R.C., Jr.; Ford, H.C.; Graham, J.A.; et al. Final Results from the Hubble Space Telescope Key Project to Measure the Hubble Constant. *Astrophys. J.* **2001**, *553*, 47–72. [CrossRef]

© 2019 by the author. Licensee MDPI, Basel, Switzerland. This article is an open access article distributed under the terms and conditions of the Creative Commons Attribution (CC BY) license (http://creativecommons.org/licenses/by/4.0/).

Review

Massive Stars in the Tarantula Nebula: A Rosetta Stone for Extragalactic Supergiant HII Regions

Paul A. Crowther

Department of Physics & Astronomy, University of Sheffield, Sheffield S3 7RH, UK; paul.crowther@sheffield.ac.uk

Received: 3 September 2019; Accepted: 6 November 2019; Published: 8 November 2019

Abstract: A review of the properties of the Tarantula Nebula (30 Doradus) in the Large Magellanic Cloud is presented, primarily from the perspective of its massive star content. The proximity of the Tarantula and its accessibility to X-ray through radio observations permit it to serve as a Rosetta Stone amongst extragalactic supergiant HII regions since one can consider both its integrated characteristics and the individual properties of individual massive stars. Recent surveys of its high mass stellar content, notably the VLT FLAMES Tarantula Survey (VFTS), are reviewed, together with VLT/MUSE observations of the central ionizing region NGC 2070 and HST/STIS spectroscopy of the young dense cluster R136, provide a near complete Hertzsprung-Russell diagram of the region, and cumulative ionizing output. Several high mass binaries are highlighted, some of which have been identified from a recent X-ray survey. Brief comparisons with the stellar content of giant HII regions in the Milky Way (NGC 3372) and Small Magellanic Cloud (NGC 346) are also made, together with Green Pea galaxies and star forming knots in high-z galaxies. Finally, the prospect of studying massive stars in metal poor galaxies is evaluated.

Keywords: galaxies; star formation–galaxies; star clusters; general–open clusters and associations; individual; 30 Doradus–stars; massive–stars; early-type

1. Introduction

The Tarantula Nebula (alias 30 Doradus) in the Large Magellanic Cloud (LMC) is the brightest supergiant HII region in the Local Group of galaxies, and serves as a local analogue to metal-poor starburst knots in high redshift galaxies [1]. Its proximity (50 kpc), and high galactic latitude (and hence low extinction) have permitted a myriad of ground-based and space-based surveys across the electromagnetic spectrum, revealing an exceptional population of massive stars ($\geq 8 M_\odot$), including the dense, young star cluster R136 that is home to some of the most massive stars known. The advent of modern highly multiplexed spectrographs coupled with large ground-based telescopes, has permitted multi-epoch optical spectroscopic surveys of the massive star population of the Tarantula Nebula for the first time. The VLT FLAMES Tarantula Survey, hereafter VFTS [2], has provided the multiplicity, rotational velocities and initial mass function of massive stars. In distant starburst regions it is not possible to resolve individual stars, such that studies rely on techniques based on their integrated properties, which themselves involve assumptions of binarity, stellar rotation and mass function.

The Tarantula Nebula is not the sole example of a supergiant HII region within the Local Group. However, studies of the richest Milky Way star-forming regions are limited by dust extinction (e.g., Westerlund 1, NGC 3603). Counterparts in other Local Group galaxies suffer from a number of limitations, including a relatively modest stellar content (NGC 346 in the Small Magellanic Cloud, SMC) or much greater distance (NGC 604 in M 33), such that the Tarantula—whose metallicity is approximately half-solar [3]—serves as the only credible Rosetta Stone for rich extragalactic star-forming regions. This review will provide a brief overview of the structural properties of the Tarantula Nebula, but will largely focus on its massive star content, drawn from results from VFTS and other optical spectroscopic surveys, supplemented by the recent deep Chandra X-ray survey 'Tarantula-Revealed by X-rays' (T-ReX). Finally, comparisons with local and high-redshift star-forming regions will be provided to put the properties of the Tarantula into a broader context. Indeed, the nebular properties of the central NGC 2070 region of the Tarantula are strikingly similar to Green Pea galaxies, exhibiting intense [O III] λ4959, 5007 emission, and its star-formation rate is comparable to intense star-forming clumps at high redshift.

2. Tarantula Nebula

The Tarantula Nebula is the most striking star-forming region in the LMC, whether viewed in the far ultraviolet (hot, luminous stars), Hα (ionized gas) or mid-IR (warm dust), extending over several hundred parsec, owing to a massive stellar content producing an ionizing output which is a thousand-fold higher than the Orion Nebula [4].

NGC 2070 is the dominant ionized region within the Tarantula Nebula, powered by a large number of hot luminous stars from R136 at its heart plus many more within its vicinity. NGC 2060, located 6 arcmin (90 pc) to the southwest, is host to a more modest population of OB stars plus an X-ray pulsar PSR J0537-6910 and its supernova remnant (SNR) N157B. Hodge 301 is located 3 arcmin (45 pc) to the north west of R136, but does not possess significant nebulosity since previous supernovae are likely to have cleared this region of gas, and its stellar population (B-type stars and red supergiants) does not possess a significant Lyman continuum output. NGC 2070 is often referred to as a cluster in the literature but it extends over tens of parsecs whereas genuine star clusters are an order of magnitude smaller, such that the only rich star clusters within the Tarantula are R136, with an age of 1–2 Myr [5] and Hodge 301 with an age of 20–30 Myr [6] with a few additional lower mass young, compact clusters (e.g., TLD1, SL 639). Table 1 compares various regions within the Tarantula Nebula, adapted from a previous review by Walborn [7]. Although the focus of the present review is on spectroscopic results for massive stars, Sabbi et al. [8] have undertaken a deep Hubble Space Telescope (HST) multi-colour photometric survey, known as the Hubble Tarantula Treasury Project (HTTP) which permits lower mass stars in the Tarantula Nebula to be studied.

Table 1. Physical scales within the Tarantula Nebula, adapted from Walborn [7]. Ionizing outputs, N(LyC) are obtained from the present work.

Region	Angular Radius (″)	Physical Radius (pc)	N(LyC) (10^{51} ph s^{-1})	Content	Reference
R136a	0.8	0.2	2	R136a1 (WN5h), R136a2 (WN5h)	[9]
R136	4	1.0	4	R136b (O4 If/WN8), R136c (WN5h+)	[5]
NGC 2070	80	20.	9	R140a (WC4 + WN6+), Mk34 (WN5h + WN5h)	[10,11]
Tarantula	600	150.	12	Hodge 301, PSR J0537-6910 (pulsar), N157B (SNR)	[12]

A number of complementary studies of the star formation history of the Tarantula Nebula have been carried out, exploiting pre-main sequence low mass stars [13,14] and massive stars [15]. Significant star formation commenced ~25 Myr ago, as witnessed by Hodge 301, and reached a peak several Myr ago, with the young massive star cluster R136 at its heart no more than 2 Myr old. It is apparent that star formation within the Tarantula Nebula has not been limited to specific parsec-scale star clusters, such as Hodge 301 or R136, but has been distributed across the entire region, akin to a super OB association. Wright et al. [16] have established from proper motion observations that star formation in the far smaller Milky Way Cygnus OB2 region did not originate in a star cluster, but involved individual sub-regions in virial equilibrium. Indeed, median ages of massive stars show little radial dependence on their projected distance from R136, with very massive stars ($\geq$100$M_\odot$) identified throughout the region [15]. Infrared and radio observations of the Tarantula reveal ongoing regions of massive star formation, to which the reader is referred to the review by Walborn [17] and a more recent study of the brightest embedded sources based on Spitzer/IRAC imaging [18]. Atacama

Large Millimetre Array (ALMA) has obtained high resolution observations of parsec-scale clumps within the Tarantula [19], with the rate of star formation in the Tarantula anticipated to decline in the future.

3. Massive Star Content

The integrated nebular properties of the Tarantula Nebula and other selected giant H II regions in the Local Group is presented in Table 2. The ionizing output of the Tarantula Nebula corresponds to the equivalent of over a thousand O7V stars, each with 10^{49} ph s^{-1}. In reality, the Tarantula hosts somewhat fewer O-type stars since the most extreme examples—early O stars and luminous Wolf-Rayet (WR) stars—each produce an order of magnitude more Lyman continuum photons. Nevertheless, this population represents an order of magnitude more O and WR stars than any Milky Way or SMC giant HII regions, and is not likely to be improved upon until extremely large telescopes are capable of resolving the massive stellar content of more extreme giant HII regions, such as NGC 5461, 5462 and 5471 in M 101 [1].

Table 2. Integrated nebular properties of nearby giant HII regions, adapted from Kennicutt [20] for an assumed O7V Lyman continuum output of 10^{49} ph s^{-1}.

Region (Galaxy)	Diameter pc	L(Hα) 10^{39} erg s^{-1}	N(O7V)
NGC 3372 (Milky Way)	200	0.8	45
NGC 346 (SMC)	220	0.8	45
NGC 3603 (Milky Way)	100	1.5	110
NGC 604 (M33)	400	4.5	320
Tarantula (LMC)	370	15	1100

Historically, there have been several photometric and spectroscopic surveys of early-type stars in the Tarantula Nebula, each of which have employed contemporary (Galactic) temperature calibrations to produce Hertzsprung-Russell diagrams. Parker [21,22] obtained the first extensive study of the entire region. Subsequently, Massey and Hunter [23] obtained high-spatial resolution HST spectroscopy of early-type stars in the central, crowded R136 region, revealing an exceptional population of very early O stars, whilst Melnick and colleagues [24,25] obtained spectroscopy for a large number of early-type stars in the NGC 2070 region. The advent of efficient multi-object spectrographs on 8–10 m telescopes, has permitted the most comprehensive optical spectroscopic survey of massive stars in the Tarantula Nebula to date through the VLT FLAMES Tarantula Survey (VFTS) [2]. Multi-object, multi-epoch spectroscopy of ~800 massive stars across the entire region, for which detailed spectroscopic analyses have been undertaken for over 500 O and early B stars, such that temperature calibrations are no longer necessary. Although this represents the most extensive of early-type stars in a single star-forming region to date, this survey is incomplete owing the fibre-placement limitations, sampling ~70% of massive stars exterior to the dense R136 cluster from comparison with photometric surveys [12].

Two additional surveys have recently provided complete optical spectroscopic observations of all bright sources within the central crowded region of the Tarantula: (a) the central 4 arcsec (1 pc) of the R136 cluster exploiting the high spatial resolution of HST/STIS [5]; (b) the central 2 × 2 arcmin (30 × 30 pc) region of NGC 2070 using the VLT/MUSE integral field spectrograph [10]. Although these lack the multi-epoch capabilities of VFTS, they are complementary since the richest stellar populations of the Tarantula are found within NGC 2070/R136, and provide extended wavelength coverage (yellow and red for MUSE, ultraviolet for STIS), albeit at reduced spectral resolution. A summary of these three contemporary surveys is provided in Table 3, together with literature results. In total, approximately 1100 early-type massive stars have been spectroscopically observed. Our discussion of the massive star content of the Tarantula will largely draw upon results from VFTS, but include others where appropriate. It should be noted that analyses based on spectroscopic fibre-fed observations of early-type stars in regions of strong, highly variable nebulosity, is inherently problematic owing to the lack of local sky subtraction. Such issues do no affect long-slit STIS or integral field MUSE observations.

Table 3. Summary of stellar content of the Tarantula Nebula from recent spectroscopic surveys (excluding sources in common, although including individual components within SB2 binaries).

Telescope/inst	Target	N(O-type)	N(B-type)	N(WR)	N(Of/WN)	N(A+)	Reference
VLT/FLAMES	30 Dor	369	436	9	6	35	[2]
HST/STIS	R136	57	..	3	2	..	[5]
VLT/MUSE	NGC 2070	115	79	..	..	1	[10]
Other	30 Dor	29	8	16	..	5	[22–24,26]
Total	30 Dor	570	523	28	8	41	

VFTS and other surveys have revealed that the Tarantula hosts extreme examples of stellar exotica, including the most massive stars known [9], a Luminous Blue Variable, R143 [27], a very massive runaway [28], the fastest rotating stars [29], a massive overcontact binary [30], and a supernova remnant N157B [31], with SN 1987A located ~300 pc to the south west (Figure 1).

Figure 1. (**left**) Optical image of the Tarantula Nebula from the MPG/ESO 2.2m WFI, with NGC 2060 and SN1987A indicated; (**centre**) Optical VLT/FORS2 image centred on NGC 2070, with Hodge 301 to the upper right; (**right**) an infrared VLT/MAD image of the central R136 region, with the massive colliding wind binary Mk 34 indicated. Credit: ESO/P. Crowther/C.J. Evans.

R136a, the central cluster, merits special consideration since it was considered by some to be a supermassive star as recently as the early 1980s [32]. Speckle interferometry resolved R136a into multiple sources [33], and it was subsequently established as a compact star cluster [34]. Massey and Hunter [23] established that dozens of the brightest sources within the central parsec were hot, early O stars. Spectroscopic studies of the brightest components R136a1, a2, a3, with nitrogen-sequence Wolf-Rayet spectral types, indicated masses of ~100 $M_\odot$ [23,35]. They established that these relatively weak-lined WN stars are luminous main-sequence stars close to their Eddington limits, rather than classical Wolf-Rayet stars. Subsequent analyses of the WN stars in R136 indicated significantly higher masses of 150–300 $M_\odot$ [9] as a result of increased spectroscopic luminosities, owing to higher stellar temperatures and IR photometry less affected by dust extinction. Indirectly inferred masses of massive stars are notoriously imprecise, and if binarity were established for individual stars their inferred luminosities and masses would be reduced. To date, faint companions to members of R136a have been detected with extreme adaptive optics imaging [36]. Melnick 34 is spectroscopically similar to the WN5-stars in R136a, and has recently been shown to be a colliding wind binary system comprising two WN5 components, with a total mass exceeding 250 $M_\odot$ [37]. Figure 2 shows that Melnick 34 is an order of magnitude brighter than R136a in X-rays, indicating that there are no colliding wind binaries comparable to Melnick 34 within R136a [38].

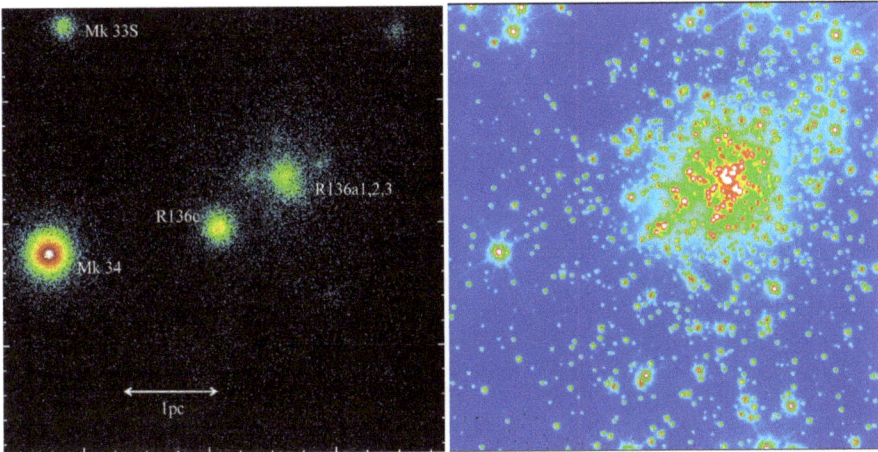

Figure 2. (**left**) Chandra ACIS X-ray logarithmic intensity image of the core of NGC 2070 from T-ReX, centred on R136c, adapted from [38], showing the relative brightness of the colliding wind binary Melnick 34 (WNh5 + WN5h) [37] to the R136a star cluster (hosting multiple WN5h stars) and R136c (WN5h+?); (**right**) HST WFC3/F555W logarithmic intensity image of the same 19 × 19 arcsec region, highlighting the rich stellar population of R136a with respect to R136c and Melnick 34.

Overall, the Tarantula hosts a remarkable number of ∼50 early-type stars with bolometric luminosities exceeding $10^6 L_\odot$. For reference, the Milky Way's Carina Nebula (NGC 3372) hosts ≈5 massive stars with such extreme properties [39]. As such, the Tarantula Nebula represents our best opportunity to study the highest mass stars known, both individually and collectively. Schneider et al. [40] analysed VFTS spectroscopic results to establish an excess of massive stars with respect to a standard Salpeter Initial Mass Function (IMF), indicating 1/3 more stars with ≥30 $M_\odot$ in 30 Doradus compared to expectations from a standard IMF. Finally, although we focus primarily on high mass early-type stars in the Tarantula Nebula, it also hosts red supergiants (RSG). Since RSG are the evolved descendants of moderately massive stars, and star formation in the Tarantula has peaked relatively recently, of order ∼10 RSG are known, most of which are associated with mature star clusters Hodge 301 and SL 639 [6].

4. Physical Properties

The determination of physical properties ($T_{\rm eff}$, $\log g$) of individual early-type stars ideally requires high S/N (≥50) intermediate resolution spectroscopy of suitable diagnostics, usually He I–II lines (Si II–IV) for temperatures of O-type (B-type) stars, plus Balmer lines for surface gravities, plus grids of model atmospheres obtained with modern codes, such as FASTWIND [41] for O stars and blue supergiants, CMFGEN [42] or PoWR [43] for emission-line stars, or TLUSTY [44] for low luminosity B stars. Analysis of late-type supergiants requires model atmosphere codes in which molecular opacities have been incorporated, such as MARCS [45]. Hot O stars require alternative temperature diagnostics to helium, with nitrogen commonly used since the blue visual spectrum of O stars includes lines of N III–V. Wolf-Rayet stars are especially problematic since photospheres are masked by the dense wind, such that gravities cannot be directly measured and temperatures usually refer to deep layers, with an optical depth of $\tau \sim 10$–20, rather than the effective temperature at $\tau = 2/3$. If stellar distances are uncertain, comparisons with evolutionary models can be made using the so-called spectroscopic Hertzsprung-Russell (sHR) diagram [46], involving temperature and $\mathscr{L} = T_{\rm eff}^4/g$, the inverse of the flux-weighted gravity, where g is the surface gravity.

The determination of stellar luminosities requires comparisons between synthetic spectral energy distributions and photometry, taking account of interstellar extinctions and distance moduli, 18.5 mag in the case of the Large Magellanic Cloud. Visual extinctions of early-type stars in the Tarantula Nebula are usually modest, although near-IR photometric comparisons usually lead to more robust luminosities since typical dust extinctions are 0.1–0.2 mag in the K-band, versus 1–2 mag in the V-band, and the lack of sensitivity of K-band extinctions to any variations in the overall extinction law. Luminosities of RSG can also be reliably estimated by integrating observed spectral energy distributions from visual to mid-IR wavelengths [47,48]. Historically, stellar estimates of masses and ages from evolutionary models involved by-eye comparisons between their position on a conventional Hertzsprung-Russell (HR) diagram and theoretical isochrones. However, additional physical information is often available, such as helium abundance or projected rotational velocities. Tools now exist which additionally take such information into account for the calculation of stellar ages and initial masses such as BONNSAI [49]. Significant discrepancies exist between current mass estimates from spectroscopic ($\log g$) and evolutionary approaches for a subset of VFTS O dwarfs [50].

Figure 3 presents the HR diagram of the Tarantula Nebula, comprising single star results from VFTS [50–54], VLT/MUSE [11], HST/STIS [55] and literature results for other stars within 160 parsec of R136, including Wolf-Rayet stars [56]. Results for binary systems have been incorporated, primarily drawn from [57] for VFTS B-type binaries, Tarantula Massive Binary Monitoring (TMBM) for VFTS O-type binaries [58] and recent literature for WR stars [37,59]. Evolutionary tracks for non-rotating, LMC metallicity massive stars up to the onset of He-burning have been included for reference [60,61]. Over 1170 massive stars have been included, revealing a well populated main sequence population up to ~200 $M_\odot$, plus classical Wolf-Rayet stars to the left of the main sequence, and evolved blue supergiants up to $\log(L/L_\odot) \sim 6$, and cool supergiants, up to $\log(L/L_\odot) \sim 5.3$ [48]. The addition of all luminous early-type stars from R136 and NGC 2070 fills in the extreme upper main sequence which is somewhat under populated from VFTS alone [40]. The overwhelming majority of the older massive stellar population—i.e., evolved stars with masses below 30 $M_\odot$—are spatially exterior to NGC 2070 (open symbols), although NGC 2070 is host to one luminous M supergiant, Melnick 9. Conversely, beyond NGC 2070, the main-sequence population at the highest stellar masses is relatively underpopulated, albeit with several WN5h stars (R146, R147) and early O stars (VFTS 16, BI 253) located 95 ± 25 parsec from R136.

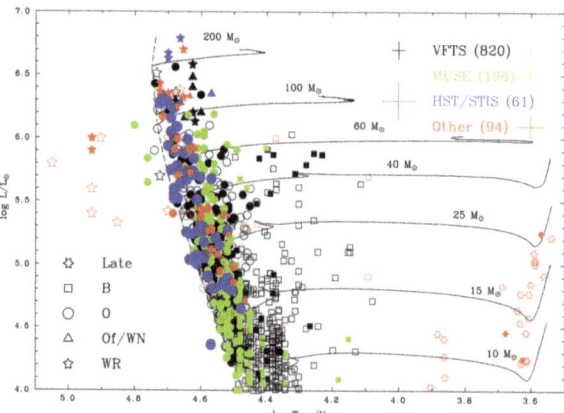

Figure 3. Hertzsprung-Russell diagram of the Tarantula Nebula, based on results from VLT FLAMES Tarantula Survey (VFTS) [50–54], MUSE [11], Hubble Space Telescope (HST)/STIS [55] and other literature results, with typical uncertainties from each survey indicated. Filled symbols are within NGC 2070, open symbols elsewhere in the Tarantula. Non-rotating tracks for 10, 15, 25, 40, 60, 100 and 200 $M_\odot$ LMC metallicity stars have been included from [60,61] which terminate at the onset of He-burning.

5. Binaries, Rotation and Runaways

Until recently, the significance of close binary evolution for massive stars was not fully recognised, in spite of a few binary "champions" [62]. The high frequency of close binaries amongst O stars in young Galactic clusters obtained from radial velocity monitoring established that only a minority of massive stars follow single stellar evolution [63]. In contrast with the majority of previous spectroscopic surveys of early-type stars, VFTS comprised multiple epochs, such that [64] were able to establish that 53% of O stars in the Tarantula Nebula inhabit a binary system with a period below 1500 days, such that binary interaction will occur. In total, 18% of O-type binaries, those with very short periods, are anticipated to merge with a companion, 27% will be stripped of their envelopes (primaries, mass donors) and 8% are predicted to be spun up (secondaries, mass gainers), as summarised in Figure 4 together with counterparts in Milky Way clusters. Broadly similar results have been obtained for VFTS B-type stars [65].

The inferred rate of envelope stripping and spin-up in the Tarantula is rather lower than [63] obtained for O stars in Milky Way clusters (Figure 4), but it is probable that the true incidence is rather higher since a subset of the current O star sample is likely to have already undergone binary evolution. Although VFTS has established the binary frequency amongst massive stars in the Tarantula, binary orbits require follow-up surveys, notably the TMBM survey [58,66].

Consequences of close binary evolution include mass gaining secondaries being spun-up, and a subset of secondaries possessing high space velocities as a result of the disruption of the binary following the core-collapse supernova (ccSN) of the original primary. Other "observables" are more challenging, including the identification of stripped primaries in close binaries which will usually be masked at visual wavelengths by mass gaining secondaries [67] which should be more common in older star clusters such as Hodge 301.

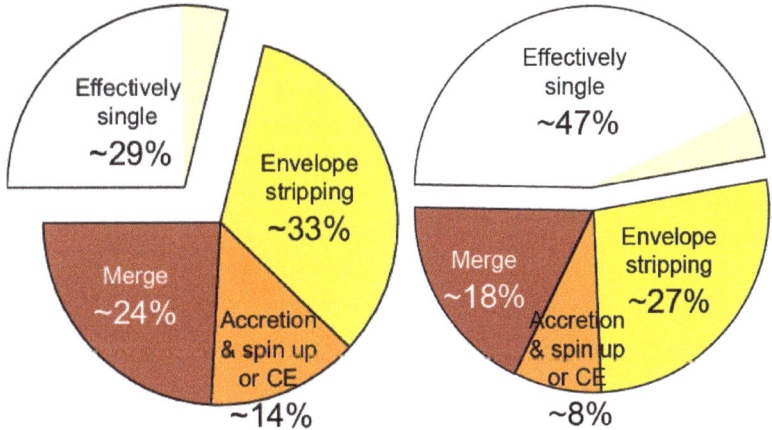

Figure 4. Pie charts, courtesy Hugues Sana and Selma de Mink, illustrating the fraction of O stars undergoing single stellar evolution versus mergers, primaries being stripped of their envelopes, and secondaries being spun up, for Milky Way young clusters (left) and VFTS O stars (right), adapted from [63,64].

Reliable measurements of projected rotation rates, $v_e \sin i$, are often problematic because strong hydrogen and helium lines are predominantly affected by pressure broadening, with an additional contribution from "macroturbulence". A Fourier Transform approach applied to metallic lines offers the most robust results, albeit requiring high resolution, high S/N spectroscopy of suitable diagnostics [68]. The lack of a spectral features originating in the hydrostatic layers of Wolf-Rayet stars prevents a direct determination of their rotational velocities. Figure 5 compares projected rotational velocities for a large sample of VFTS O and B-type stars. Typical rotational velocities of single O stars are modest, with $v_e \sin i \sim 100$ km s^{-1}, albeit with 10% exceeding 300 km s^{-1} [69], of which some examples are rotating close to their critical rates [29]. This high velocity tail is suspected of being spun-up mass gainers in former close binaries. The lack of fast rotators amongst O stars in VFTS binary systems supports this interpretation [70]. Figure 5 illustrates that rotational velocities of VFTS B-type stars are higher, with $v_e \sin i \sim 200$ km s^{-1} on average, and 20% exceeding 300 km s^{-1} [71].

Of particular interest is the rotational velocity distribution of massive stars within the young R136 star cluster, whose severe crowding prevented inclusion in VFTS. Bestenlehner et al. [55] utilised HST/STIS spectroscopy to reveal 150 km s^{-1} on average for a sample of 55 massive stars within the central parsec of R136, with no examples exceeding 250 km s^{-1}, although the low S/N of these datasets prevented distinguishing between rotational and macroturbulence, adding to the interpretation that rapid rotators originate from close binary evolution. Wolff et al. [72] obtained somewhat higher rotational velocities for OB stars in the periphery of R136, where contamination from the field population is significant.

The Tarantula hosts a number of candidate early-type runaway stars from their measured (radial and/or tangential) velocities with respect to the average for their environment, although radial velocity outliers may be unresolved binaries. Runaways can originate either from disrupted secondaries following the core collapse of primaries in close binaries, or following the dynamical ejection of stars from young star-forming regions. Platais et al. [73] have investigated high proper motion stars in the Tarantula from HST imaging obtained 3 years apart, revealing a number of potential stars ejected from R136, while Lennon et al. [28] have exploited Gaia DR2 proper motions to conclude that VFTS 16 (O2 III) [74] was likely ejected from R136 during its formation 1–2 Myr ago. Renzo et al. [75] discuss the origin of the candidate 'walkaway' very massive star VFTS 682 (WN5h).

The Tarantula hosts several notable massive binary systems, whose physical and orbital properties have been obtained from spectroscopic monitoring, with searches for massive binaries also greatly benefitting from the recent Chandra T-ReX survey (PI Leisa Townsley) which monitored the Tarantula in X-rays for almost 2 years with a total integration time of 2 Ms. Single hot, luminous stars tend to produce (thermal) X-rays due to shocks in the winds, but these are generally soft X-ray emitters with $L_X/L_{Bol} \sim 10^{-7}$ [76]. Massive stars in binary systems may lead to excess X-ray emission arising from wind-wind collisions, usually relatively hard, providing the separations are not too small (low wind velocities) or too large (low wind densities) [77]. A close binary comprising an early-type star and compact remnant (neutron star or black hole) will be extremely X-ray bright if the accretion disk of the remnant is being fed by the wind of the massive star or via Roche Lobe overflow.

A number of eclipsing binaries in the proximity of R136 have been identified [78], including # 38 from [34] comprising an O3 V + O6 V in a circular 3.4 day orbit, with component masses 57 and 23 $M_\odot$. This represented the first robust stellar mass determination for an O3 star in the LMC. VFTS revealed a large number of binaries within the Tarantula, many of which have been followed-up with TMBM. Most notably R139 has been established as an eccentric system comprising a pair of mid O supergiants in a 154 day orbit [79] with lower limits of $\sim$66 + 78 $M_\odot$ for individual component, recently revised downward to 54 + 69 $M_\odot$ [58]. R139 is amongst the brightest X-ray sources in the Tarantula in T-ReX with $L_{X,corr} \sim 5 \times 10^{33}$ erg s^{-1} and an enhanced $L_{X,corr}/L_{Bol} \sim 9 \times 10^{-7}$ based on TMBM bolometric luminosities [58].

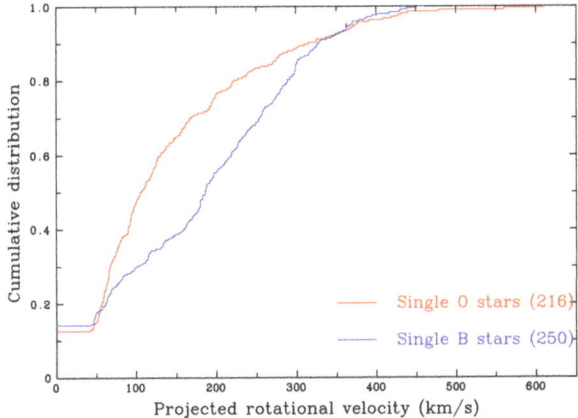

Figure 5. Cumulative distribution of rotational rates for single VFTS O (red) and B stars (blue) [69,71].

The most remarkable X-ray source in the Tarantula is VFTS 399 with $L_{X,corr} \sim 5 \times 10^{34}$ erg s^{-1} despite being associated with a low luminosity O9 giant, implying $L_{X,corr}/L_{Bol} \sim 2 \times 10^{-4}$. Clark et al. [80] conclude that VFTS 399 is a high-mass X-ray binary hosting a neutron star remnant, with the O giant known to be a rapid rotator ($v_e \sin i = 324$ km s^{-1}, according to [53]), as one would expect for a mass gainer in a close binary system.

Two point sources in the Tarantula are even brighter in X-rays than VFTS 399. Of these, R140a is a compact group of stars including two WR stars, so likely hosts one or more colliding wind binaries, while X-ray variability for Melnick 34 reveals a 155 day period, peaking at $L_{X,corr} = 3.2 \times 10^{35}$ erg s^{-1}, exceeding η Carina at X-ray maximum [38]. Melnick 34 has been confirmed to be a colliding wind binary system in a 155 day eccentric orbit, with minimum masses of 60–65 $M_\odot$ for each of the WN5h components [37]. Individual masses of 130–140 $M_\odot$ are favoured from spectroscopic analysis, such that Melnick 34 is likely to be the most massive binary known to date, with $L_{X,corr}/L_{bol} = 1.7 \times 10^{-5}$ at X-ray maximum. This arises from the collision of dense, fast moving winds at a minimum separation of $\sim$1.2 AU or 13 stellar radii for an assumed orbital inclination of $i = 50°$. Almeida et al. [30] identify the overcontact binary VFTS 352 as a prototype of systems which, at low metallicity, are plausible black hole-black hole merger progenitors.

6. Wind Properties

The underlying theory responsible for outflows by hot luminous stars has been known for several decades [81], with a metallicity dependence primarily arising from the variation in iron-peak elemental abundances [82,83]. Individual wind properties of O stars or blue supergiants usually rely on spectroscopic fits to Hα, as parameterised by the wind strength parameter, Q [41], from which the mass-loss rate requires knowledge of the physical radius (from T_{eff}, $\log L$) and measured or adopted wind velocity. Wind velocities of OB stars cannot be measured from optical spectroscopy, so usually spectral type calibrations are adopted based on measured velocities from UV P Cygni profiles of C IV, N V or Si IV [84]. Until recently, high S/N, high quality UV spectroscopy of OB stars in the Large Magellanic Cloud has been in short supply, but the situation has improved via the HST Large Program METAL (GO 14675, PI Julia Roman-Duval) [85] and upcoming ULLYSES initiative [1].

---

[1] http://www.stsci.edu/stsci-research/research-topics-and-programs/ullyses.

Specifically for the Tarantula, low resolution UV spectroscopy of the R136 star cluster has added a significant number of wind measurements for early O stars [5]. Wind velocities exceed 3000 km s^{-1} at the earliest subtypes (O2–3), reducing to ~1500 km s^{-1} for late O-types. Mass-loss rates of emission line stars typically rely on an alternative wind scaling relation, namely the transformed radius, R_t [86], which also requires knowledge of wind velocities, although these can be estimated from optical spectroscopy for Wolf-Rayet stars. An added complication arises because radiatively-driven winds are known to be inherently unstable, leading to clumped winds. Wolf-Rayet winds have been known to be clumped for 30 years [87,88], but the degree of clumping for O stars via the Hα diagnostic remains unclear, with conflicting results obtained from UV resonance lines [89] unless the wind comprises a mixture of optically thin and thick clumps within a much lower density inter-clump medium [90].

Figure 6 presents unclumped mass-loss rates of O, Of/WN and Wolf-Rayet stars in the Tarantula Nebula, obtained from VFTS [50,51,53], HST/STIS [55] and other literature results. Uncertainties have been included wherever possible. Since the primary wind diagnostic in the majority of instances presented here is Hα, it is apparent that uncertainties are large for those stars with weak stellar winds. In addition, mass-loss rates for Of/WN and Wolf-Rayet stars are anticipated to be reduced significantly owing to wind clumping, as indicated with downward arrows. If volume filling factors are ~10%, mass-loss rates will be reduced by a factor of $\sqrt{10}$.

Theoretical mass-loss rates [83] for zero-age main sequence stars at LMC composition [60,61] are included in Figure 6. At face value it would appear that the theoretical mass-loss rates of LMC O stars are supported by theory. However, the following should be borne in mind. It is not clear how significantly wind clumping affects the inferred mass-loss rates of normal O stars, although Hα results for supergiants are likely to be sensitive to wind clumping. In addition, the vast majority of mass-loss rates of VFTS O stars shown here have been inferred by adopting wind velocities from an assumed scaling relation involving escape velocities [91,92], which are themselves dependent upon spectroscopic gravities. Exceptions are HST/STIS results for early-type stars in R136 which are based on measured UV wind velocities, and span dwarfs, giants, supergiants and main sequence WN stars. In order to verify predictions for lower luminosity ($\log L/L_\odot < 5.5$) O stars, more sensitive diagnostics would need to be employed, such as UV P Cygni lines, providing complications such as porosity are accounted for [90].

It is clear from Figure 6 that rates for the highest luminosity main-sequence Of/WN and WN stars significantly exceed theoretical predictions. This discrepancy is partially addressed through wind clumping, but very massive stars close to their Eddington limits are observed to exhibit enhanced mass-loss rates which are not taken into account in standard theoretical predictions [93,94]. Unsurprisingly, classical Wolf-Rayet stars with $\log(L/L_\odot) = 5.5 - 6$ possess the strongest winds amongst early-type stars in 30 Doradus, with clumping-corrected wind densities an order of magnitude higher than O stars with similar luminosities. It is well known that the wind momenta of WR stars, $\dot{M}v_\infty$, exceeds the momentum provided by their radiation field, L/c, owing to multiple photon absorption and re-emission within their optically thick winds, permitting $\dot{M}v_\infty/(L/c) > 1$ [95].

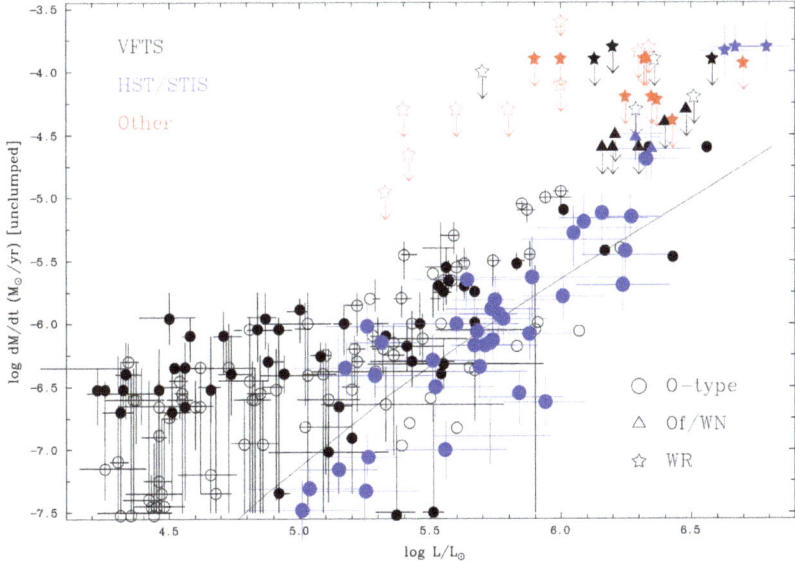

Figure 6. Unclumped mass-loss rates of O-type, Of/WN and Wolf-Rayet stars in the Tarantula Nebula (based on results from VFTS [50,51,53], HST/STIS [55] and other surveys [37,56,59] for WR stars). Filled symbols are within NGC 2070, open symbols elsewhere in the Tarantula. Theoretical mass-loss rates for zero age main sequence massive stars at the Large Magellanic Cloud (LMC) metallicity [83] are included (solid line), based on LMC metallicity evolutionary models [60,61].

7. Fate of Massive Stars in the Tarantula

The conventional picture of massive star evolution close to solar metallicity is that those with initial masses of 8–25 $M_\odot$ will end their lives as red supergiants (RSG), undergo a H-rich core collapse supernova, leaving behind a neutron star remnant, while higher mass counterparts will either circumvent the RSG phase or subsequently proceed to a Wolf-Rayet stage prior to undergoing core collapse, leading to a H-deficient supernova (neutron star remnant) or faint/failed supernova (black hole remnant) [96]. If one considers the global WR vs. RSG population in the LMC, the lower boundary to the luminosity of WR stars is $\log(L/L_\odot)$ = 5.3, while the upper luminosity of RSG is $\log(L/L_\odot)$ = 5.5 [48], supporting a transition from RSG to WR for higher mass progenitors at ~25 $M_\odot$.

Close binary evolution severely complicates this scenario, since primaries below 25 $M_\odot$ can be stripped of their hydrogen envelope, leading to a type IIb or Ib/c instead of a H-rich supernova, while secondaries will be rejuvenated, spun-up, with the potential for a core-collapse supernova for secondaries whose initial masses fall below 8 $M_\odot$. To date, there are no unambiguous cases of pre-supernova close binaries in the Tarantula hosting stripped (Wolf-Rayet or helium) stars, although it has been suggested that the WN3 binary BAT99-49 elsewhere in the LMC is the product of close binary evolution [59]. Rapid rotation of the bright O giant component of the high mass X-ray binary VFTS 399 is consistent with this evolutionary scenario. The absence of low luminosity Wolf-Rayet stars in the Tarantula does not exclude the binary channel since low-mass stripped stars would be unlikely to exhibit a Wolf-Rayet spectral appearance [67].

Initially very close binaries may merge on the main sequence, prior to following a relatively conventional evolution, albeit with unusually high rotation rates, which would lead to increased luminosities and potentially evolve blueward off the main sequence [60,61]. Extremely rapid rotation in some VFTS OB stars favours close binary evolution or stellar mergers. Very massive stars in the Tarantula up to ~300 $M_\odot$ are expected to lead to 30–50 $M_\odot$ CO cores and black hole fates, unlikely to produce any associated supernova [97], such that a subset of binary VMS are plausible progenitors of LIGO black hole binary mergers, although their exact fate crucially depends on their mass-loss properties, which remain uncertain.

8. Integrated Properties and Comparison With Star-Forming Regions, Near And Far

The Tarantula Nebula would subtend little more than one arcsec if it were located at a distance of 50 Mpc, and so provides us with a unique opportunity to compare the individual spatially-resolved properties of an intensively star-forming region and its aggregate characteristics. Using integrated Hα observations of the Tarantula Nebula [98], an age of ∼3.5 Myr would be inferred from a comparison between the inferred Hα equivalent width of 1100 Å and population synthesis models for a coeval population at LMC metallicity [12]. This is in reasonable agreement with the typical age of massive stars, albeit failing to reflect the complexity in its star-formation history (recall Figure 3).

An analysis of the integrated UV spectrum of NGC 2070 supports a young (≤3 Myr) starburst episode [99], while the high spatial resolution of HST/STIS has permitted a comparison between the individual and integrated UV spectroscopic appearance of the central R136 cluster [5]. Very massive stars contributed a significant fraction of its far UV continuum flux, and completely dominate the strong, broad He II λ1640 emission. The integrated UV spectroscopic appearance of R136 closely resembles some star clusters in star-forming galaxies at Mpc distances, such as NGC5253-5 [100], suggesting the presence of VMS in these other young massive clusters. From comparison with the predictions of standard population synthesis models, both Starburst99 [101] and BPASS [102] models fail to predict any significant emission prior to the conventional Wolf-Rayet phase (Figure 7), owing to the use of inadequate wind theory for VMS [93,94]. Consequently, neither Starburst99 nor BPASS accounts for the powerful winds of very massive stars in R136, and the adopted mass function follows a Salpeter slope, rather than the top heavy IMF identified by [40] for the Tarantula region as a whole.

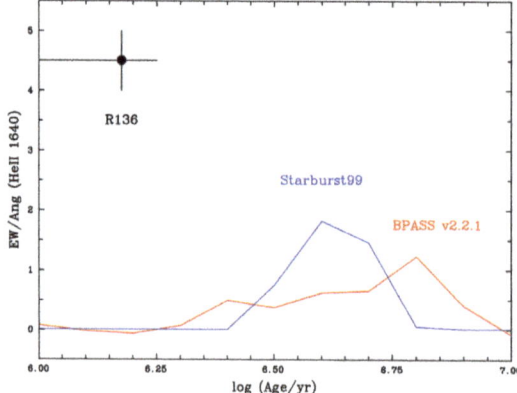

Figure 7. Comparison between observed He II λ1640 emission equivalent widths in R136 [5] versus predicted emission from BPASS (v.2.2.1, red) and Starburst99 (blue) population synthesis models (absorption lines are shown as negative values).

An estimate of the cumulative ionizing and mechanical feedback from massive stars within the Tarantula has revealed a major contribution from VMS towards the collective ionizing output and a dominant role from WR stars to the mechanical feedback [12]. However this analysis relied on calibrations and estimates of spectral types for a significant subset of the massive star content, so we are able to provide updates from recent spectroscopic observations (e.g., VLT/MUSE, HST/STIS) and analyses. We present the updated cumulative ionizing output from 1170 massive stars in the Tarantula Nebula in Figure 8, indicating a total Lyman continuum ionizing output of 1.2×10^{52} ph s^{-1} within 150 pc of R136a. A quarter of the total ionizing radiation originates from the R136a cluster, while members of NGC 2070 produce three quarters of the global feedback. Ten systems alone, listed in Table 4, collectively contribute a quarter of the ionizing budget of the Tarantula Nebula, comprising main sequence and classical WR stars, plus early O supergiants. Indeed, half of the global ionizing output originates from 40 early O stars, main sequence WN stars and classical WR stars, with 1130 stars contributing the remaining 50%. The collective bolometric luminosity of these stars is $10^{8.4}$ $L_\odot$, of which the 40 UV-bright stars contribute $10^{8.0}$ $L_\odot$.

Table 4. Top ten stellar systems contributing to the Lyman continuum output of the Tarantula Nebula, comprising very massive early O stars and WN5 stars, and classical Wolf-Rayet stars, updated from [12].

Star (Alias)	Sp Type	log $L/L_\odot$	log N(LyC)	Reference
R136a1 (BAT99-108)	WN5h	6.8	50.6	[55]
Mk 34 (BAT99-116)	WN5h + WN5h	6.4 + 6.4	50.6	[37]
R136a2 (BAT99-109)	WN5h	6.7	50.5	[55]
R144 (BAT99-118)	WN5-6 + WN6-7	6.7	50.5	[56,64]
R136a3 (BAT99-106)	WN5h	6.6	50.5	[55]
Mk 49 (BAT99-98)	WN6(h)	6.7	50.5	[56]
R145 (BAT99-119)	WN6h + O3.5If/WN7	6.3+6.3	50.4	[59]
Mk 42 (BAT99-105)	O2 If	6.6	50.4	[51]
VFTS 682	WN5h	6.5	50.4	[51]
R136c (BAT99-112)	WN5h +?	6.6	50.4	[51]

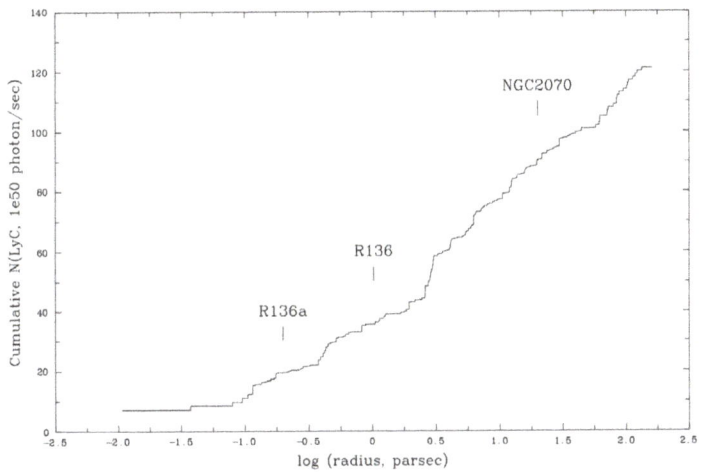

Figure 8. Cumulative ionizing output (10^{50} ph/s) from spectroscopically classified early-type stars in the Tarantula, obtained from VFTS [50,51,53], VLT/MUSE [11], HST/STIS [55] and literature results [56], updated from [12]. Specific regions within 30 Dor are indicated from Table 1.

Recalling Table 2, the highest mass stars and evolved high mass stars in other giant H II regions in the Local Group dominate their radiative and mechanical feedback, including NGC 3372 (Carina Nebula) in the Milky Way, N206 in the LMC and NGC 346 in the SMC. By way of example, Smith [39] established that only a handful of early O-type stars and H-rich WN stars contribute the majority of the Lyman continuum flux of the Carina Nebula, while η Car, four Wolf-Rayet stars and two early O supergiants completely dominate the stellar mechanical luminosity. The central ionizing cluster of the Galactic NGC 3603 star-forming region is host to a stellar content analogous to R136a, including a number of early O stars, nitrogen-sequence Wolf-Rayet stars [103,104]. Weak main-sequence wind properties of metal-poor massive stars conspire to even fewer massive stars

(HD 5980, Sk 80) dominating the cumulative stellar feedback in NGC 346. Similar conclusions were reached by Ramachandran et al. [105] for the supergiant shell in the wing of the SMC.

Although the Tarantula Nebula is the most extreme giant H II region in the Local Group, how does it rank against star-forming regions of galaxies in the near universe or knots at high redshift? Figure 9 compares the star-formation rate versus size of regions spanning $z = 0$ to 3.4, adapted from [106], indicating that the Tarantula (red square) is forming stars more vigorously than typical low-redshift counterparts, resembling some star-forming regions at high redshift. Indeed, 80% of the cumulative ionizing radiation originates from NGC 2070, such that this region corresponds closely with typical clumps in the lensed galaxy SDSS J1110 + 6459 at $z = 2.5$ (green circles).

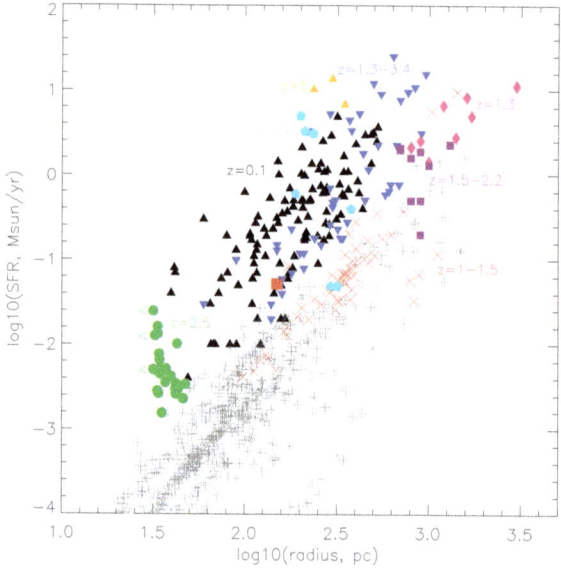

Figure 9. Comparison between the integrated star-formation rate versus size of the Tarantula (filled red square) and star-forming knots from galaxies spanning a range of redshifts, adapted from [106].

In addition to its unusually high star formation rate, the Tarantula also possesses high ionization parameter nebular properties with respect to star-forming galaxies in the local universe. Figure 10 presents a Baldwin, Philipps and Terlevich [107] (BPT) diagnostic diagram of Sloan Digital Sky Survey (SDSS) galaxies, in which the Tarantula (red square) has been indicated, along with Green Pea galaxies from Micheva et al. [108] which are low-metallicity, intensively star-forming galaxies exhibiting unusually strong [O III] $\lambda 5007$ emission. Steidel et al. [109] showed that $z = 2$–3 star forming galaxies share similar extreme nebular properties, and a subset of Green Pea galaxies have been established as Lyman continuum leakers [110,111]. Focusing again on NGC 2070, this sits amongst the extreme Green Pea galaxies in the BPT diagram.

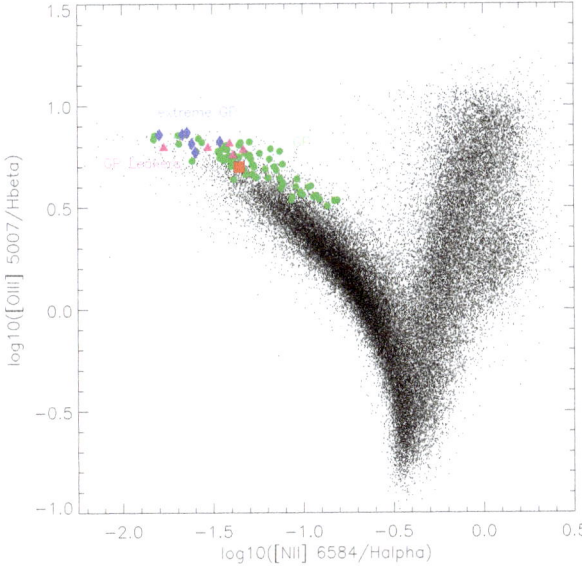

Figure 10. BPT diagram illustrating the similarity in integrated strengths between the Tarantula Nebula (red square), Green Pea (green circles), extreme Green Peas (blue diamonds), Lyman-continuum leaking Green Peak (pink triangles), updated from [108], plus SDSS star-forming galaxies.

9. Summary and Outlook

Recent comprehensive spectroscopic and imaging surveys have revealed that the Tarantula Nebula hosts the most exceptional massive star population within the Local Group of galaxies, including the most massive stars identified to date, the fastest rotating early-type stars, and the X-ray brightest colliding wind system. In particular, the VFTS survey has revealed an excess of massive stars with respect to a Salpeter IMF [40] and added support from previous results for the importance of close binary evolution in the evolution of massive stars [64]. As such, the Tarantula Nebula is the closest analogue to the Hubble Deep Field for the community interested in the evolution of massive stars since its richness provides us with a huge breadth of extreme stars.

The integrated appearance of R136 resembles young extragalactic star clusters, while the integrated nebular properties of NGC 2070 is analogous to extreme Green Pea galaxies at low redshift and star-forming knots in high-redshift galaxies. Typical metallicities of Green Pea galaxies are lower than the LMC, as measured from oxygen nebular lines, while the oxygen content of high-z star forming galaxies tends to be similar to those of the Magellanic Clouds. However, α/Fe abundances of high redshift galaxies are likely to be higher than young populations in the Milky Way, such that winds from OB stars at high redshift are anticipated to be weaker than in the LMC or SMC [112].

The LMC metallicity is only a factor of two below that of the Solar neighbourhood [3], so ideally we would like to supplement the extensive survey of the Tarantula Nebula with counterparts at significantly lower metallicity. The SMC (1/5 solar) represents our best opportunity to study the formation and evolution of massive stars at a metallicity significantly below that of the LMC. Alas, it does not host as rich a massive star-forming region as the Tarantula, but cumulatively does host a substantial number of O stars so is key towards our improved understanding of massive stars at low metallicity, especially as it can be studied in exquisite detail with current instrumentation.

Since we need to look beyond the SMC to study metal-poor counterparts to the Tarantula Nebula, Figure 11 compares the star-formation rate, metallicity, and distance modulus of Local Group dwarf galaxies. Rates of star formation in metal-poor (≤20% of solar oxygen content) galaxies are significantly lower than the Magellanic Clouds, so there are no rich metal-poor massive star populations elsewhere in the Local Group, and those few that are present are much more distant. The absence of nearby metal-poor counterparts to the Tarantula implies that we currently have to rely on the interpretation of integrated populations in order to understand massive stellar evolution at low metallicity, notably extremely metal-poor dwarf star-forming galaxies I Zw 18 and SBS-0335.

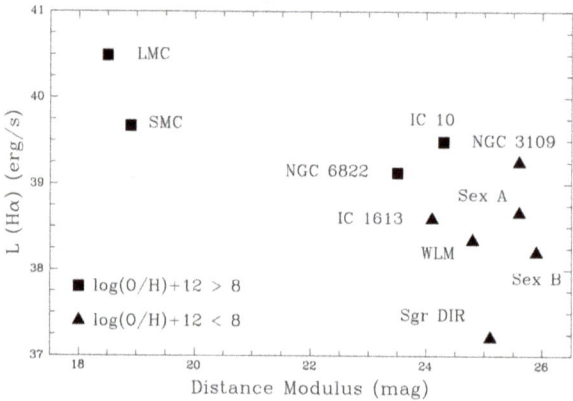

Figure 11. Comparison between present-day star formation rates, as measured by Hα luminosity [113], distance modulus (mag), and oxygen metal content (squares: ≥20% of solar value, triangles: <20% of solar value, for Local Group dwarf galaxies. Metal-poor galaxies possess low star-formation rates, so host small numbers of OB stars, and these are ≥6 magnitudes fainter than Magellanic Cloud counterparts.

In order to test predictions of the metallicity dependence of massive star winds, it is necessary to measure mass-loss properties across a wide range of metallicities. Although theory has been qualitatively supported from the observed wind properties of Milky Way, LMC and SMC early-type stars [82], some issues remain, including weak winds in low luminosity OB stars. Our only opportunity to study individual massive stars below 1/10th of the solar oxygen content is to observe O stars in Sextans A and B with 7% solar [114] or the Sagittarius Dwarf Irregular Galaxy (SgrDIR) with 5% solar [115]. Stellar winds from early-type stars at such low metallicities are anticipated to be much weaker than in metal-rich populations, which has been confirmed by UV spectroscopy. By way of example, Figure 12 compares the far UV spectrum of ζ Per (O7.5 III) with a counterpart in Sextans A [116], revealing negligible wind signatures in the latter (e.g., C IV λ1550).

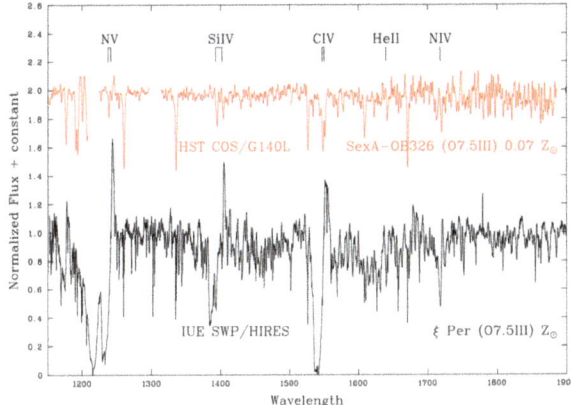

Figure 12. Comparison between far UV spectroscopy of mid O giants in metal-rich [117] and metal-deficient [116] environments, illustrating the extreme differences in wind features (e.g., N v λ1240, Si iv λ1400, C iv λ1550) and the iron forest (Fe iv-v).

Funding: This research received no external funding.

Acknowledgments: This review is dedicated to the memory of Nolan Walborn, from whom the author learnt a great deal about the Tarantula Nebula. Thanks to Leisa Townsley and Patrick Broos for access to T-ReX point source results prior to publication, Miriam Garcia for the UV spectrum of Sextans A OB 326, Selma De Mink for the VFTS pie chart, Heloise Stevance for the BPASS predictions, and Joachim Bestenlehner for converting Q wind density results into mass-loss rates. Feedback from Fabian Schneider, Chris Evans, Joachim Bestenlehner, Andy Pollock, Roberta Humphreys and external referees on an earlier draft is greatly appreciated.

Conflicts of Interest: The author declares no conflict of interest.

References

1. Kennicutt, R.C., Jr. Star Formation in Galaxies Along the Hubble Sequence. *Ann. Rev. Astron. Astrophys.* **1998**, *36*, 189–232. [CrossRef]
2. Evans, C.J.; Taylor, W.D.; Hénault-Brunet, V.; Sana, H.; de Koter, A.; Simón-Díaz, S.; Carraro, G.; Bagnoli, T.; Bastian, N.; Bestenlehner, J.M.; et al. The VLT-FLAMES Tarantula Survey. I. Introduction and observational overview. *Astron. Astrophys.* **2011**, *530*, A108. [CrossRef]
3. Tsamis, Y.G.; Péquignot, D. A photoionization-modelling study of 30 Doradus: The case for small-scale chemical inhomogeneity. *Mon. Not. R. Astron. Soc.* **2005**, *364*, 687–704. [CrossRef]
4. Baldwin, J.A.; Ferland, G.J.; Martin, P.G.; Corbin, M.R.; Cota, S.A.; Peterson, B.M.; Slettebak, A. Physical conditions in the Orion Nebula and an assessment of its helium abundance. *Astrophys. J.* **1991**, *374*, 580–609. [CrossRef]
5. Crowther, P.A.; Caballero-Nieves, S.M.; Bostroem, K.A.; Maíz Apellániz, J.; Schneider, F.R.N.; Walborn, N.R.; Angus, C.R.; Brott, I.; Bonanos, A.; de Koter, A.; et al. The R136 star cluster dissected with Hubble Space Telescope/STIS. I. Far-ultraviolet spectroscopic census and the origin of He II λ1640 in young star clusters. *Mon. Not. R. Astron. Soc.* **2016**, *458*, 624–659. [CrossRef]
6. Britavskiy, N.; Lennon, D.J.; Patrick, L.R.; Evans, C.J.; Herrero, A.; Langer, N.; van Loon, J.T.; Clark, J.S.; Schneider, F.R.N.; Almeida, L.A.; et al. The VLT-FLAMES Tarantula Survey. XXX. Red stragglers in the clusters Hodge 301 and SL 639. *Astron. Astrophys.* **2019**, *624*, A128. [CrossRef]
7. Walborn, N.R. The Starburst Region 30 Doradus. In *The Magellanic Clouds: Proceedings of the 148th Symposium of the International Astronomical UnionThe Magellanic Clouds*; Haynes, R., Milne, D., Eds.; Kluwer Academic Publishers: Dordrecht, The Netherland, 1991; Volume 148, p. 145.
8. Sabbi, E.; Anderson, J.; Lennon, D.J.; van der Marel, R.P.; Aloisi, A.; Boyer, M.L.; Cignoni, M.; de Marchi, G.; de Mink, S.E.; Evans, C.J.; et al. Hubble Tarantula Treasury Project: Unraveling Tarantula's Web. I. Observational Overview and First Results. *Astron. J.* **2013**, *146*, 53. [CrossRef]

9. Crowther, P.A.; Schnurr, O.; Hirschi, R.; Yusof, N.; Parker, R.J.; Goodwin, S.P.; Kassim, H.A. The R136 star cluster hosts several stars whose individual masses greatly exceed the accepted 150M$_{solar}$ stellar mass limit. *Mon. Not. R. Astron. Soc.* **2010**, *408*, 731–751. [CrossRef]
10. Castro, N.; Crowther, P.A.; Evans, C.J.; Mackey, J.; Castro-Rodriguez, N.; Vink, J.S.; Melnick, J.; Selman, F. Mapping the core of the Tarantula Nebula with VLT-MUSE. I. Spectral and nebular content around R136. *Astron. Astrophys.* **2018**, *614*, A147. [CrossRef]
11. Castro, N.; Crowther, P.A.; Evans, C.J.; Mackey, J.; Castro-Rodriguez, N.; Vink, J.S.; Melnick, J.; Selman, F. Mapping the core of the Tarantula Nebula with VLT-MUSE. III. The spectroscopic Hertzsprung-Russell diagram of OB stars in NGC2070. *Astron. Astrophys.* in preparation.
12. Doran, E.I.; Crowther, P.A.; de Koter, A.; Evans, C.J.; McEvoy, C.; Walborn, N.R.; Bastian, N.; Bestenlehner, J.M.; Gräfener, G.; Herrero, A.; et al. The VLT-FLAMES Tarantula Survey. XI. A census of the hot luminous stars and their feedback in 30 Doradus. *Astron. Astrophys.* **2013**, *558*, A134. [CrossRef]
13. De Marchi, G.; Paresce, F.; Panagia, N.; Beccari, G.; Spezzi, L.; Sirianni, M.; Andersen, M.; Mutchler, M.; Balick, B.; Dopita, M.A.; et al. Star Formation in 30 Doradus. *Astrophys. J.* **2011**, *739*, 27. [CrossRef]
14. Sabbi, E.; Lennon, D.J.; Anderson, J.; Cignoni, M.; van der Marel, R.P.; Zaritsky, D.; De Marchi, G.; Panagia, N.; Gouliermis, D.A.; Grebel, E.K.; et al. Hubble Tarantula Treasury Project. III. Photometric Catalog and Resulting Constraints on the Progression of Star Formation in the 30 Doradus Region. *Astrophys. J.* **2016**, *222*, 11. [CrossRef]
15. Schneider, F.R.N.; Ramírez-Agudelo, O.H.; Tramper, F.; Bestenlehner, J.M.; Castro, N.; Sana, H.; Evans, C.J.; Sabín-Sanjulián, C.; Simón-Díaz, S.; Langer, N.; et al. The VLT-FLAMES Tarantula Survey. XXIX. Massive star formation in the local 30 Doradus starburst. *Astron. Astrophys.* **2018**, *618*, A73. [CrossRef]
16. Wright, N.J.; Bouy, H.; Drew, J.E.; Sarro, L.M.; Bertin, E.; Cuillandre, J.C.; Barrado, D. Cygnus OB2 DANCe: A high-precision proper motion study of the Cygnus OB2 association. *Mon. Not. R. Astron. Soc.* **2016**, *460*, 2593–2610. [CrossRef]
17. Walborn, N.R. The Pillars of the Second Generation. In *Hot Star Workshop III: The Earliest Phases of Massive Star Birth*; Astronomical Society of the Pacific Conference Series; Crowther, P.A., Ed.; Astronomical Society of the Pacific: San Francisco, CA, USA, 2002; Volume 267, p. 111.
18. Walborn, N.R.; Barbá, R.H.; Sewiło, M.M. The Top 10 Spitzer Young Stellar Objects in 30 Doradus. *Astron. J.* **2013**, *145*, 98. [CrossRef]
19. Indebetouw, R.; Brogan, C.; Chen, C.H.R.; Leroy, A.; Johnson, K.; Muller, E.; Madden, S.; Cormier, D.; Galliano, F.; Hughes, A.; et al. ALMA Resolves 30 Doradus: Sub-parsec Molecular Cloud Structure near the Closest Super Star Cluster. *Astrophys. J.* **2013**, *774*, 73. [CrossRef]
20. Kennicutt, R.C., Jr. Structural properties of giant H II regions in nearby galaxies. *Astrophys. J.* **1984**, *287*, 116–130. [CrossRef]
21. Parker, J.W. The OB associations of 30 Doradus in the Large Magellanic Cloud. I—Stellar observations and data reductions. *Astron. J.* **1993**, *106*, 560–577. [CrossRef]
22. Parker, J.W.; Garmany, C.D. The OB associations of 30 Doradus in the Large Magellanic Cloud. II—Stellar content and initial mass function. *Astron. J.* **1993**, *106*, 1471–1483. [CrossRef]
23. Massey, P.; Hunter, D.A. Star Formation in R136: A Cluster of O3 Stars Revealed by Hubble Space Telescope Spectroscopy. *Astrophys. J.* **1998**, *493*, 180–194. [CrossRef]
24. Bosch, G.; Terlevich, R.; Melnick, J.; Selman, F. The ionising cluster of 30 Doradus. II. Spectral classification for 175 stars. *Astron. Astrophys.* **1999**, *137*, 21–41. [CrossRef]
25. Selman, F.; Melnick, J.; Bosch, G.; Terlevich, R. The ionizing cluster of 30 Doradus. III. Star-formation history and initial mass function. *Astron. Astrophys.* **1999**, *347*, 532–549.
26. Breysacher, J.; Azzopardi, M.; Testor, G. The fourth catalogue of Population I Wolf-Rayet stars in the Large Magellanic Cloud *Astron. Astrophys.* **1999** *137*, 117–145.
27. Walborn, N.R.; Gamen, R.C.; Morrell, N.I.; Barbá, R.H.; Fernández Lajús, E.; Angeloni, R. Active Luminous Blue Variables in the Large Magellanic Cloud. *Astron. J.* **2017**, *154*, 15. [CrossRef]
28. Lennon, D.J.; Evans, C.J.; van der Marel, R.P.; Anderson, J.; Platais, I.; Herrero, A.; de Mink, S.E.; Sana, H.; Sabbi, E.; Bedin, L.R.; et al. Gaia DR2 reveals a very massive runaway star ejected from R136. *Astron. Astrophys.* **2018**, *619*, A78. [CrossRef]

29. Dufton, P.L.; Dunstall, P.R.; Evans, C.J.; Brott, I.; Cantiello, M.; de Koter, A.; de Mink, S.E.; Fraser, M.; Hénault-Brunet, V.; Howarth, I.D.; et al. The VLT-FLAMES Tarantula Survey: The Fastest Rotating O-type Star and Shortest Period LMC Pulsar: Remnants of a Supernova Disrupted Binary? *Astrophys. J. Lett.* **2011**, *743*, L22. [CrossRef]
30. Almeida, L.A.; Sana, H.; de Mink, S.E.; Tramper, F.; Soszyński, I.; Langer, N.; Barbá, R.H.; Cantiello, M.; Damineli, A.; de Koter, A.; et al. Discovery of the Massive Overcontact Binary VFTS352: Evidence for Enhanced Internal Mixing. *Astrophys. J.* **2015**, *812*, 102. [CrossRef]
31. Chen, Y.; Wang, Q.D.; Gotthelf, E.V.; Jiang, B.; Chu, Y.H.; Gruendl, R. Chandra ACIS Spectroscopy of N157B: A Young Composite Supernova Remnant in a Superbubble. *Astrophys. J.* **2006**, *651*, 237–249. [CrossRef]
32. Savage, B.D.; Fitzpatrick, E.L.; Cassinelli, J.P.; Ebbets, D.C. The nature of R136a, the superluminous central object of the 30 Doradus nebula. *Astrophys. J.* **1983**, *273*, 597–623. [CrossRef]
33. Weigelt, G.; Baier, G. R136a in the 30 Doradus nebula resolved by holographic speckle interferometry. *Astron. Astrophys.* **1985**, *150*, L18–L20.
34. Hunter, D.A.; Shaya, E.J.; Holtzman, J.A.; Light, R.M.; O'Neil, E.J., Jr.; Lynds, R. The Intermediate Stellar Mass Population in R136 Determined from Hubble Space Telescope Planetary Camera 2 Images. *Astrophys. J.* **1995**, *448*, 179. [CrossRef]
35. de Koter, A.; Heap, S.R.; Hubeny, I. On the Evolutionary Phase and Mass Loss of the Wolf-Rayet–like Stars in R136a. *Astrophys. J.* **1997**, *477*, 792. [CrossRef]
36. Khorrami, Z.; Vakili, F.; Lanz, T.; Langlois, M.; Lagadec, E.; Meyer, M.R.; Robbe-Dubois, S.; Abe, L.; Avenhaus, H.; Beuzit, J.L.; et al. Uncrowding R 136 from VLT/SPHERE extreme adaptive optics. *Astron. Astrophys.* **2017**, *602*, A56. [CrossRef]
37. Tehrani, K.A.; Crowther, P.A.; Bestenlehner, J.M.; Littlefair, S.P.; Pollock, A.M.T.; Parker, R.J.; Schnurr, O. Weighing Melnick 34: The most massive binary system known. *Mon. Not. R. Astron. Soc.* **2019**, *484*, 2692–2710. [CrossRef]
38. Pollock, A.M.T.; Crowther, P.A.; Tehrani, K.; Broos, P.S.; Townsley, L.K. The 155-day X-ray cycle of the very massive Wolf-Rayet star Melnick 34 in the Large Magellanic Cloud. *Mon. Not. R. Astron. Soc.* **2018**, *474*, 3228–3236. [CrossRef]
39. Smith, N. A census of the Carina Nebula - I. Cumulative energy input from massive stars. *Mon. Not. R. Astron. Soc.* **2006**, *367*, 763–772. [CrossRef]
40. Schneider, F.R.N.; Sana, H.; Evans, C.J.; Bestenlehner, J.M.; Castro, N.; Fossati, L.; Gräfener, G.; Langer, N.; Ramírez-Agudelo, O.H.; Sabín-Sanjulián, C.; et al. An excess of massive stars in the local 30 Doradus starburst. *Science* **2018**, *359*, 69–71. [CrossRef]
41. Puls, J.; Urbaneja, M.A.; Venero, R.; Repolust, T.; Springmann, U.; Jokuthy, A.; Mokiem, M.R. Atmospheric NLTE-models for the spectroscopic analysis of blue stars with winds. II. Line-blanketed models. *Astron. Astrophys.* **2005**, *435*, 669–698. [CrossRef]
42. Hillier, D.J.; Miller, D.L. The Treatment of Non-LTE Line Blanketing in Spherically Expanding Outflows. *Astrophys. J.* **1998**, *496*, 407–427. [CrossRef]
43. Gräfener, G.; Koesterke, L.; Hamann, W.R. Line-blanketed model atmospheres for WR stars. *Astron. Astrophys.* **2002**, *387*, 244–257. [CrossRef]
44. Lanz, T.; Hubeny, I. A Grid of NLTE Line-blanketed Model Atmospheres of Early B-Type Stars. *Astrophys. J. Suppl.* **2007**, *169*, 83–104. [CrossRef]
45. Gustafsson, B.; Edvardsson, B.; Eriksson, K.; Jørgensen, U.G.; Nordlund, Å.; Plez, B. A grid of MARCS model atmospheres for late-type stars. I. Methods and general properties. *Astron. Astrophys.* **2008**, *486*, 951–970. [CrossRef]
46. Langer, N.; Kudritzki, R.P. The spectroscopic Hertzsprung-Russell diagram. *Astron. Astrophys.* **2014**, *564*, A52. [CrossRef]
47. Gordon, M.S.; Humphreys, R.M.; Jones, T.J. Luminous and Variable Stars in M31 and M33. III. The Yellow and Red Supergiants and Post-red Supergiant Evolution. *Astrophys. J.* **2016**, *825*, 50. [CrossRef]
48. Davies, B.; Crowther, P.A.; Beasor, E.R. The luminosities of cool supergiants in the Magellanic Clouds, and the Humphreys-Davidson limit revisited. *Mon. Not. R. Astron. Soc.* **2018**, *478*, 3138–3148. [CrossRef]
49. Schneider, F.R.N.; Langer, N.; de Koter, A.; Brott, I.; Izzard, R.G.; Lau, H.H.B. Bonnsai: A Bayesian tool for comparing stars with stellar evolution models. *Astron. Astrophys.* **2014**, *570*, A66. [CrossRef]

50. Sabín-Sanjulián, C.; Simón-Díaz, S.; Herrero, A.; Puls, J.; Schneider, F.R.N.; Evans, C.J.; Garcia, M.; Najarro, F.; Brott, I.; Castro, N.; et al. The VLT-FLAMES Tarantula Survey. XXVI. Properties of the O-dwarf population in 30 Doradus. *Astron. Astrophys.* **2017**, *601*, A79. [CrossRef]
51. Bestenlehner, J.M.; Gräfener, G.; Vink, J.S.; Najarro, F.; de Koter, A.; Sana, H.; Evans, C.J.; Crowther, P.A.; Hénault-Brunet, V.; Herrero, A.; et al. The VLT-FLAMES Tarantula Survey. XVII. Physical and wind properties of massive stars at the top of the main sequence. *Astron. Astrophys.* **2014**, *570*, A38. [CrossRef]
52. McEvoy, C.M.; Dufton, P.L.; Evans, C.J.; Kalari, V.M.; Markova, N.; Simón-Díaz, S.; Vink, J.S.; Walborn, N.R.; Crowther, P.A.; de Koter, A.; et al. The VLT-FLAMES Tarantula Survey. XIX. B-type supergiants: Atmospheric parameters and nitrogen abundances to investigate the role of binarity and the width of the main sequence. *Astron. Astrophys.* **2015**, *575*, A70. [CrossRef]
53. Ramírez-Agudelo, O.H.; Sana, H.; de Koter, A.; Tramper, F.; Grin, N.J.; Schneider, F.R.N.; Langer, N.; Puls, J.; Markova, N.; Bestenlehner, J.M.; et al. The VLT-FLAMES Tarantula Survey. XXIV. Stellar properties of the O-type giants and supergiants in 30 Doradus. *Astron. Astrophys.* **2017**, *600*, A81. [CrossRef]
54. Dufton, P.L.; Thompson, A.; Crowther, P.A.; Evans, C.J.; Schneider, F.R.N.; de Koter, A.; de Mink, S.E.; Garland, R.; Langer, N.; Lennon, D.J.; et al. The VLT-FLAMES Tarantula Survey. XXVIII. Nitrogen abundances for apparently single dwarf and giant B-type stars with small projected rotational velocities. *Astron. Astrophys.* **2018**, *615*, A101. [CrossRef]
55. Bestenlehner, J.M.; Crowther, P.; Caballero-Nieves, S.M.; Simón-Díaz, S.; Schneider, F. The R136 star cluster dissected with Hubble Space Telescope.III Physical properties of the most massive stars in R136. *Mon. Not. R. Astron. Soc.* in preparation.
56. Hainich, R.; Rühling, U.; Todt, H.; Oskinova, L.M.; Liermann, A.; Gräfener, G.; Foellmi, C.; Schnurr, O.; Hamann, W.R. The Wolf-Rayet stars in the Large Magellanic Cloud. A comprehensive analysis of the WN class. *Astron. Astrophys.* **2014**, *565*, A27. [CrossRef]
57. Garland, R.; Dufton, P.L.; Evans, C.J.; Crowther, P.A.; Howarth, I.D.; de Koter, A.; de Mink, S.E.; Grin, N.J.; Langer, N.; Lennon, D.J.; et al. The VLT-FLAMES Tarantula Survey. XXVII. Physical parameters of B-type main-sequence binary systems in the Tarantula nebula. *Astron. Astrophys.* **2017**, *603*, A91. [CrossRef]
58. Mahy, L.; Sana, H.; Abdul-Masih, M.; Almeida, L.A.; Langer, N.; Shenar, T.; de Koter, A.; de Mink, S.; de Wit, S.; Grin, N.; et al. The Tarantula Massive Binary Monitoring: III. Atmosphere analysis of double-lined spectroscopic systems. *Astron. Astrophys.* submitted.
59. Shenar, T.; Sablowski, D.P.; Hainich, R.; Todt, H.; Moffat, A.F.J.; Oskinova, L.M.; Ramachandran, V.; Sana, H.; Sander, A.A.C.; Schnurr, O.; et al. The Wolf-Rayet binaries of the nitrogen sequence in the Large Magellanic Cloud. Spectroscopy, orbital analysis, formation, and evolution. *Astron. Astrophys.* **2019**, *627*, A151. [CrossRef]
60. Brott, I.; de Mink, S.E.; Cantiello, M.; Langer, N.; de Koter, A.; Evans, C.J.; Hunter, I.; Trundle, C.; Vink, J.S. Rotating massive main-sequence stars. I. Grids of evolutionary models and isochrones. *Astron. Astrophys.* **2011**, *530*, A115. [CrossRef]
61. Köhler, K.; Langer, N.; de Koter, A.; de Mink, S.E.; Crowther, P.A.; Evans, C.J.; Gräfener, G.; Sana, H.; Sanyal, D.; Schneider, F.R.N.; et al. The evolution of rotating very massive stars with LMC composition. *Astron. Astrophys.* **2015**, *573*, A71. [CrossRef]
62. Vanbeveren, D.; De Loore, C.; Van Rensbergen, W. Massive stars. *Astron. Astrophys. Rev.* **1998**, *9*, 63–152. [CrossRef]
63. Sana, H.; de Mink, S.E.; de Koter, A.; Langer, N.; Evans, C.J.; Gieles, M.; Gosset, E.; Izzard, R.G.; Le Bouquin, J.B.; Schneider, F.R.N. Binary Interaction Dominates the Evolution of Massive Stars. *Science* **2012**, *337*, 444. [CrossRef] [PubMed]
64. Sana, H.; de Koter, A.; de Mink, S.E.; Dunstall, P.R.; Evans, C.J.; Hénault-Brunet, V.; Maíz Apellániz, J.; Ramírez-Agudelo, O.H.; Taylor, W.D.; Walborn, N.R.; et al. The VLT-FLAMES Tarantula Survey. VIII. Multiplicity properties of the O-type star population. *Astron. Astrophys.* **2013**, *550*, A107. [CrossRef]
65. Dunstall, P.R.; Dufton, P.L.; Sana, H.; Evans, C.J.; Howarth, I.D.; Simón-Díaz, S.; de Mink, S.E.; Langer, N.; Maíz Apellániz, J.; Taylor, W.D. The VLT-FLAMES Tarantula Survey. XXII. Multiplicity properties of the B-type stars. *Astron. Astrophys.* **2015**, *580*, A93. [CrossRef]
66. Almeida, L.A.; Sana, H.; Taylor, W.; Barbá, R.; Bonanos, A.Z.; Crowther, P.; Damineli, A.; de Koter, A.; de Mink, S.E.; Evans, C.J.; et al. The Tarantula Massive Binary Monitoring. I. Observational campaign and OB-type spectroscopic binaries. *Astron. Astrophys.* **2017**, *598*, A84. [CrossRef]

67. Götberg, Y.; de Mink, S.E.; Groh, J.H.; Kupfer, T.; Crowther, P.A.; Zapartas, E.; Renzo, M. Spectral models for binary products: Unifying subdwarfs and Wolf-Rayet stars as a sequence of stripped-envelope stars. *Astron. Astrophys.* **2018**, *615*, A78. [CrossRef]
68. Simón-Díaz, S.; Herrero, A. The IACOB project. I. Rotational velocities in northern Galactic O- and early B-type stars revisited. The impact of other sources of line-broadening. *Astron. Astrophys.* **2014**, *562*, A135. [CrossRef]
69. Ramírez-Agudelo, O.H.; Simón-Díaz, S.; Sana, H.; de Koter, A.; Sabín-Sanjulían, C.; de Mink, S.E.; Dufton, P.L.; Gräfener, G.; Evans, C.J.; Herrero, A.; et al. The VLT-FLAMES Tarantula Survey. XII. Rotational velocities of the single O-type stars. *Astron. Astrophys.* **2013**, *560*, A29. [CrossRef]
70. Ramírez-Agudelo, O.H.; Sana, H.; de Mink, S.E.; Hénault-Brunet, V.; de Koter, A.; Langer, N.; Tramper, F.; Gräfener, G.; Evans, C.J.; Vink, J.S.; et al. The VLT-FLAMES Tarantula Survey. XXI. Stellar spin rates of O-type spectroscopic binaries. *Astron. Astrophys.* **2015**, *580*, A92. [CrossRef]
71. Dufton, P.L.; Langer, N.; Dunstall, P.R.; Evans, C.J.; Brott, I.; de Mink, S.E.; Howarth, I.D.; Kennedy, M.; McEvoy, C.; Potter, A.T.; et al. The VLT-FLAMES Tarantula Survey. X. Evidence for a bimodal distribution of rotational velocities for the single early B-type stars. *Astron. Astrophys.* **2013**, *550*, A109. [CrossRef]
72. Wolff, S.C.; Strom, S.E.; Cunha, K.; Daflon, S.; Olsen, K.; Dror, D. Rotational Velocities for Early-Type Stars in the Young Large Magellanic Cloud Cluster R136: Further Study of the Relationship Between Rotation Speed and Density in Star-Forming Regions. *Astron. J.* **2008**, *136*, 1049–1060. [CrossRef]
73. Platais, I.; Lennon, D.J.; van der Marel, R.P.; Bellini, A.; Sabbi, E.; Watkins, L.L.; Sohn, S.T.; Walborn, N.R.; Bedin, L.R.; Evans, C.J.; et al. HST Astrometry in the 30 Doradus Region. II. Runaway Stars from New Proper Motions in the Large Magellanic Cloud. *Astron. J.* **2018**, *156*, 98. [CrossRef]
74. Evans, C.J.; Walborn, N.R.; Crowther, P.A.; Hénault-Brunet, V.; Massa, D.; Taylor, W.D.; Howarth, I.D.; Sana, H.; Lennon, D.J.; van Loon, J.T. A Massive Runaway Star from 30 Doradus. *Astrophys. J. Lett.* **2010**, *715*, L74–L79. [CrossRef]
75. Renzo, M.; de Mink, S.E.; Lennon, D.J.; Platais, I.; van der Marel, R.P.; Laplace, E.; Bestenlehner, J.M.; Evans, C.J.; Hénault-Brunet, V.; Justham, S.; et al. Space astrometry of the very massive 150 $M_\odot$ candidate runaway star VFTS682. *Mon. Not. R. Astron. Soc.* **2019**, *482*, L102–L106. [CrossRef]
76. Pallavicini, R.; Golub, L.; Rosner, R.; Vaiana, G.S.; Ayres, T.; Linsky, J.L. Relations among stellar X-ray emission observed from Einstein, stellar rotation and bolometric luminosity. *Astrophys. J.* **1981**, *248*, 279–290. [CrossRef]
77. Stevens, I.R.; Blondin, J.M.; Pollock, A.M.T. Colliding winds from early-type stars in binary systems. *Astrophys. J.* **1992**, *386*, 265–287. [CrossRef]
78. Massey, P.; Penny, L.R.; Vukovich, J. Orbits of Four Very Massive Binaries in the R136 Cluster. *Astrophys. J.* **2002**, *565*, 982–993. [CrossRef]
79. Taylor, W.D.; Evans, C.J.; Sana, H.; Walborn, N.R.; de Mink, S.E.; Stroud, V.E.; Alvarez-Candal, A.; Barbá, R.H.; Bestenlehner, J.M.; Bonanos, A.Z.; et al. The VLT-FLAMES Tarantula Survey. II. R139 revealed as a massive binary system. *Astron. Astrophys.* **2011**, *530*, L10. [CrossRef]
80. Clark, J.S.; Bartlett, E.S.; Broos, P.S.; Townsley, L.K.; Taylor, W.D.; Walborn, N.R.; Bird, A.J.; Sana, H.; de Mink, S.E.; Dufton, P.L.; et al. The VLT-FLAMES Tarantula survey. XX. The nature of the X-ray bright emission-line star VFTS 399. *Astron. Astrophys.* **2015**, *579*, A131. [CrossRef]
81. Castor, J.I.; Abbott, D.C.; Klein, R.I. Radiation-driven winds in Of stars. *Astrophys. J.* **1975**, *195*, 157–174. [CrossRef]
82. Mokiem, M.R.; de Koter, A.; Vink, J.S.; Puls, J.; Evans, C.J.; Smartt, S.J.; Crowther, P.A.; Herrero, A.; Langer, N.; Lennon, D.J.; et al. The empirical metallicity dependence of the mass-loss rate of O- and early B-type stars. *Astron. Astrophys.* **2007**, *473*, 603–614. [CrossRef]
83. Vink, J.S.; de Koter, A.; Lamers, H.J.G.L.M. Mass-loss predictions for O and B stars as a function of metallicity. *Astron. Astrophys.* **2001**, *369*, 574–588. [CrossRef]
84. Prinja, R.K.; Barlow, M.J.; Howarth, I.D. Terminal velocities for a large sample of O stars, B supergiants, and Wolf-Rayet stars. *Astrophys. J.* **1990**, *361*, 607–620. [CrossRef]
85. Roman-Duval, J.; Jenkins, E.B.; Williams, B.; Tchernyshyov, K.; Gordon, K.; Meixner, M.; Hagen, L.; Peek, J.; Sandstrom, K.; Werk, J.; et al. METAL: The Metal Evolution, Transport, and Abundance in the Large Magellanic Cloud Hubble Program. I. Overview and Initial Results. *Astrophys. J.* **2019**, *871*, 151. [CrossRef]

86. Schmutz, W.; Hamann, W.R.; Wessolowski, U. Spectral analysis of 30 Wolf-Rayet stars. *Astron. Astrophys.* **1989**, *210*, 236–248.
87. St.-Louis, N.; Moffat, A.F.J.; Drissen, L.; Bastien, P.; Robert, C. Polarization variability among Wolf-Rayet stars. III - A new way to derive mass-loss rates for Wolf-Rayet stars in binary systems. *Astrophys. J.* **1988**, *330*, 286–304. [CrossRef]
88. Hillier, D.J. The effects of electron scattering and wind clumping for early emission line stars. *Astron. Astrophys.* **1991**, *247*, 455–468.
89. Fullerton, A.W.; Massa, D.L.; Prinja, R.K. The Discordance of Mass-Loss Estimates for Galactic O-Type Stars. *Astrophys. J.* **2006**, *637*, 1025–1039. [CrossRef]
90. Sundqvist, J.O.; Puls, J. Atmospheric NLTE models for the spectroscopic analysis of blue stars with winds. IV. Porosity in physical and velocity space. *Astron. Astrophys.* **2018**, *619*, A59. [CrossRef]
91. Lamers, H.J.G.L.M.; Snow, T.P.; Lindholm, D.M. Terminal Velocities and the Bistability of Stellar Winds. *Astrophys. J.* **1995**, *455*, 269. [CrossRef]
92. Kudritzki, R.P.; Puls, J. Winds from Hot Stars. *Ann. Rev. Astron. Astrophys.* **2000**, *38*, 613–666. [CrossRef]
93. Gräfener, G.; Vink, J.S.; de Koter, A.; Langer, N. The Eddington factor as the key to understand the winds of the most massive stars. Evidence for a Γ-dependence of Wolf-Rayet type mass loss. *Astron. Astrophys.* **2011**, *535*, A56. [CrossRef]
94. Bestenlehner, J.M. Mass loss and the Eddington factor: An updated stellar wind theory for hot massive stars. *Mon. Not. R. Astron. Soc.* in preparation.
95. Puls, J.; Vink, J.S.; Najarro, F. Mass loss from hot massive stars. *Astron. Astrophys. Rev.* **2008**, *16*, 209–325. [CrossRef]
96. Langer, N. Presupernova Evolution of Massive Single and Binary Stars. *Ann. Rev. Astron. Astrophys.* **2012**, *50*, 107–164. [CrossRef]
97. Yusof, N.; Hirschi, R.; Meynet, G.; Crowther, P.A.; Ekström, S.; Frischknecht, U.; Georgy, C.; Abu Kassim, H.; Schnurr, O. Evolution and fate of very massive stars. *Mon. Not. R. Astron. Soc.* **2013**, *433*, 1114–1132. [CrossRef]
98. Pellegrini, E.W.; Baldwin, J.A.; Ferland, G.J. Structure and Feedback in 30 Doradus. I. Observations. *Astrophys. J. Suppl.* **2010**, *191*, 160–178. [CrossRef]
99. Vacca, W.D.; Robert, C.; Leitherer, C.; Conti, P.S. The stellar content of 30 doradus derived from spatially integrated ultraviolet spectra: A test of spectral synthesis models. *Astrophys. J.* **1995**, *444*, 647–662. [CrossRef]
100. Smith, L.J.; Crowther, P.A.; Calzetti, D.; Sidoli, F. The Very Massive Star Content of the Nuclear Star Clusters in NGC 5253. *Astrophys. J.* **2016**, *823*, 38. [CrossRef]
101. Levesque, E.M.; Leitherer, C.; Ekstrom, S.; Meynet, G.; Schaerer, D. The Effects of Stellar Rotation. I. Impact on the Ionizing Spectra and Integrated Properties of Stellar Populations. *Astrophys. J.* **2012**, *751*, 67. [CrossRef]
102. Eldridge, J.J.; Stanway, E.R.; Xiao, L.; McClelland, L.A.S.; Taylor, G.; Ng, M.; Greis, S.M.L.; Bray, J.C. Binary Population and Spectral Synthesis Version 2.1: Construction, Observational Verification, and New Results. *Pub. Astron. Soc. Austr.* **2017**, *34*, e058. [CrossRef]
103. Moffat, A.F.J.; Corcoran, M.F.; Stevens, I.R.; Skalkowski, G.; Marchenko, S.V.; Mücke, A.; Ptak, A.; Koribalski, B.S.; Brenneman, L.; Mushotzky, R.; et al. Galactic Starburst NGC 3603 from X-Rays to Radio. *Astrophys. J.* **2002**, *573*, 191–198. [CrossRef]
104. Melena, N.W.; Massey, P.; Morrell, N.I.; Zangari, A.M. The Massive Star Content of NGC 3603. *Astron. J.* **2008**, *135*, 878–891. [CrossRef]
105. Ramachandran, V.; Hamann, W.R.; Oskinova, L.M.; Gallagher, J.S.; Hainich, R.; Shenar, T.; Sander, A.A.C.; Todt, H.; Fulmer, L. Testing massive star evolution, star formation history, and feedback at low metallicity. Spectroscopic analysis of OB stars in the SMC Wing. *Astron. Astrophys.* **2019**, *625*, A104. [CrossRef]
106. Johnson, T.L.; Rigby, J.R.; Sharon, K.; Gladders, M.D.; Florian, M.; Bayliss, M.B.; Wuyts, E.; Whitaker, K.E.; Livermore, R.; Murray, K.T. Star Formation at z = 2.481 in the Lensed Galaxy SDSS J1110+6459: Star Formation Down to 30 pc Scales. *Astrophys. J. Lett.* **2017**, *843*, L21. [CrossRef] [PubMed]
107. Baldwin, J.A.; Phillips, M.M.; Terlevich, R. Classification parameters for the emission-line spectra of extragalactic objects. *Pub. Astron. Soc. Pac.* **1981**, *93*, 5–19. [CrossRef]
108. Micheva, G.; Oey, M.S.; Jaskot, A.E.; James, B.L. Mrk 71/NGC 2366: The Nearest Green Pea Analog. *Astrophys. J.* **2017**, *845*, 165. [CrossRef]

109. Steidel, C.C.; Rudie, G.C.; Strom, A.L.; Pettini, M.; Reddy, N.A.; Shapley, A.E.; Trainor, R.F.; Erb, D.K.; Turner, M.L.; Konidaris, N.P.; et al. Strong Nebular Line Ratios in the Spectra of z ~ 2-3 Star Forming Galaxies: First Results from KBSS-MOSFIRE. *Astrophys. J.* **2014**, *795*, 165. [CrossRef]
110. Izotov, Y.I.; Orlitová, I.; Schaerer, D.; Thuan, T.X.; Verhamme, A.; Guseva, N.G.; Worseck, G. Eight per cent leakage of Lyman continuum photons from a compact, star-forming dwarf galaxy. *Nature* **2016**, *529*, 178–180. [CrossRef]
111. Izotov, Y.I.; Schaerer, D.; Thuan, T.X.; Worseck, G.; Guseva, N.G.; Orlitová, I.; Verhamme, A. Detection of high Lyman continuum leakage from four low-redshift compact star-forming galaxies. *Mon. Not. R. Astron. Soc.* **2016**, *461*, 3683–3701. [CrossRef]
112. Steidel, C.C.; Strom, A.L.; Pettini, M.; Rudie, G.C.; Reddy, N.A.; Trainor, R.F. Reconciling the Stellar and Nebular Spectra of High-redshift Galaxies. *Astrophys. J.* **2016**, *826*, 159. [CrossRef]
113. Kennicutt, R.C., Jr.; Lee, J.C.; Funes, J.G.; Sakai, S.; Akiyama, S. An Hα Imaging Survey of Galaxies in the Local 11 Mpc Volume. *Astrophys. J. Suppl.* **2008**, *178*, 247–279. [CrossRef]
114. van Zee, L.; Skillman, E.D.; Haynes, M.P. Oxygen and Nitrogen in Leo A and GR 8. *Astrophys. J.* **2006**, *637*, 269–282. [CrossRef]
115. Saviane, I.; Rizzi, L.; Held, E.V.; Bresolin, F.; Momany, Y. New abundance measurements in UKS 1927-177, a very metal-poor galaxy in the Local Group. *Astron. Astrophys.* **2002**, *390*, 59–64. [CrossRef]
116. Garcia, M.; Herrero, A.; Najarro, F.; Camacho, I.; Lorenzo, M. Ongoing star formation at the outskirts of Sextans A: Spectroscopic detection of early O-type stars. *Mon. Not. R. Astron. Soc.* **2019**, *484*, 422–430. [CrossRef]
117. Walborn, N.R.; Nichols-Bohlin, J.; Panek, R.J. *International Ultraviolet Explorer Atlas of O-type Spectra from 1200 to 1900 Å*; Technical Report NASA-RP-1155, NAS 1.61:1155; NASA Goddard Space Flight Center: Greenbelt, MD, USA, 1 December 1985; p. 56.

© 2019 by the authors. Licensee MDPI, Basel, Switzerland. This article is an open access article distributed under the terms and conditions of the Creative Commons Attribution (CC BY) license (http://creativecommons.org/licenses/by/4.0/).

Review

Red Supergiants, Yellow Hypergiants, and Post-RSG Evolution

Michael S. Gordon [1,*,†] **and Roberta M. Humphreys** [2]

1. SOFIA Science Center, NASA Ames Research Center, Mountain View, CA 94035-0001, USA
2. Minnesota Institute for Astrophysics, School of Physics and Astronomy, 116 Church St. SE, University of Minnesota, Minneapolis, MN 55455, USA; roberta@umn.edu
* Correspondence: mgordon@sofia.usra.edu
† Current address: SOFIA Science Center, NASA Ames Research Center, Moffett Field, CA 940353, USA.

Received: 14 October 2019; Accepted: 26 November 2019; Published: 3 December 2019

Abstract: How massive stars end their lives remains an open question in the field of star evolution. While the majority of stars above $\gtrsim 9$ $M_\odot$ will become red supergiants (RSGs), the terminal state of these massive stars can be heavily influenced by their mass-loss histories. Periods of enhanced circumstellar wind activity can drive stars off the RSG branch of the HR Diagram. This phase, known as post-RSG evolution, may well be tied to high mass-loss events or eruptions as seen in the Luminous Blue Variables (LBVs) and other massive stars. This article highlights some of the recent observational and modeling studies that seek to characterize this unique class of stars, the post-RSGs and link them to other massive objects on the HR Diagram such as LBVs, Yellow Hypergiants and dusty RSGs.

Keywords: evolved stars; Yellow Hypergiants; Red Supergiants; stellar mass loss

1. Introduction

The standard model of massive star evolution follows a rapid progression from the main-sequence, through a blue supergiant (BSG) phase, to the red supergiant (RSG) branch, to terminal supernova (SN) explosion. However, surveys of the brightest supergiants revealed an empirical upper luminosity limit to stars on the Hertzprung-Russell (HR) diagram [1]. This limit suggests that stars above some initial zero-age main-sequence (ZAMS) mass ($\approx$30–40 $M_\odot$) do not evolve to the RSG branch on the HR Diagram and therefore follow an alternative evolutionary pathway. Since the 1980s, both observational and modeling studies have attempted to describe and constrain the stellar populations and instabilities at the upper luminosity boundary, as well as explore the local environments that influence these massive stars during both their main- and post-main-sequence lives.

It has long been established that massive stars at any stage of evolution provide a favorable environment for enhanced mass-loss in their stellar winds due to low surface gravity (g) in their outer atmospheres. The outer circumstellar (CS) material is only tenuously gravitationally bound to the star itself. Indeed, mass-loss rates for RSGs range from 10^{-6} $M_\odot$ yr^{-1} [2,3] to as high as 10^{-4} $M_\odot$ yr^{-1} in extreme supergiant stars like VY CMa [4]—mass-loss rates that represent a significant fraction of a star's initial mass being shed during its post-main-sequence lifetime. The evolution and terminal state of a massive star is ultimately governed not just by its ZAMS mass but also by these drastic changes in total stellar mass and outer envelope conditions. We refer to these changes in stellar mass through ejection of CS material as the "mass-loss history." In this chapter, we summarize some of the literature on the mass-loss histories of evolved supergiant stars and the evidence for post-red supergiant evolution both in observational studies of the circumstellar ejecta and in evolutionary models that predict the effect of various mass-loss mechanisms on massive star evolution.

2. Context—The Red Supergiant Problem

Further context for much of the observational exploration of the last decade comes from a recently-identified "red supergiant problem" [5–7]. A survey of Type II-P supernova progenitors using optical and near-IR pre-explosion archival images revealed an upper limit of only 16–17 $M_\odot$ for the initial stellar masses of their likely red supergiant (RSG) progenitors [6]. Type II-P SN remnants are a useful laboratory for population statistics of RSGs since they represent the most abundant class of CCSNe (~70% of hydrogen-rich SNe; see [8] for a discussion of SN rate estimates).

From the notable lack of II-P SN progenitors above ~17 $M_\odot$ (Figure 1), Smartt et al. [6] suggested two possible scenarios:

1. Systematic underestimation of progenitor mass due to improper extinction correction.
2. Red supergiants greater than 17 $M_\odot$ have another terminal state besides II-P CCSNe.

We explore these two scenarios below.

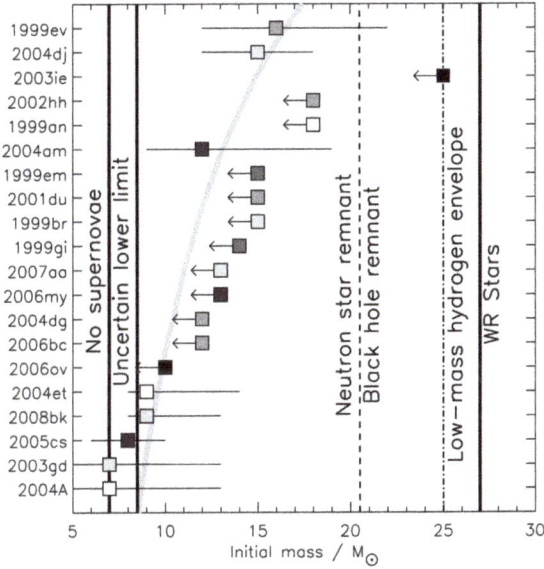

Figure 1. Initial masses of observed supernova (SN) Type II-P progenitors in the Smartt et al. [6] survey. Labels indicate theoretical limits for types of compact remnants. Darker shading is higher metallicity. The thick gray line represents a cumulative frequency distribution of a Salpeter IMF with $\Gamma = -1.35$. Figure reproduced from Smartt et al. [6].

Though no Type II-P SN progenitors appear to exist much above ~16 $M_\odot$, red supergiants have certainly been observed with masses far greater than this. For examples, see the recent HR Diagrams for massive evolved stars in the Galaxy [9], the Magellanic Clouds [10] and in M31 and M33 [11–13] A comparison with the evolutionary models [14], illustrates that many stars are present on the RSG branch above 20 $M_\odot$ in Local Group galaxies.

One potential caveat for these RSG surveys is that the derived masses require accurate measurement of bolometric luminosity. As Smartt et al. [6] suggest in scenario 1. above, underestimating RSG masses could be a potential solution to the RSG problem. Extra intrinsic extinction due to dust close to the RSG progenitor would yield lower luminosities and their estimated masses [15,16]. Indeed, mass-loss rates and dust ejecta masses scale with luminosity [3,13,17]. Additionally, mid-IR interferometry around evolved stars has revealed dust close enough in to the central source [18] that the dust grains could potentially be destroyed by the star's SN explosion.

Walmswell and Eldridge [16] examined the effect of dust on derived RSG masses by applying the Cambridge STARS code [19,20] combined with various mass-loss schema [21–23] to simulate SEDs throughout a massive star's evolution. They model a series of dust shells and estimate a simulated extinction to measure an average A_V of around 1 mag from the circumstellar ejecta. The resulting model SEDs do indeed yield lower deduced stellar masses than what would have been observed with proper extinction-correction for the model dust shells, with an underestimate of as much as 5 $M_\odot$ for supergiants in the ~20–25 $M_\odot$ range. Still, the authors note that even a change in several solar masses worth of dust material does not solve the red supergiant problem.

Beasor and Davies [24] explored this missing mass problem combining mid-IR WISE and Spitzer/IRAC photometry with circumstellar dust shell models from DUSTY [25]. The authors apply their model analysis to a co-eval population of RSG cluster stars (NGC 2100), which allows for studying stars with similar initial conditions—mass, metallicity, local environment and so forth. If the cluster stars all have roughly the same initial masses (within a few tenths of a solar mass), then the evolutionary pathway should be the same and any differences in luminosity should be due to the slightly more massive stars evolving faster on the HR Diagram. This allowed the authors to use luminosity as a proxy for evolution. Based on their models and estimated mass-loss rates, they find an increase in mass-loss rates along the RSG branch as high as a factor of 40 over the post-main-sequence lifetime of the star, which appears to be consistent with the de Jager et al. [21] mass-loss prescription. If the increased mass loss is translated into an intrinsic extinction, they argue that the increased reddening may substantially increase derived masses of Type II-P SN RSG progenitors. As an example, the authors show that similar dust extinction conditions on SNe 1999gi, 2001du and 2012ec could revise initial mass estimates by as many as 10 $M_\odot$.

Kilpatrick and Foley [26], however, argue that while circumstellar dust can alter the observed SEDs of supergiant progenitors, several studies of the circumstellar environments around SN progenitors suggest that there cannot be enough material around at least some SN Type II progenitor systems to hide an underlying high-mass RSG. They note that the total dust mass in progenitor systems is independently constrained by radio and X-ray observations. For example, X-ray light curves of CCSNe have been used to estimate stellar wind parameters and the density structure of the CS medium [27,28]. Many of these studies, though, have broad wavelength coverage of the SN progenitor SED. It is possible that for some RSG progenitors with less constrained SEDs and sparse pre-SN imaging/photometry, the "missing mass" scenario from CS dust may indeed be biasing RSG mass statistics. However, as IR photometry exists for many SN RSG progenitors, this argument is only a partial solution to the red supergiant problem.

As for scenario 2. above from Smartt et al. [6]—that high-mass red supergiant progenitors simply do not exist—there are two possible explanations: first, that higher mass RSGs collapse directly to black holes; and second, that stellar evolution to the warmer, blue side of the HR Diagram produces stellar end products other than Type II-P SNe. The subject of black hole formation, either through direct collapse or fall back, merits a longer discussion that is beyond the scope of this work. For a review of some of the work surrounding black hole formation in massive stars, see the annual review by Smartt [29].

One realm of exploration in the literature is the idea of failed supernovae, or "unnova"—stars that collapse to black holes with little or no energy released (e.g., References [30–33]). Such events may have no significant transient and thus may be almost impossible to observe [34]. However, models by Lovegrove and Woosley [35] and Piro [36] find that RSGs in the 15–25 $M_\odot$ range can lose so much energy in neutrinos during collapse that the resulting shock in the stellar envelope is expected to create an optical signature. This can be as bright as $L_{bol} \sim 10^6 - 10^7 \, L_\odot$, though perhaps lasting for only a few days [36].

These results suggest that an optical transient of this type from a failed SN would have only a small observable window, thereby decreasing the likelihood of detection. Nonetheless, surveys like that on the LBT [30,33,37] have potentially found one such source, N6946-BH1, which brightened to $\gtrsim 10^6 \, L_\odot$ in March 2009 before fading below its pre-outburst luminosity [33]. SED modeling constrained the mass of the RSG progenitor to ~25 $M_\odot$ [37], above the apparent Smartt et al. [6] SN Type II-P progenitor limit. Based on near-IR IRAC photometry and an unusually high dust temperature, Humphreys [38] suggests that the progenitor may have actually been a warm hypergiant in a post-RSG state, rather than an M-type RSG. Despite a decade of monitoring, objects like this remain exceedingly rare. While failed SNe may indeed represent some high-mass RSG population that is as of yet undiscovered, for the moment this does not seem to solve missing high-mass SN progenitors.

In this chapter, we focus on another population of transient objects—the yellow supergiants (YSGs) and hypergiants and evidence for post-red supergiant evolution.

3. The Milky Way Hypergiants and Post-RSG Evolution

Many years before attention was drawn to the red supergiant problem, a small group of high luminosity evolved supergiants was recognized with a range of intermediate to cool temperatures, high mass loss and unstable atmospheres (see several papers cited in Reference [39]). These stars, now referred to as yellow or red hypergiants, defined the empirical upper luminosity boundary in the HR Diagram for evolved massive stars [1]. The well-studied members of this elite group are Galactic members with a few examples in the Magellanic Clouds and M31 and M33 (Section 5). The visibly bright Galactic stars all exhibit spectroscopic and photometric variability, high mass loss and several show dusty ejecta. The evolutionary state of the warm or yellow hypergiants was not obvious; they could be evolving toward the red supergiant region or on a blue-loop back to warmer temperatures. The instability and brief high mass-loss events exhibited by ρ Cas, for example, Reference [40], during which it developed TiO bands and the increasing evidence for episodic high mass-loss events especially visible in the ejecta of IRC +10420 (Section 4) favored a post-RSG evolved state for these stars.

Other warm or yellow hypergiants include the Galactic stars HR 8752 and HR 5171A [41–43]. These hypergiants are visually bright and relatively nearby, which has made them important laboratories for study of late-stage evolution. Interestingly, de Jager [44] suggests that all of the yellow hypergiants are post-red supergiants. During blueward evolution, their atmospheres contract, the atmospheric opacity increases and their rotation increases. Having shed a sizable fraction of their mass on the RSG branch, these stars are now closer to the Eddington Limit for their ZAMS mass. The stars thus enter a temperature range (6000–9000 K) of increased dynamical instability, that de Jager called the "yellow void," where high mass-loss episodes occur. Figure 2 is a schematic HR Diagram showing the positions of some of the better-studied Galactic yellow hypergiants plus Var A in M33 with respect to the critical temperature region.

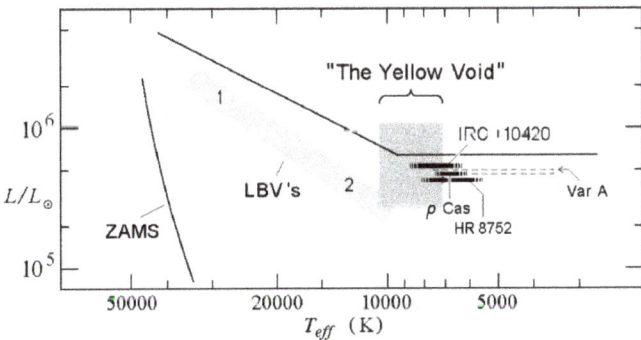

Figure 2. Schematic HRD of Galactic warm hypergiants (and Var A in M33) illustrating the location of these massive stars relative to the Humphreys-Davidson limit [1] and the "yellow void"—a temperature and luminosity band region for increased dynamical instability. The location of the LBV instability strip is also shown with the classical (LBV 1) and less-luminous (LBV 2) LBVs in their quiescent state.

4. IRC +10420 and Var A in M33—Clues to Post-RSG Evolution

The luminosities and apparent temperatures of the two evolved yellow supergiants IRC +10420 and Var A place them at the upper luminosity boundary for evolved stars in the HR Diagram. Both stars exhibit a history of photometric and spectroscopic variability with high mass loss episodes and dusty circumstellar ejecta making them excellent candidates for post-red supergiant evolution.

At its initial discovery, IRC +10420 (V1302 Aql) was quickly recognized as remarkable with its very large infrared excess and late F-type high luminosity spectrum [45]. It was soon identified as a powerful maser source and is one of the warmest known OH/IR stars [46].

IRC +10420 is a Galactic star and because of its relative proximity, its circumstellar ejecta is easily resolved by HST imaging [47] which revealed a complex environment. The color image in Figure 3 shows the spatial extent of the ejecta, more that 5 arcsec across. Numerous features are visible within two arcsec of the embedded star including condensations arrayed in jet-like structures, rays and an intriguing group of small, nearly spherical shells or arcs apparently at the ends of some of the jet-like features. One or more distant reflection shells at 5 to 6 arcsec from the star are visible in the longer exposure images.

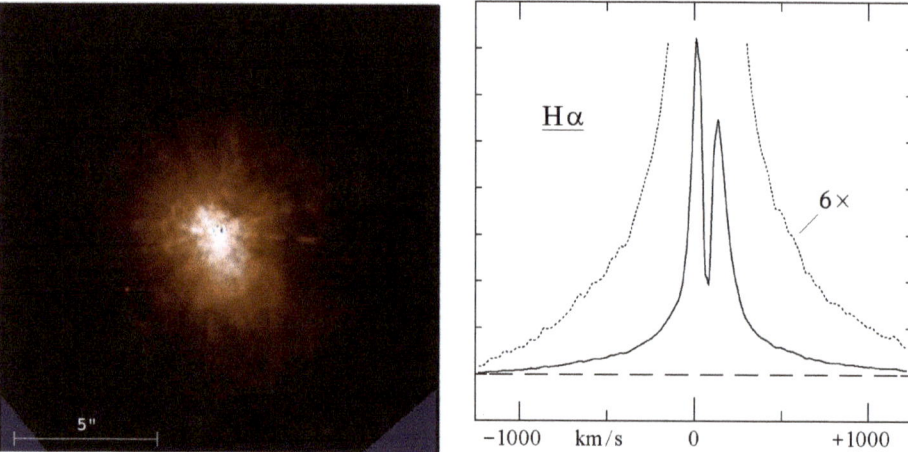

Figure 3. Left: The combined color image of IRC +10420 from HST/WFPC2 [47]. **Right**: Profile of the Hα emission line showing the broad electron scattering wings and the split profile (adapted from Reference [48]).

While its actual distance is somewhat uncertain, numerous arguments (e.g., Reference [49]) clearly demonstrated that it was not a post-AGB star and therefore above the AGB-limit at $M_{Bol} \approx -7$ mag. The reddening of its optical spectral energy distribution, infrared polarization and its radial velocity [50] suggested a distance of 4–6 kpc and a luminosity of $\approx -9.6 \pm 0.5$ mag (at 5 kpc), which places IRC +10420 at the upper luminosity boundary in the HRD for evolved stars. Jones et al. [50] therefore proposed that IRC +10420 may be evolving from a red supergiant across the HR Diagram to warmer temperatures and in a phase of its evolution analogous to the post-AGB lower mass giants evolving to the planetary nebula phase but at much higher luminosities.

The early photographic image-tube spectra [45] showed late F-type absorption features, however, 23 years later, Oudmaijer [51] identified H lines and other absorption features typical of a warmer A-type supergiant implying a significant change in its apparent temperature. HST/STIS spectra a few years later were consistent with the higher temperature [48]. The spectrum is dominated by a strong Hα stellar wind split emission line (Figure 3, right) due either to a bi-polar outflow or an equatorial disk. Strong Ca II triplet emission lines, also with a split profiles and the [Ca II] doublet in emission formed in the extended low density ejecta plus numerous Fe II emission lines typical of a stellar wind are present. Humphreys et al. [48] demonstrated that the wind was optically thick. Thus, observed variations in the apparent spectral type and the inferred temperature are due to changes in the wind and not to changes in the interior, that is, evolution, of the star on such a short timescale. Subsequent spectroscopic monitoring by Klochkova et al. [52] and by our group do not show any further increase in its apparent temperature suggesting that the blueward motion of IRC +10420 on the HR Diagram has slowed.

The morphology of IRC +10420's circumstellar ejecta had always been elusive, with suggestions of a bipolar outflow or a circumstellar disk with different orientations ranging from edge-on to an inclined disk at different angles. To investigate its three-dimensional morphology, Tiffany et al. [53] combined the transverse velocities for several knots and condensations in the inner ejecta, measured from second-epoch HST imaging, with their Doppler velocities. The resulting total space motions and direction of the outflows showed that these knots were ejected at different times and in different directions over the last ≈400 years, a relatively recent period of asymmetric mass loss. Interestingly, they are all moving within a few degrees of the plane of the sky. Thus we are viewing IRC +10420 nearly pole-on and are looking nearly directly down onto its equatorial plane. This orientation is confirmed by both the highly polarized 2.2 µm emission around the star, which places the scattering dust in the plane of the sky [54] and high resolution near-infrared interferometry [55]. The more distant reflection shells were ejected about 3000 years ago, suggesting more than one epoch of high mass loss.

To explore IRC +10420's mass loss history, Shenoy et al. [56] used far-infrared imaging from SOFIA/FORCAST at 11–37 µm to probe the extended cold dust plus high-resolution adaptive optics imaging at 8–12 µm. They found evidence for two distinct periods of high mass loss, an earlier episode from 6000 to about 2000 years ago with a high rate of 2×10^{-3} $M_\odot$ yr^{-1}, followed by an order of magnitude decrease with a current rate of $\approx 10^{-4}$ $M_\odot$ yr^{-1}, consistent with other recent measurements. This change is additional evidence for IRC +10420's evolution from the red supergiant stage and its transition to a warmer state.

Var A in M33 is significant since it has actually been observed to transition to a red supergiant and back to its presumably normal state as a high luminosity F-type supergiant within the last century. This color and spectral change, however, was not due to interior evolution but to a high mass-loss episode that produced a dense, cooler wind. Var A provides additional evidence for the highly unstable state of evolved stars near the upper limit in the HR Diagram. This supergiant has the important advantage that its distance and therefore its intrinisic luminosity, $M_{Bol} \approx -9.5$, are known. In M33, however, it is too distant for direct imaging of its ejecta.

Var A is one of the original Hubble-Sandage (H-S) variables [57]. However, unlike the other H-S variables that have been subsequently identified as evolved hot stars with episodes of high mass loss—the LBVs—Var A's quiescent state is a high luminosity yellow or intermediate temperature supergiant. Its historic light curve (Figure 6 in Reference [57]) is remarkable. At maximum light it was one of the visually-brightest stars in M33 but then in 1951 its luminosity rapidly declined by 3.5 mag, becoming faint and red after what had been a slow increase in brightness during the previous 50 years. Spectra from 1985 and 1986 revealed an M-type supergiant with prominent TiO bands [58]. Its spectral energy distribution not only showed the shift to cooler temperatures but a large mid-infrared excess due to extensive circumstellar dust and Var A was as luminous at 10 µm as at its visual maximum. The star had experienced a high mass-loss event that had produced an optically thick, cooler wind—a "false" or "pseudo" photosphere—that resembled a red supergiant.

Subsequent spectra, not observed until 2003–2004, revealed that its "eruption," which had begun ∼1951, had indeed ended, having lasted ≈45 years [59]. The spectrum showed that the star or its dense wind was now in a much warmer state with absorption lines consistent with an F-type supergiant and emission lines of Ca II, [Ca II] and K I, similar to IRC +10420, in addition to strong H emission formed in it surrounding low-density gas. The optical photometry shows the transition to bluer, warmer colors but Var A remained visually faint and was still obscured by circumstellar dust. The spectra from

1985 and 2004 are shown in Figure 4 and its light curve and SED in Figure 5. Its 10 μm flux shows an unexpected decline, which implies an unexpected decrease in the star's total luminosity. The most likely explanation is that the radiation is escaping in some direction other than our line-of-sight. This possibility is supported by recent spectra of small clumps and knots in the inner ejecta of the red hypergiant VY CMa (Reference [60], see Section 6.1 below), which require a clear line of sight to the star and therefore imply large, low density regions even holes in the circumstellar material which may also be the case for Var A.

Thus, Var A and IRC +10420 are not only probable post-red supergiants but their shared characteristics of photometric and spectroscopic variability, surface instability and stellar winds, high mass loss and a history of enhanced mass-loss episodes are clues to understanding the evolution of stars near the upper luminosity boundary and their transit across the HRD from red to blue.

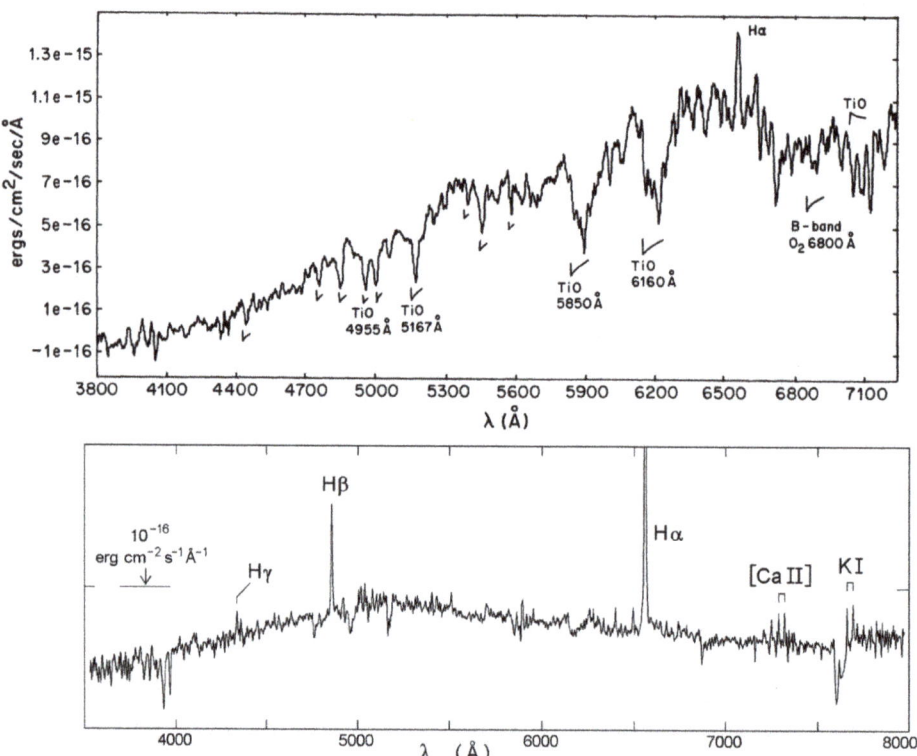

Figure 4. Top: Optical spectrum of Var A from 1985 [58] showing Hα emission and TiO absorption bands. **Bottom**: Optical spectrum of Var A from 2004 [59] with the strongest emission lines marked.

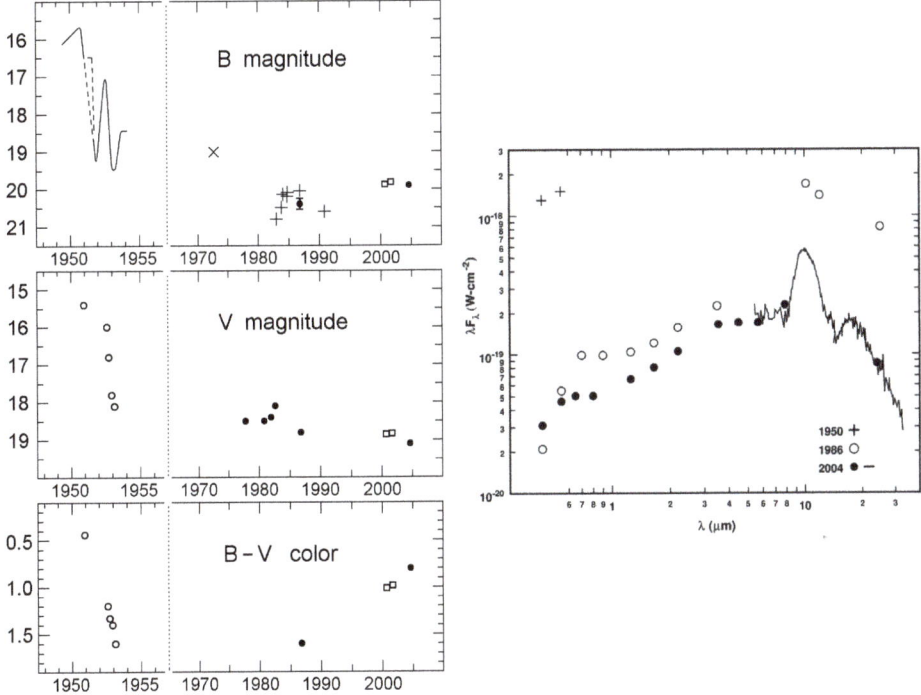

Figure 5. Left: Light curve of Var A from 1950 to the present. The top panel shows the photographic and B-band magnitudes. The middle and bottom panels show the variability in the V-band and the B–V color. See Reference [59]. **Right**: Spectral energy distribution of Var A from 1986 [58,59]. The plus signs show its apparent magnitudes at maximum light from Reference [57].

5. YSGs and Post-RSG Candidates in M31 and M33

Other than Var A, IRC +10420 and the Galactic hypergiant candidates, what fraction of known evolved supergiants may be in a post-RSG state? These statistics, as well as the physical characteristics of candidate post-RSGs and their locations on the HR Diagram, are crucial to our understanding the final stages of the majority of massive stars.

Due in part to their position on the HR Diagram, few post-RSGs are known. They occupy a relatively brief, transient state between the blue and red supergiants and may either be evolving from the main sequence to cooler temperatures, or back to warmer temperatures from the RSG stage. In the Galaxy, the warm or yellow hypergiants, close to the upper luminosity boundary in the HRD with high mass-loss rates, enhanced abundances and dusty CS environments, are excellent candidates for post-RSG evolution. These stars contrast with the intermediate-type yellow supergiants which have normal spectra and long-wavelength SEDs—that is, no evidence for circumstellar dust or mass loss in their spectra. Considering how few objects of this type are known locally, many studies have pursued observations of supergiants outside of the Galaxy.

As part of a larger program on the luminous and variable emission-line stars in M31 and M33, Humphreys et al. [4] recognized a few high luminosity, A- to F-type stars in each galaxy with spectroscopic evidence for high mass loss and extensive gaseous and dusty circumstellar ejecta revealed in their spectra and SEDs; characteristics shared with the warm hypergiants IRC +10420 and the peculiar Var A also in M33. They demonstrated that these stars were indeed evolved, intermediate temperature supergiants with strong winds and mass loss and like IRC +10420 and Var A, they were candidates for post-RSG evolution. Based on their luminosities, their initial masses would be greater than 20 M$_\odot$ or more. One possible exception was B324, one of the visually brightest stars in M33. Its SED showed strong free-free emission in the near-infrared but lacked the cooler dust expected in a post-RSG star. B324, just at the upper luminosity boundary, with high mass loss, could be approaching the limit to its redward evolution and therefore a candidate for future high mass episodes. Humphreys et al. [4] had identified a few candidates but was not a comprehensive survey for post-red supergiants.

Gordon et al. [13] conducted a survey of the yellow and red supergiants to search for post-RSG candidates. The targets were primarily selected from the published surveys of M31 and M33 for yellow and red supergiants [11,12,61] chosen from the Local Group Galaxies Survey (LGGS; [62]). Post-RSG candidates were identified based on spectroscopic evidence for mass loss and the presence of circumstellar dust in their SEDs. In that work, Gordon et al. [13] spectroscopically confirmed 75 YSGs in M31, 30 of which (40%) are likely in a post-RSG state based on spectroscopic and photometric markers for dusty wind. For M33, 27 of the observed 86 YSGs (31%) were determined to be post-RSG candidates. Further discussion of this work and its methodologies is included below. We note that a similar survey was conducted by Kourniotis et al. [63], which flagged yellow super and hypergiant candidates based on photometric criteria for follow-up spectroscopy.

The greatest challenge in photometric surveys of supergiants is distinguishing extragalactic sources from foreground disk dwarfs as well as halo giants in the Milky Way. Humphreys and Sandage [64] highlighted the magnitude of this issue in a survey of the brightest blue and red supergiants in M33. There is significant contamination of foreground K and M dwarfs in the red supergiant region of the M33 color-magnitude diagram (CMD), which presents some observational challenges. Since there is little star formation in the Milky Way halo, there is essentially no foreground contamination in the "blue plume" of the CMD. Massey et al. [65] applied the Bahcall and Soneira [66] model to estimate that almost 80% of red stars ($1.2 < B - V < 1.8$) fainter than $V \sim 16$ seen toward M31 will be foreground stars. The central portion of the CMD, representing the yellow supergiant population, is similarly affected by foreground contamination. Drout et al. [61] and later Massey et al. [67] apply the Besançon model [68] of the Milky Way (two disks + halo) to illustrate that over 70% of bright stars redward of the blue plume ($B - V > 0.4$) could be foreground contamination.

Massey et al. [11], Massey [69] and Drout et al. [12] demonstrated that color criteria could be used as an effective metric for distinguishing foreground contaminants in the RSG surveys in M31 and M33 but few such two-color discriminants have been used for YSGs, except for Bonanos et al. [70,71], who defined color ranges for a variety of massive star types in the Magellanic Clouds using 2MASS and Spitzer/IRAC photometry. In general, however, spectra are needed to determine both extragalactic membership and evolutionary state.

5.1. Spectral Types and Luminosity Classification

Drout et al. [12,61] use radial velocities from spectral-line features to generate a catalog of extragalactic YSG candidates, whereas both Gordon et al. [13] and Massey et al. [67] classified the stars based on the spectral type and luminosity criteria in their absorption-line spectra. For example, the blends of Ti II and Fe II at $\lambda\lambda 4172$-8 and $\lambda\lambda 4395$-4400 are valuable luminosity criteria in the blue when compared against Fe I lines, which show little luminosity sensitivity such as $\lambda 4046$ and $\lambda 4271$. The O I $\lambda 7774$ triplet in the red spectra—also used in Drout et al. [12] as part of their classification scheme—is also a particularly strong luminosity indicator in A- to F-type supergiants.

Using these and several other classifiers, Gordon et al. [13] confirmed extragalactic membership of $\sim$150 yellow supergiants in M31 and M33. Thirty, or $\sim$20%, of the observed YSGs in each galaxy showed evidence for stellar winds in their ejecta and enhanced mass-loss, not shared with the other YSGs and therefore possible post-RSG evolution. The notable spectral features include P Cygni profiles in hydrogen emission, broad wings in Hα or Hβ emission indicative of Thomson scattering and [Ca II]/Ca II triplet emission from circumstellar gas. If mass-loss markers in the YSG SEDs are included (discussed below), the fraction of YSGs likely in a post-RSG state increases to ~40% of the observed sources.

5.2. Photometric Evidence of Mass Loss

Gordon et al. [13] also examined the SEDs of the YSG and RSG populations in M31 and M33 to identify what fraction of the evolved supergiants have circumstellar dust and are in a mass-losing state. The RSGs currently experiencing episodes of high mass loss may eventually evolve to become post-RSG warm supergiants, LBVs, or WR stars.

The defining signature of mass loss in RSGs is the presence of circumstellar dust, usually revealed as excess radiation in their IR SEDs from the silicate emission features at 9.8 µm and 18 µm, corresponding to the Si–O vibrational [72] and O–Si–O bending modes [73], respectively. The strength of the silicate emission feature is (to first order) correlated with the luminosity and apparent temperature as revealed by the spectral type; that is, the higher the luminosity and cooler the star, the stronger the silicate emission and the larger the IR excess, as well as a larger mass of ejected dust.

In the YSGs, the presence of excess radiation due to circumstellar warm dust and/or free-free emission in the near and mid-infrared wavelengths is evidence for mass loss. This additional radiation is apparent in their SEDs if the flux in the near-IR bands exceeds the expected Rayleigh-Jeans tail of the stellar component. For example, an infrared excess in the 1–2 µm 2MASS bands is a well-known characteristic of free-free emission in stellar winds, while the 3.6 to 8 µm Spitzer/IRAC data provides evidence for warm CS dust. Free-free emission is generally identified as constant F_ν in the near-infrared, often extending out to 5 µm. Examples are shown in Figure 6 for two warm hypergiants in M31. Beyond being useful for identifying mass loss, this IR excess in the stellar SED is crucial for accurately calculating the bolometric luminosity. The CS dust will re-radiate the central star's optical flux into the infrared and this processed radiation can contribute significantly to the total bolometric luminosity of the star + ejecta system. There are various methods for fitting models to stellar SEDs to account for this and an example can be found in Kourniotis et al. [63], who fit an ATLAS9 stellar atmosphere model [74,75] for the stellar component and up to three distinct blackbodies for the warm and cool dust components of their YSG SEDs.

Gordon et al. [13] find ~50–60% of the observed RSGs in M31 and M33 show evidence for an IR excess in their near- to mid-IR SEDs. The IRAC 8 µm photometry is used in Gordon et al. [13] to provide an estimate of the total dust mass lost over a timescale of about a century and estimate that the RSGs in both galaxies tend to have dusty ejecta of the order of 10^{-3}–10^{-2} $M_\odot$, assuming a warm dust component of ~350 K. Consistent with the de Jager et al. [21] prescription, mass loss correlates with luminosity along the RSG branch. If more than 50% of RSGs are indeed experiencing sufficient mass loss to produce CS dusty ejecta, a large fraction of stars along the RSG branch may evolve back toward the blue to become the warm post-RSG stars before their terminal state as SNe or black holes.

We note that the target selection from Gordon et al. [13] was derived from optical surveys. Thus, it may be likely that our surveys of the most luminous stars in M31 and M33 do not necessarily include some supergiant populations that are heavily obscured. Since the most luminous warm and cool supergiant populations are likely to have the highest mass-loss rates, it is probable that some will be obscured in the optical by their own CS ejecta in the optical surveys. To complete the upper portion of the HRD would require a further search in the IR to find the brightest infrared sources. There are several IR surveys of M31 and M33 with Spitzer/IRAC (References [76–78], for example) that have specifically targeted the bright and/or variable stellar populations in the Local Group. These surveys have already revealed many unique supergiant stars that were obscured in high-resolution optical surveys—for example, the discovery of optically-obscured η Carinae analogs by Khan et al. [79].

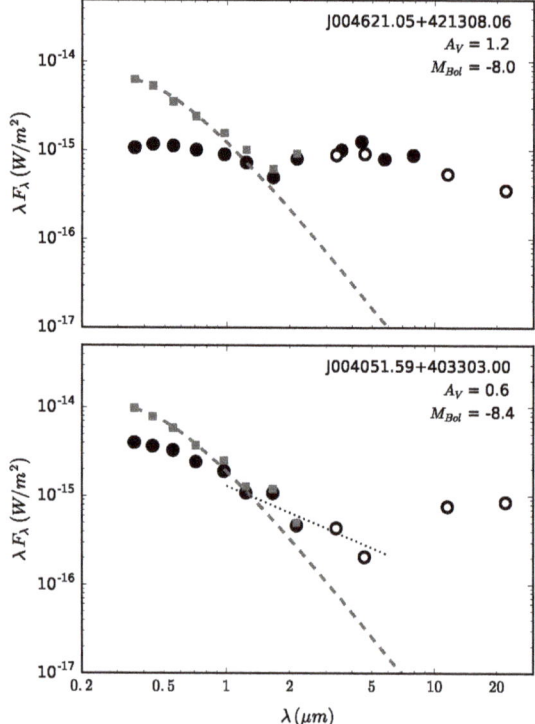

Figure 6. SEDs of warm hypergiant candidates in M31. The observed visual, 2MASS and IRAC magnitudes are shown as filled circles and WISE data as open circles. The extinction-corrected photometry is plotted as filled squares, with the measured line-of-sight A_V specified in each legend. The SED of J004621.05+421308.06 (**top**) reveals a prominent CS dust envelope in the IRAC and WISE bands. The WISE photometry of J004051.59+403303.00 (**bottom**) is suggestive of silicate dust emission but is most likely due to contamination from a nearby H II region and nebulosity. The dotted line is a curve of constant F_ν, which is evidence for free-free emission in wind. Figure adapted from Reference [13].

5.3. The Post-RSG Candidates, the HR Diagram and Comparison with Evolutionary Models

The HR Diagrams for the observed YSGs and RSGs in M31 and M33 Gordon et al. [13] are reproduced in Figure 7. For the YSGs with observed optical spectra, effective temperatures can be derived through comparison to intrinsic colors of the stars' identified spectral types. However, for sources without observed spectra/spectral-type, several photometric temperature scales exist in the literature. For example, Massey et al. [11] compare the $(V - K)$ colors of their M31 RSGs to MARCS atmosphere synthetic photometric colors and Drout et al. [12] adopt the $(V - R)$ color transformations from LMC sources [10] for their observed RSGs in M33. We note that in the absence of spectral types, photometric temperature scales can be somewhat uncertain.

In both M31 and M33, the post-RSG candidates—flagged in Gordon et al. [13] based on their spectroscopic and/or photometric mass-loss indicators—are preferentially more abundant at higher luminosities. Also shown in Figure 7 are Geneva Group [80,81] evolutionary tracks for different ZAMS mass models. The higher mass models (M $\gtrsim$ 20 M$_\odot$) loop through the YSG region of the HRD, perhaps even in multiple passes, before terminating on the RSG branch. These stars are those supergiants undergoing post-RSG evolution and are sometimes referred to in the literature as "group

2 blue supergiants", (e.g, References [82,83]). We loosely define the YSG region as ~4000 to 12,000 K and this evolution across the HRD can occur over timescales of just a few Myr.

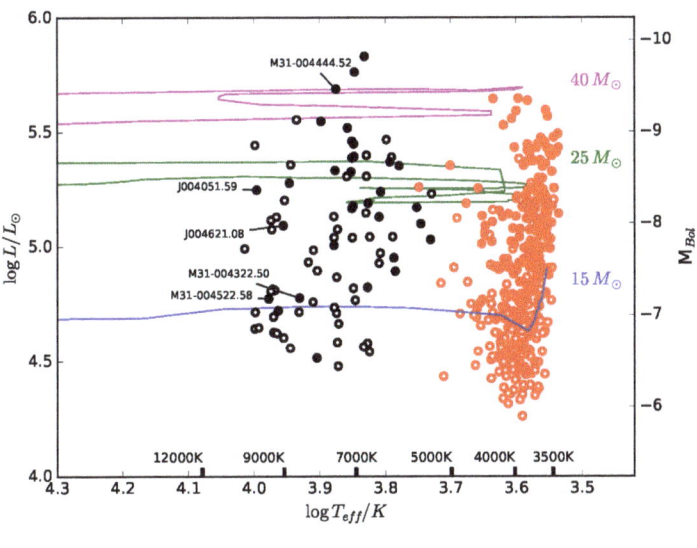

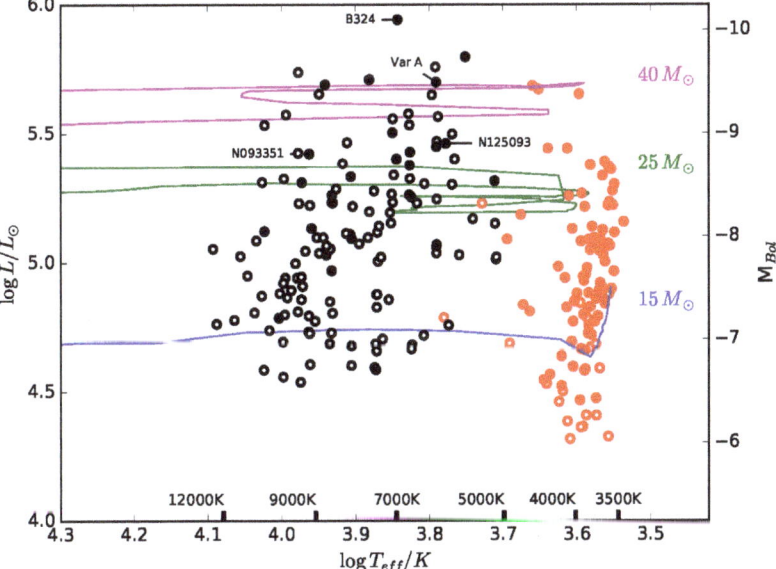

Figure 7. HR Diagrams of M31 (**top**) and M33 (**bottom**). Red circles represent the RSG sample from Gordon et al. [13], black circles are the YSGs. Closed symbols are sources with evidence of mass loss, either in their spectra (for the YSGs) or their SEDs (for both the YSGs and RSGs). Non-rotating stellar evolution tracks for three mass bins from Ekström et al. [80] are shown for comparison. The stars with mass loss, the post-RSG candidates, appear to dominate the upper portion of the HR diagram. Figures adapted from Reference [13].

Comparison with the evolutionary tracks suggests that most of the progenitor main-sequence stars have masses $\gtrsim 20~M_\odot$. Likewise, the dusty RSGs dominate the higher luminosities. This is not surprising considering results from Mauron and Josselin [3] (Figure 8) and others that $\dot{M}$ and total mass lost in the RSGs correlates with luminosity.

HR Diagrams of massive stars in the Local Group like those in Gordon et al. [13] and others, such as References [10–12,63,67,84,85], suggest that the mass-losing post-RSG candidates are more common at luminosities above $\sim 10^5~L_\odot$. Most appear to have initial masses of 20–40 $M_\odot$ and may be the evolutionary descendants of the more massive RSGs that do not explode as supernovae (i.e., the "missing" RSGs from Smartt et al. [6]). The eventual fate of these stars may be either as "less-luminous" LBVs or WR stars before their terminal explosion.

6. Mass-Loss in the Yellow and Red Supergiants

For many YSG and RSG stars, the thermal excess flux is fairly constant across the mid-infrared, which implies that the dust is emitting over a range of temperatures and distances from the central star. With some assumptions on dust temperature, grain size distributions, silicate grain chemistry and gas-to-dust ratio, near- to mid-infrared photometry can be used directly to estimate the total mass of the CS ejecta around each supergiant star. With some additional measurements and/or assumptions on timescales—such as the stellar wind velocity [4,13] or the dust condensation timescale—estimates on mass-loss rates can be extracted from the mid-infrared flux alone. For example, Mauron and Josselin [3] apply the de Jager et al. [21] mass-loss prescription to Galactic RSGs to estimate an average mass-loss rate of $\sim 10^{-6}~M_\odot~\mathrm{yr}^{-1}$ from IRAS 60 µm flux. Figure 8 from Mauron and Josselin [3] illustrates the de Jager et al. [21] prediction of increasing mass-loss rate with increasing luminosity for a handful of Galactic RSGs. Similar figures exist in Gordon et al. [13], Meynet et al. [84] and others for Galactic and extragalactic RSGs (see Figure 9 below which illustrates a similar trend for total ejecta mass lost).

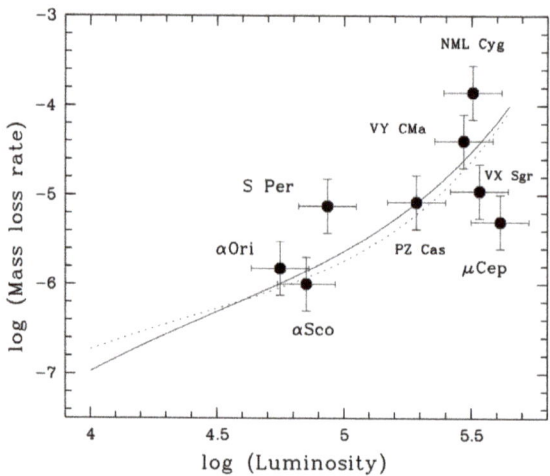

Figure 8. Mass-loss rates vs. luminosity for Galactic RSGs. The solid line represents the de Jager et al. [21] model for stellar $T_{\mathrm{eff}} = 4000$ K and the dotted line for $T_{\mathrm{eff}} = 3500$ K. Figure adapted from Reference [3].

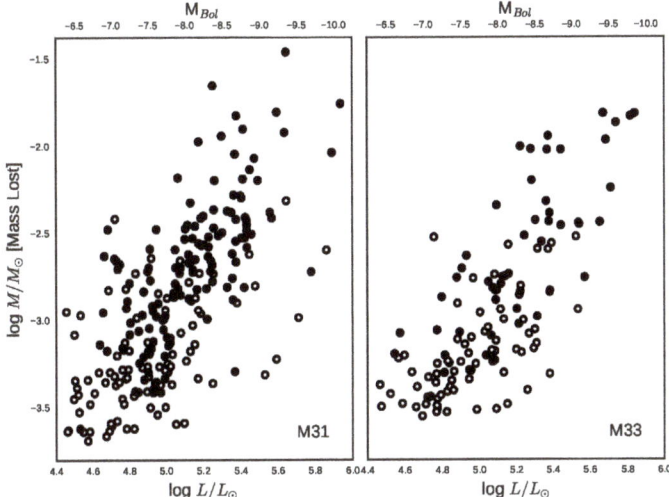

Figure 9. Bolometric luminosity vs. total mass lost based on dust measurements for RSG candidates in M31 and M33. Closed circles are those with clear evidence for mass loss in their SEDs. Open circles are the less certain mass losers. We note that the RSGs with higher luminosity tend to have lost more mass, consistent with the prescription of de Jager et al. [21] for mass loss in RSGs. Figure adapted from Reference [13].

The DUSTY radiative transfer code [25] is now often used to derive mass-loss rates or total ejected mass. DUSTY solves the radiative transfer equation for a spherically-symmetric dust distribution around a central source. Input parameters include the spectrum of the illuminating source, the optical properties and size distribution of the dust grains, the dust temperature at the inner boundary of the shell and a functional form for the radial profile of the dust density throughout the shell. The primary output is the resulting SED of the modeled system. This code has been recently been applied to different populations of RSGs and their ejecta to derive $\dot{M}$-luminosity relations from the IR SED fitting [24,56,86,87].

Shenoy et al. [56] and Gordon et al. [87] used DUSTY to generate radial profiles for a variety of model dust density profiles to test whether the mass-loss rates of the target RSGs are constant and smooth over time (e.g., $\rho_{dust} \propto r^{-2}$) or if the circumstellar ejecta can be better modeled by one (or more) discrete, high-mass ejecta events. This methodology, however, requires high-resolution imaging both to trace the ejecta close to the central star and also to resolve the thermal emission above the PSF of the telescope/instrument used for the observations. These studies demonstrated that a spherically-symmetric shell model with constant mass loss over time does not adequately explain the morphology of the circumstellar ejecta in many yellow and red supergiants. In fact, variable mass loss over time is required to build up the multiple dust shells observed around several Galactic RSGs.

6.1. Mass-Loss Mechanisms and High Mass-Loss Events

Both ground and space-based high-resolution imaging and interferometry of evolved massive stars are transforming our view of mass loss and the mass-loss mechanism in evolved stars. The precise mass-loss mechanism for red supergiants is not fully understood. The leading processes have included radiation pressure on grains, pulsation and convection. The discovery of large-scale surface asymmetries or hot spots on the surfaces of red supergiants ([88–90] and more recently [91–93]), which vary on short timescales of months or years, supports the important role of convection and surface activity.

Pulsation and dust-driven winds have been successful at explaining mass loss in Miras and AGB stars, which are fundamental-mode pulsators. However, less variable RSGs with extended, low-density atmospheres are quite different environments than their lower-mass counterparts. Pulsation may be important for the YSGs, which are at the upper-luminosity limit of the Cepheid instability strip. For example, the light and velocity curves for ρ Cas [40] support a pulsational instability as the origin of its three brief, high mass-loss episodes. Yet, as discussed in Section 4, the peculiar M33 Var A's 45+ year high mass-loss episode [59], during which it resembled an M supergiant, required some high mass-loss mechanism lasting decades. Additionally, there exists significant dispersion in the measured mass-loss rates for stars of a given luminosity class. For example, Mauron and Josselin [3] compiled mass-loss rates for LMC RSGs from several data sets [17,94,95] to demonstrate that for stars around 10^5 $L_\odot$, a rather wide range of mass-loss rates between 10^{-6} and 10^{-4} $M_\odot$ yr^{-1} have been measured (see their Figure 5).

This dispersion may well be due to observational bias or different measurement techniques or may indeed be a manifestation of whatever physical mass-loss mechanism is at play. One approach to mitigate systematics in this mass-loss rate dispersion is to study individual RSG cluster populations. Beasor and Davies [86] compared RSGs within NGC 7419 and χ Per, whose stars are of similar ages, ~14 Myr; [96,97]. With a focus on these coeval populations, the effects of age, metallicity and environment on $\dot{M}$ are removed and they find a tight correlation of mass-loss rate with luminosity.

Optical and near-IR imaging of the extreme OH/IR supergiant VY CMa and the post-RSG IRC +10420 (Section 4) have yielded surprising results about the circumstellar environments around massive stars. VY CMa has an extensive, highly structured nebula consisting of multiple knots and arcs ejected within the last 1000 years [47,53,98–100]. The numerous knots, arcs and loops visible in scattered light from the dust in their ejecta are structurally and kinematically distinct from the surrounding diffuse ejecta (see, for example, References [60,100]). These features were each ejected at different times over several hundred years, presumably by localized processes from different regions on the star. Estimates of the mass in some of the arcs and clumps in VY CMa's ejecta from surface photometry in the HST images and from the near-IR imaging of the southwest clump feature [101,102], yield minimum masses of 3–5×10^{-3} $M_\odot$ implying short term, high mass-loss events. These discrete ejecta events hint at a very different ejecta mechanism than the slow, spherical shell paradigm. The presence of magnetic fields from Zeeman splitting and polarization of the OH/water masers has been detected in the circumstellar ejecta of VY CMa and other OH/IR supergiants such as VX Sgr, NML Cyg, and S Per [103–105]. These results suggest that enhanced surface convective activity (e.g.,in α Orionis; [92,93,106]) together with magnetic activity may be important for these high mass ejection events.

Recently, HST/STIS spectra revealed TiO and VO molecular emission discrete ejecta close to the central star in VY CMa [60]. These molecules, previously believed to form in low-density dusty CS shells, instead appear concentrated in small clumps and knots. Coupled with extremely strong K I emission (4 $L_\odot$ in just two narrow doublet lines; [60,98]), the emission features imply a dust-free environment between the knots and the star. These localized sources of atomic and molecular emission imply major gaps or holes in the star's envelope or outflow structure perhaps formed by large-scale surface activity.

7. Related Work

Two related fields of study not yet discussed are the effects of stellar pulsation and binarity in RSG/YSG stars. Stellar pulsations may serve as an effective observational discriminant between post-RSG stars and main-sequence OB-stars migrating towards the RSG branch for the first time. Saio et al. [82] modeled the pulsation periods of supergiants with the Geneva stellar evolution code and found that most non-radial pulsations are only excited after significant mass loss on the RSG branch. In particular, the models appear to predict the pulsation periods of the α Cygni variables, suggesting these stars are in a He-burning, post-RSG phase. Further, the models presented in that work and its follow-up [83] suggest that CNO surface abundances should be different in the two stellar populations, owing to increased dredge up along the RSG branch. Convective cells and their effect on both surface abundance and pulsational properties remains largely unexplored on the observational side of stellar astrophysics for post-RSG stars but may well be a useful diagnostic for evolutionary state.

One final topic to consider is the role of binarity in massive star systems. More than 70% of O- and B-type stars have a binary companion [107,108] suggesting that their evolved counterparts—typically RSG+B binary pairs—should also be numerous. Recent observations by Neugent et al. [109] show that optical spectroscopy alone can be used to detect RSG+B star binaries and studying these systems prior to a SN explosion is critical, since Kochanek et al. [110] demonstrates that only ~5% of SN remnants contain a surviving star plus remnant binary. Eldridge et al. [111] suggest that binary interactions may well be the cause of SN Type Ibc explosions, as the interaction strips much of the hydrogen from the stellar surface before the terminal explosion. However, the modeling from that work does not necessarily implicate the higher mass RSG/YSG pairs in the statistics for Type II-P progenitor systems. That said, earlier work by Eldridge et al. [5] suggest that enhanced mass transfer, colliding winds, gravitational distortion, and other binary interaction effects may indeed hasten an RSG towards core-collapse. The field of binary interactions is an entire genre of astrophysics which we cannot hope to summarize in this work but we note that binarity can significantly alter the HR diagram for high-mass/high-luminosity objects. A great summary of binarity and multiplicity in stellar systems as it relates to RSGs in particular can be found in Chapter 5 in Levesque [112].

Many of the luminous warm and cool hypergiants have extensive CS ejecta and evidence for high mass-loss events. The yellow hypergiants and many of the yellow supergiants are candidates for post red supergiant evolution. IRC +10420, Var A and the extreme red supergiant VY CMa may be the special cases that provide the clues to understanding evolution near the top of the HR Diagram. These stars represent short-lived, unstable states that signal the last stages in RSG evolution and the brief post-RSG transition as the star returns to warmer temperatures. This class of post-RSG stars with complex mass-loss histories may be the missing piece on the HR Diagram and the solution to the red supergiant problem.

Funding: Support for this research was funded in part by the Universities Space Research Association (USRA) for the author's post-doctoral research position at the SOFIA Science Center.

Conflicts of Interest: The authors declare no conflict of interest.

Abbreviations

The following abbreviations are used in this manuscript:

2MASS	Two-Micron All Sky Survey
BSG	Blue Supergiant
CCSNe	Core-collapse supernovae
CMD	Color-magnitude diagram
CS	Circumstellar [ejecta]
HRD	Hertzsprung-Russell diagram
HST	Hubble Space Telescope
IR	Infrared
IRAC	Infrared Array Camera (Spitzer)
IRAS	Infrared Astronomical Satellite
LBV	Luminous Blue Variable
LGGS	Local Group Galaxy Survey
LMC	Large Magellanic Cloud

MIRAC	Mid-Infrared Array Camera (MMT)
PSF	Point spread function
RSG	Red Supergiant
SED	Spectral energy distribution
SN[e]	Supernova[e]
STIS	Space Telescope Imaging Spectrograph (HST)
WISE	Wide-field Infrared Survey Explorer
WR	Wolf-Rayet
YSG	Yellow Supergiant
ZAMS	Zero-age main-sequence

References

1. Humphreys, R.M.; Davidson, K. Studies of luminous stars in nearby galaxies. III—Comments on the evolution of the most massive stars in the Milky Way and the Large Magellanic Cloud. *Astrophys. J.* **1979**, *232*, 409–420, doi:10.1086/157301. [CrossRef]
2. Gehrz, R.D.; Woolf, N.J. Mass Loss from M Stars. *Astrophys. J.* **1971**, *165*, 285, doi:10.1086/150897. [CrossRef]
3. Mauron, N.; Josselin, E. The mass-loss rates of red supergiants and the de Jager prescription. *Astron. Astrophys.* **2011**, *526*, A156. [CrossRef]
4. Humphreys, R.M.; Davidson, K.; Grammer, S.; Kneeland, N.; Martin, J.C.; Weis, K.; Burggraf, B. Luminous and Variable Stars in M31 and M33. I. The Warm Hypergiants and Post-red Supergiant Evolution. *Astrophys. J.* **2013**, *773*, 46. [CrossRef]
5. Eldridge, J.J.; Izzard, R.G.; Tout, C.A. The effect of massive binaries on stellar populations and supernova progenitors. *Mon. Not. R. Astron. Soc.* **2008**, *384*, 1109–1118. [CrossRef]
6. Smartt, S.J.; Eldridge, J.J.; Crockett, R.M.; Maund, J.R. The death of massive stars—I. Observational constraints on the progenitors of Type II-P supernovae. *Mon. Not. R. Astron. Soc.* **2009**, *395*, 1409–1437. [CrossRef]
7. Smartt, S.J. Progenitors of Core-Collapse Supernovae. *Annu. Rev. Astron. Astrophys.* **2009**, *47*, 63–106. [CrossRef]
8. Li, W.; Leaman, J.; Chornock, R.; Filippenko, A.V.; Poznanski, D.; Ganeshalingam, M.; Wang, X.; Modjaz, M.; Jha, S.; Foley, R.J.; et al. Nearby supernova rates from the Lick Observatory Supernova Search—II. The observed luminosity functions and fractions of supernovae in a complete sample. *Mon. Not. R. Astron. Soc.* **2011**, *412*, 1441–1472. [CrossRef]
9. Levesque, E.M.; Massey, P.; Olsen, K.A.G.; Plez, B.; Josselin, E.; Maeder, A.; Meynet, G. The Effective Temperature Scale of Galactic Red Supergiants: Cool, but Not As Cool As We Thought. *Astrophys. J.* **2005**, *628*, 973–985. [CrossRef]
10. Levesque, E.M.; Massey, P.; Olsen, K.A.G.; Plez, B.; Meynet, G.; Maeder, A. The Effective Temperatures and Physical Properties of Magellanic Cloud Red Supergiants: The Effects of Metallicity. *Astrophys. J.* **2006**, *645*, 1102–1117. [CrossRef]
11. Massey, P.; Silva, D.R.; Levesque, E.M.; Plez, B.; Olsen, K.A.G.; Clayton, G.C.; Meynet, G.; Maeder, A. Red Supergiants in the Andromeda Galaxy (M31). *Astrophys. J.* **2009**, *703*, 420–440. [CrossRef]
12. Drout, M.R.; Massey, P.; Meynet, G. The Yellow and Red Supergiants of M33. *Astrophys. J.* **2012**, *750*, 97. [CrossRef]
13. Gordon, M.S.; Humphreys, R.M.; Jones, T.J. Luminous and Variable Stars in M31 and M33. III. The Yellow and Red Supergiants and Post-red Supergiant Evolution. *Astrophys. J.* **2016**, *825*, 50. doi:10.3847/0004-637X/825/1/50. [CrossRef]
14. Meynet, G.; Maeder, A. Stellar evolution with rotation. X. Wolf-Rayet star populations at solar metallicity. *Astron. Astrophys.* **2003**, *404*, 975–990. [CrossRef]
15. Smith, N.; Li, W.; Filippenko, A.V.; Chornock, R. Observed fractions of core-collapse supernova types and initial masses of their single and binary progenitor stars. *Mon. Not. R. Astron. Soc.* **2011**, *412*, 1522–1538. [CrossRef]
16. Walmswell, J.J.; Eldridge, J.J. Circumstellar dust as a solution to the red supergiant supernova progenitor problem. *Mon. Not. R. Astron. Soc.* **2012**, *419*, 2054–2062. [CrossRef]

17. van Loon, J.T.; Cioni, M.R.L.; Zijlstra, A.A.; Loup, C. An empirical formula for the mass-loss rates of dust-enshrouded red supergiants and oxygen-rich Asymptotic Giant Branch stars. *Astron. Astrophys.* **2005**, *438*, 273–289. [CrossRef]
18. Danchi, W.C.; Bester, M.; Degiacomi, C.G.; Greenhill, L.J.; Townes, C.H. Characteristics of dust shells around 13 late-type stars. *Astron. J.* **1994**, *107*, 1469–1513, doi:10.1086/116960. [CrossRef]
19. Eggleton, P.P. The evolution of low mass stars. *Mon. Not. R. Astron. Soc.* **1971**, *151*, 351, doi:10.1093/mnras/151.3.351. [CrossRef]
20. Stancliffe, R.J.; Eldridge, J.J. Modelling the binary progenitor of Supernova 1993J. *Mon. Not. R. Astron. Soc.* **2009**, *396*, 1699–1708. [CrossRef]
21. De Jager, C.; Nieuwenhuijzen, H.; van der Hucht, K.A. Mass loss rates in the Hertzsprung-Russell diagram. *Astron. Astrophys.* **1988**, *72*, 259–289.
22. Vink, J.S.; de Koter, A.; Lamers, H.J.G.L.M. Mass-loss predictions for O and B stars as a function of metallicity. *Astron. Astrophys.* **2001**, *369*, 574–588. [CrossRef]
23. Eldridge, J.J.; Tout, C.A. The progenitors of core-collapse supernovae. *Mon. Not. R. Astron. Soc.* **2004**, *353*, 87–97. [CrossRef]
24. Beasor, E.R.; Davies, B. The evolution of red supergiants to supernova in NGC 2100. *Mon. Not. R. Astron. Soc.* **2016**, *463*, 1269–1283. [CrossRef]
25. Ivezic, Z.; Nenkova, M.; Elitzur, M. *DUSTY: Radiation Transport in a Dusty Environment*; Astrophysics Source Code Library: Houghton, MI, USA, 1999.
26. Kilpatrick, C.D.; Foley, R.J. The dusty progenitor star of the Type II supernova 2017eaw. *Mon. Not. R. Astron. Soc.* **2018**, *481*, 2536–2547. [CrossRef]
27. Chevalier, R.A.; Fransson, C. Supernova Interaction with a Circumstellar Medium. In *Supernovae and Gamma-Ray Bursters*; Weiler, K., Ed.; Springer: Berlin, Germany, 2003; Volume 598, pp. 171–194. [CrossRef]
28. Dwarkadas, V.V.; Gruszko, J. What are published X-ray light curves telling us about young supernova expansion? *Mon. Not. R. Astron. Soc.* **2012**, *419*, 1515–1524. [CrossRef]
29. Smartt, S.J. Observational Constraints on the Progenitors of Core-Collapse Supernovae: The Case for Missing High-Mass Stars. *PASA* **2015**, *32*, e016. [CrossRef]
30. Kochanek, C.S.; Beacom, J.F.; Kistler, M.D.; Prieto, J.L.; Stanek, K.Z.; Thompson, T.A.; Yüksel, H. A Survey About Nothing: Monitoring a Million Supergiants for Failed Supernovae. *Astrophys. J.* **2008**, *684*, 1336–1342. [CrossRef]
31. Ugliano, M.; Janka, H.T.; Marek, A.; Arcones, A. Progenitor-explosion Connection and Remnant Birth Masses for Neutrino-driven Supernovae of Iron-core Progenitors. *Astrophys. J.* **2012**, *757*, 69. [CrossRef]
32. Pejcha, O.; Thompson, T.A. The Landscape of the Neutrino Mechanism of Core-collapse Supernovae: Neutron Star and Black Hole Mass Functions, Explosion Energies, and Nickel Yields. *Astrophys. J.* **2015**, *801*, 90. [CrossRef]
33. Gerke, J.R.; Kochanek, C.S.; Stanek, K.Z. The search for failed supernovae with the Large Binocular Telescope: first candidates. *Mon. Not. R. Astron. Soc.* **2015**, *450*, 3289–3305. [CrossRef]
34. Woosley, S.E.; Heger, A. Long Gamma-Ray Transients from Collapsars. *Astrophys. J.* **2012**, *752*, 32. [CrossRef]
35. Lovegrove, E.; Woosley, S.E. Very Low Energy Supernovae from Neutrino Mass Loss. *Astrophys. J.* **2013**, *769*, 109. [CrossRef]
36. Piro, A.L. Taking the "Un" out of "Unnovae". *Astrophys. J.* **2013**, *768*, L14. [CrossRef]
37. Adams, S.M.; Kochanek, C.S.; Gerke, J.R.; Stanek, K.Z.; Dai, X. The search for failed supernovae with the Large Binocular Telescope: confirmation of a disappearing star. *Mon. Not. R. Astron. Soc.* **2017**, *468*, 4968–4981. [CrossRef]
38. Humphreys, R.M. Comments on the Progenitor of NGC 6946-BH1. *Res. Notes Am. Astron. Soc.* **2019**, *3*, 164. [CrossRef]
39. De Jager, C.; Nieuwenhuijzen, H. (Eds.) *Instabilities in Evolved Super- and Hypergiants*; Royal Netherlands Academy of Arts and Sciences: Amsterdam, The Netherlands, 1992.
40. Lobel, A.; Dupree, A.K.; Stefanik, R.P.; Torres, G.; Israelian, G.; Morrison, N.; de Jager, C.; Nieuwenhuijzen, H.; Ilyin, I.; Musaev, F. High-Resolution Spectroscopy of the Yellow Hypergiant ρ Cassiopeiae from 1993 through the Outburst of 2000-2001. *Astrophys. J.* **2003**, *583*, 923–954. [CrossRef]

41. Nieuwenhuijzen, H.; De Jager, C.; Kolka, I.; Israelian, G.; Lobel, A.; Zsoldos, E.; Maeder, A.; Meynet, G. The hypergiant HR 8752 evolving through the yellow evolutionary void. *Astron. Astrophys.* **2012**, *546*, A105, doi:10.1051/0004-6361/201117166. [CrossRef]
42. Lobel, A.; de Jager, C.; Nieuwenhuijzen, H.; van Genderen, A.M.; Oudmaijer, R. Yellow Hypergiants: A Comparative Study of HR 5171A, Rho Cas, and HR 8752. *EAS Publ. Ser.* **2015**, *71*, 279–280, doi:10.1051/eas/1571062. [CrossRef]
43. Chesneau, O.; Meilland, A.; Chapellier, E.; Millour, F.; van Genderen, A.M.; Nazé, Y.; Smith, N.; Spang, A.; Smoker, J.V.; Dessart, L.; et al. The yellow hypergiant HR 5171 A: Resolving a massive interacting binary in the common envelope phase. *Astron. Astrophys.* **2014**, *563*, A71. [CrossRef]
44. De Jager, C. The yellow hypergiants. *Astron. Astrophys.* **1998**, *8*, 145–180, doi:10.1007/s001590050009. [CrossRef]
45. Humphreys, R.M.; Strecker, D.W.; Murdock, T.L.; Low, F.J. IRC+10420—Another Eta Carinae? *Astrophys. J.* **1973**, *179*, L49, doi:10.1086/181114. [CrossRef]
46. Giguere, P.T.; Woolf, N.J.; Webber, J.C. IRC +10 420—A hot supergiant maser. *Astrophys. J.* **1976**, *207*, L195–L198, doi:10.1086/182211. [CrossRef]
47. Humphreys, R.M.; Smith, N.; Davidson, K.; Jones, T.J.; Gehrz, R.T.; Mason, C.G.; Hayward, T.L.; Houck, J.R.; Krautter, J. HST and Infrared Images of the Circumstellar Environment of the Cool Hypergiant IRC + 10420. *Astron. J.* **1997**, *114*, 2778, doi:10.1086/118686. [CrossRef]
48. Humphreys, R.M.; Davidson, K.; Smith, N. Crossing the Yellow Void: Spatially Resolved Spectroscopy of the Post-Red Supergiant IRC +10420 and Its Circumstellar Ejecta. *Astron. J.* **2002**, *124*, 1026–1044. doi: 10.1086/341380. [CrossRef]
49. Oudmaijer, R.D.; Groenewegen, M.A.T.; Matthews, H.E.; Blommaert, J.A.D.L.; Sahu, K.C. The spectral energy distribution and mass-loss history of IRC+10420. *Mon. Not. R. Astron. Soc.* **1996**, *280*, 1062–1070, doi:10.1093/mnras/280.4.1062. [CrossRef]
50. Jones, T.J.; Humphreys, R.M.; Gehrz, R.D.; Lawrence, G.F.; Zickgraf, F.J.; Moseley, H.; Casey, S.; Glaccum, W.J.; Koch, C.J.; Pina, R.; et al. IRC +10420—A cool hypergiant near the top of the H-R diagram. *Astrophys. J.* **1993**, *411*, 323–335, doi:10.1086/172832. [CrossRef]
51. Oudmaijer, R.D. High resolution spectroscopy of the post-red supergiant IRC+10420. I. The data. *Astron. Astrophys.s* **1998**, *129*, 541–552, doi:10.1051/aas:1998404. [CrossRef]
52. Klochkova, V.G.; Chentsov, E.L.; Miroshnichenko, A.S.; Panchuk, V.E.; Yushkin, M.V. High-resolution optical spectroscopy of the yellow hypergiant V1302 Aql (=IRC+10420) in 2001-2014. *Mon. Not. R. Astron. Soc.* **2016**, *459*, 4183–4190. doi: 10.1093/mnras/stw902. [CrossRef]
53. Tiffany, C.; Humphreys, R.M.; Jones, T.J.; Davidson, K. The Morphology of IRC+10420's Circumstellar Ejecta. *Astron. J.* **2010**, *140*, 339–349. doi: 10.1088/0004-6256/140/2/339. [CrossRef]
54. Shenoy, D.P.; Jones, T.J.; Packham, C.; Lopez-Rodriguez, E. Probing Hypergiant Mass Loss with Adaptive Optics Imaging and Polarimetry in the Infrared: MMT-Pol and LMIRCam Observations of IRC +10420 and VY Canis Majoris. *Astron. J.* **2015**, *150*, 15. doi: 10.1088/0004-6256/150/1/15. [CrossRef]
55. De Wit, W.J.; Oudmaijer, R.D.; Fujiyoshi, T.; Hoare, M.G.; Honda, M.; Kataza, H.; Miyata, T.; Okamoto, Y.K.; Onaka, T.; Sako, S.; et al. A Red Supergiant Nebula at 25 μm: Arcsecond-Scale Mass-Loss Asymmetries of μ Cephei. *Astrophys. J.* **2008**, *685*, L75. doi: 10.1086/592384. [CrossRef]
56. Shenoy, D.; Humphreys, R.M.; Jones, T.J.; Marengo, M.; Gehrz, R.D.; Helton, L.A.; Hoffmann, W.F.; Skemer, A.J.; Hinz, P.M. Searching for Cool Dust in the Mid-to-far Infrared: The Mass-loss Histories of the Hypergiants μ Cep, VY CMa, IRC+10420, and ρ Cas. *Astron. J.* **2016**, *151*, 51. doi: 10.3847/0004-6256/151/3/51. [CrossRef]
57. Hubble, E.; Sandage, A. The Brightest Variable Stars in Extragalactic Nebulae. I. M31 and M33. *Astrophys. J.* **1953**, *118*, 353, doi:10.1086/145764. [CrossRef]
58. Humphreys, R.M.; Jones, T.J.; Gehrz, R.D. The enigmatic object variable A in M33. *Astron. J.* **1987**, *94*, 315–323, doi:10.1086/114473. [CrossRef]
59. Humphreys, R.M.; Jones, T.J.; Polomski, E.; Koppelman, M.; Helton, A.; McQuinn, K.; Gehrz, R.D.; Woodward, C.E.; Wagner, R.M.; Gordon, K.; et al. M33's Variable A: A Hypergiant Star More Than 35 YEARS in Eruption. *Astron. J.* **2006**, *131*, 2105–2113. doi: 10.1086/500811. [CrossRef]
60. Humphreys, R.M.; Ziurys, L.M.; Bernal, J.J.; Gordon, M.S.; Helton, L.A.; Ishibashi, K.; Jones, T.J.; Richards, A.M.S.; Vlemmings, W. The Unexpected Spectrum of the Innermost Ejecta of the Red Hypergiant VY CMa. *Astrophys. J.* **2019**, *874*, L26. doi: 10.3847/2041-8213/ab11e5. [CrossRef]

61. Drout, M.R.; Massey, P.; Meynet, G.; Tokarz, S.; Caldwell, N. Yellow Supergiants in the Andromeda Galaxy (M31). *Astrophys. J.* **2009**, *703*, 441–460. doi: 10.1088/0004-637X/703/1/441. [CrossRef]
62. Massey, P.; Olsen, K.A.G.; Hodge, P.W.; Strong, S.B.; Jacoby, G.H.; Schlingman, W.; Smith, R.C. A Survey of Local Group Galaxies Currently Forming Stars. I. UBVRI Photometry of Stars in M31 and M33. *Astron. J.* **2006**, *131*, 2478–2496. doi: 10.1086/503256. [CrossRef]
63. Kourniotis, M.; Bonanos, A.Z.; Yuan, W.; Macri, L.M.; Garcia-Alvarez, D.; Lee, C.H. Monitoring luminous yellow massive stars in M 33: New yellow hypergiant candidates. *Astron. Astrophys.* **2017**, *601*, A76. doi: 10.1051/0004-6361/201629146. [CrossRef]
64. Humphreys, R.M.; Sandage, A. On the stellar content and structure of the spiral Galaxy M33. *Astrophys. J.* **1980**, *44*, 319–381, doi:10.1086/190696. [CrossRef]
65. Massey, P.; Olsen, K.A.G.; Hodge, P.W.; Jacoby, G.H.; McNeill, R.T.; Smith, R.C.; Strong, S.B. A Survey of Local Group Galaxies Currently Forming Stars. II. UBVRI Photometry of Stars in Seven Dwarfs and a Comparison of the Entire Sample. *Astron. J.* **2007**, *133*, 2393–2417. doi: 10.1086/513319. [CrossRef]
66. Bahcall, J.N.; Soneira, R.M. The universe at faint magnitudes. I—Models for the galaxy and the predicted star counts. *Astrophys. J.s* **1980**, *44*, 73–110, doi:10.1086/190685. [CrossRef]
67. Massey, P.; Neugent, K.F.; Smart, B.M. A Spectroscopic Survey of Massive Stars in M31 and M33. *Astron. J.* **2016**, *152*, 62. doi: 10.3847/0004-6256/152/3/62. [CrossRef]
68. Robin, A.C.; Reylé, C.; Derrière, S.; Picaud, S. A synthetic view on structure and evolution of the Milky Way. *Astron. Astrophys.* **2003**, *409*, 523–540, doi:10.1051/0004-6361:20031117. [CrossRef]
69. Massey, P. Evolved Massive Stars in the Local Group. I. Identification of Red Supergiants in NGC 6822, M31, and M33. *Astrophys. J.* **1998**, *501*, 153–174, doi:10.1086/305818. [CrossRef]
70. Bonanos, A.Z.; Massa, D.L.; Sewilo, M.; Lennon, D.J.; Panagia, N.; Smith, L.J.; Meixner, M.; Babler, B.L.; Bracker, S.; Meade, M.R.; et al. Spitzer SAGE Infrared Photometry of Massive Stars in the Large Magellanic Cloud. *Astron. J.* **2009**, *138*, 1003–1021. doi: 10.1088/0004-6256/138/4/1003. [CrossRef]
71. Bonanos, A.Z.; Lennon, D.J.; Köhlinger, F.; van Loon, J.T.; Massa, D.L.; Sewilo, M.; Evans, C.J.; Panagia, N.; Babler, B.L.; Block, M.; et al. Spitzer SAGE-SMC Infrared Photometry of Massive Stars in the Small Magellanic Cloud. *Astron. J.* **2010**, *140*, 416–429. doi: 10.1088/0004-6256/140/2/416. [CrossRef]
72. Woolf, N.J.; Ney, E.P. Circumstellar Infrared Emission from Cool Stars. *Astrophys. J.* **1969**, *155*, L181, doi:10.1086/180331. [CrossRef]
73. Treffers, R.; Cohen, M. High-resolution spectra of cool stars in the 10- and 20-micron regions. *Astrophys. J.* **1974**, *188*, 545–552, doi:10.1086/152746. [CrossRef]
74. Kurucz, R.L. Model atmospheres for G, F, A, B, and O stars. *Astrophys. J.* **1979**, *40*, 1–340, doi:10.1086/190589. [CrossRef]
75. Howarth, I.D. New limb-darkening coefficients and synthetic photometry for model-atmosphere grids at Galactic, LMC and SMC abundances. *Mon. Not. R. Astron. Soc.* **2011**, *413*, 1515–1523. doi: 10.1111/j.1365-2966.2011.18122.x. [CrossRef]
76. McQuinn, K.B.W.; Woodward, C.E.; Willner, S.P.; Polomski, E.F.; Gehrz, R.D.; Humphreys, R.M.; van Loon, J.T.; Ashby, M.L.N.; Eicher, K.; Fazio, G.G. The M33 Variable Star Population Revealed by Spitzer. *Astrophys. J.* **2007**, *664*, 850–861. doi: 10.1086/519068. [CrossRef]
77. Mould, J.; Barmby, P.; Gordon, K.; Willner, S.P.; Ashby, M.L.N.; Gehrz, R.D.; Humphreys, R.; Woodward, C.E. A Point-Source Survey of M31 with the Spitzer Space Telescope. *Astrophys. J.* **2008**, *687*, 230–241. doi: 10.1086/591844. [CrossRef]
78. Khan, R.; Stanek, K.Z.; Kochanek, C.S.; Sonneborn, G. Spitzer Point-source Catalogs of ∼300,000 Stars in Seven Nearby Galaxies. *Astrophys. J.* **2015**, *219*, 42. doi: 10.1088/0067-0049/219/2/42. [CrossRef]
79. Khan, R.; Kochanek, C.S.; Stanek, K.Z.; Gerke, J. Finding η Car Analogs in Nearby Galaxies Using Spitzer. II. Identification of An Emerging Class of Extragalactic Self-Obscured Stars. *Astrophys. J.* **2015**, *799*, 187. doi: 10.1088/0004-637X/799/2/187. [CrossRef]
80. Ekström, S.; Georgy, C.; Eggenberger, P.; Meynet, G.; Mowlavi, N.; Wyttenbach, A.; Granada, A.; Decressin, T.; Hirschi, R.; Frischknecht, U.; et al. Grids of stellar models with rotation - I. Models from 0.8 to 120 M solar metallicity (Z = 0.014). *Astron. Astrophys.* **2012**, *537*, A146. doi: 10.1051/0004-6361/201117751. [CrossRef]

81. Meynet, G.; Kudritzki, R.P.; Georgy, C. The flux-weighted gravity-luminosity relationship of blue supergiant stars as a constraint for stellar evolution. *Astron. Astrophys.* **2015**, *581*, A36. doi:10.1051/0004-6361/201526035. [CrossRef]
82. Saio, H.; Georgy, C.; Meynet, G. Evolution of blue supergiants and α Cygni variables: Puzzling CNO surface abundances. *Mon. Not. R. Astron. Soc.* **2013**, *433*, 1246–1257. doi:10.1093/mnras/stt796. [CrossRef]
83. Georgy, C.; Saio, H.; Meynet, G. The puzzle of the CNO abundances of α Cygni variables resolved by the Ledoux criterion. *Mon. Not. R. Astron. Soc.* **2014**, *439*, L6–L10. doi:10.1093/mnrasl/slt165. [CrossRef]
84. Meynet, G.; Chomienne, V.; Ekström, S.; Georgy, C.; Granada, A.; Groh, J.; Maeder, A.; Eggenberger, P.; Levesque, E.; Massey, P. Impact of mass-loss on the evolution and pre-supernova properties of red supergiants. *Astron. Astrophys.* **2015**, *575*, A60. doi:10.1051/0004-6361/201424671. [CrossRef]
85. Kourniotis, M.; Kraus, M.; Arias, M.L.; Cidale, L.; Torres, A.F. On the evolutionary state of massive stars in transition phases in M33. *Mon. Not. R. Astron. Soc.* **2018**, *480*, 3706–3717. doi:10.1093/mnras/sty2087. [CrossRef]
86. Beasor, E.R.; Davies, B. The evolution of red supergiant mass-loss rates. *Mon. Not. R. Astron. Soc.* **2018**, *475*, 55–62. doi:10.1093/mnras/stx3174. [CrossRef]
87. Gordon, M.S.; Humphreys, R.M.; Jones, T.J.; Shenoy, D.; Gehrz, R.D.; Helton, L.A.; Marengo, M.; Hinz, P.M.; Hoffmann, W.F. Searching for Cool Dust. II. Infrared Imaging of The OH/IR Supergiants, NML Cyg, VX Sgr, S Per, and the Normal Red Supergiants RS Per and T Per. *Astron. J.* **2018**, *155*, 212. doi:10.3847/1538-3881/aab961. [CrossRef]
88. Gilliland, R.L.; Dupree, A.K. First Image of the Surface of a Star with the Hubble Space Telescope. *Astrophys. J.l* **1996**, *463*, L29, doi:10.1086/310043. [CrossRef]
89. Tuthill, P.G.; Haniff, C.A.; Baldwin, J.E. Hotspots on late-type supergiants. *Mon. Not. R. Astron. Soc.* **1997**, *285*, 529–539, doi:10.1093/mnras/285.3.529. [CrossRef]
90. Monnier, J.D.; Millan-Gabet, R.; Tuthill, P.G.; Traub, W.A.; Carleton, N.P.; Coudé du Foresto, V.; Danchi, W.C.; Lacasse, M.G.; Morel, S.; Perrin, G.; et al. High-Resolution Imaging of Dust Shells by Using Keck Aperture Masking and the IOTA Interferometer. *Astrophys. J.* **2004**, *605*, 436–461. doi:10.1086/382218. [CrossRef]
91. Montargès, M.; Norris, R.; Chiavassa, A.; Tessore, B.; Lèbre, A.; Baron, F. The convective photosphere of the red supergiant CE Tauri. I. VLTI/PIONIER H-band interferometric imaging. *Astron. Astrophys.* **2018**, *614*, A12. doi:10.1051/0004-6361/201731471. [CrossRef]
92. Nance, S.; Sullivan, J.M.; Diaz, M.; Wheeler, J.C. The Betelgeuse Project II: asteroseismology. *Mon. Not. R. Astron. Soc.* **2018**, *479*, 251–261. doi:10.1093/mnras/sty1418. [CrossRef]
93. López Ariste, A.; Mathias, P.; Tessore, B.; Lèbre, A.; Aurière, M.; Petit, P.; Ikhenache, N.; Josselin, E.; Morin, J.; Montargès, M. Convective cells in Betelgeuse: imaging through spectropolarimetry. *Astron. Astrophys.* **2018**, *620*, A199. doi:10.1051/0004-6361/201834178. [CrossRef]
94. Reid, N.; Tinney, C.; Mould, J. Luminous asymptotic giant branch stars in the Large Magellanic Cloud. *Astrophys. J.* **1990**, *348*, 98–119, doi:10.1086/168217. [CrossRef]
95. Groenewegen, M.A.T.; Sloan, G.C.; Soszyński, I.; Petersen, E.A. Luminosities and mass-loss rates of SMC and LMC AGB stars and red supergiants. *Astron. Astrophys.* **2009**, *506*, 1277–1296. doi:10.1051/0004-6361/200912678. [CrossRef]
96. Currie, T.; Hernandez, J.; Irwin, J.; Kenyon, S.J.; Tokarz, S.; Balog, Z.; Bragg, A.; Berlind, P.; Calkins, M. The Stellar Population of h and χ Persei: Cluster Properties, Membership, and the Intrinsic Colors and Temperatures of Stars. *Astrophys. J.* **2010**, *186*, 191–221. doi:10.1088/0067-0049/186/2/191. [CrossRef]
97. Marco, A.; Negueruela, I. NGC 7419 as a template for red supergiant clusters. *Astron. Astrophys.* **2013**, *552*, A92. doi:10.1051/0004-6361/201220750. [CrossRef]
98. Humphreys, R.M.; Davidson, K.; Ruch, G.; Wallerstein, G. High-Resolution, Long-Slit Spectroscopy of VY Canis Majoris: The Evidence for Localized High Mass Loss Events. *Astron. J.* **2005**, *129*, 492–510. doi:10.1086/426565. [CrossRef]
99. Humphreys, R.M.; Helton, L.A.; Jones, T.J. The Three-Dimensional Morphology of VY Canis Majoris. I. The Kinematics of the Ejecta. *Astron. J.* **2007**, *133*, 2716–2729. doi:10.1086/517609. [CrossRef]
100. Smith, N.; Humphreys, R.M.; Davidson, K.; Gehrz, R.D.; Schuster, M.T.; Krautter, J. The Asymmetric Nebula Surrounding the Extreme Red Supergiant VY Canis Majoris. *Astron. J.* **2001**, *121*, 1111–1125, doi:10.1086/318748. [CrossRef]

101. Shenoy, D.P.; Jones, T.J.; Humphreys, R.M.; Marengo, M.; Leisenring, J.M.; Nelson, M.J.; Wilson, J.C.; Skrutskie, M.F.; Hinz, P.M.; Hoffmann, W.F.; et al. Adaptive Optics Imaging of VY Canis Majoris at 2-5 μm with LBT/LMIRCam. *Astron. J.* **2013**, *146*, 90. doi: 10.1088/0004-6256/146/4/90. [CrossRef]
102. Gordon, M.S.; Jones, T.J.; Humphreys, R.M.; Ertel, S.; Hinz, P.M.; Hoffmann, W.F.; Stone, J.; Spalding, E.; Vaz, A. Thermal Emission in the Southwest Clump of VY CMa. *Astron. J.* **2019**, *157*, 57. doi: 10.3847/1538-3881/aaf5cb. [CrossRef]
103. Vlemmings, W.H.T.; Diamond, P.J.; van Langevelde, H.J. Circular polarization of water masers in the circumstellar envelopes of late type stars. *Astron. Astrophys.* **2002**, *394*, 589–602. doi: 10.1051/0004-6361:20021166. [CrossRef]
104. Vlemmings, W.H.T.; van Langevelde, H.J.; Diamond, P.J.; Habing, H.J.; Schilizzi, R.T. VLBI astrometry of circumstellar OH masers: Proper motions and parallaxes of four AGB stars. *Astron. Astrophys.* **2003**, *407*, 213–224. doi: 10.1051/0004-6361:20030766. [CrossRef]
105. Vlemmings, W.H.T.; van Langevelde, H.J.; Diamond, P.J. The magnetic field around late-type stars revealed by the circumstellar H_2O masers. *Astron. Astrophys.* **2005**, *434*, 1029–1038. doi: 10.1051/0004-6361:20042488. [CrossRef]
106. Kervella, P.; Lagadec, E.; Montargès, M.; Ridgway, S.T.; Chiavassa, A.; Haubois, X.; Schmid, H.M.; Langlois, M.; Gallenne, A.; Perrin, G. The close circumstellar environment of Betelgeuse. III. SPHERE/ZIMPOL imaging polarimetry in the visible. *Astron. Astrophys.* **2016**, *585*, A28. doi: 10.1051/0004-6361/201527134. [CrossRef]
107. Gies, D.R. Binaries in Massive Star Formation. In *Massive Star Formation: Observations Confront Theory*; ASP Conference Series; Beuther, H., Linz, H., Henning, T., Eds.; Astronomical Society of the Pacific: San Francisco, CA, USA, 2008; Volume 387, p. 93.
108. Sana, H.; de Mink, S.E.; de Koter, A.; Langer, N.; Evans, C.J.; Gieles, M.; Gosset, E.; Izzard, R.G.; Le Bouquin, J.B.; Schneider, F.R.N. Binary Interaction Dominates the Evolution of Massive Stars. *Science* **2012**, *337*, 444. doi: 10.1126/science.1223344. [CrossRef]
109. Neugent, K.F.; Levesque, E.M.; Massey, P. Binary Red Supergiants: A New Method for Detecting B-type Companions. *Astron. J.* **2018**, *156*, 225. doi: 10.3847/1538-3881/aae4e0. [CrossRef]
110. Kochanek, C.S.; Auchettl, K.; Belczynski, K. Stellar binaries that survive supernovae. *Mon. Not. R. Astron. Soc.* **2019**, *485*, 5394–5410. doi: 10.1093/mnras/stz717. [CrossRef]
111. Eldridge, J.J.; Fraser, M.; Smartt, S.J.; Maund, J.R.; Crockett, R.M. The death of massive stars—II. Observational constraints on the progenitors of Type Ibc supernovae. *Mon. Not. R. Astron. Soc.* **2013**, *436*, 774–795. doi: 10.1093/mnras/stt1612. [CrossRef]
112. Levesque, E.M. *Astrophysics of Red Supergiants*; IOP Publishing: Bristol, UK, 2017, doi:10.1088/978-0-7503-1329-2. [CrossRef]

© 2019 by the authors. Licensee MDPI, Basel, Switzerland. This article is an open access article distributed under the terms and conditions of the Creative Commons Attribution (CC BY) license (http://creativecommons.org/licenses/by/4.0/).

Review

A Census of B[e] Supergiants

Michaela Kraus

Astronomical Institute, Czech Academy of Sciences, Fričova 298, 251 65 Ondřejov, Czech Republic; michaela.kraus@asu.cas.cz

Received: 7 August 2019; Accepted: 25 September 2019; Published: 29 September 2019

Abstract: Stellar evolution theory is most uncertain for massive stars. For reliable predictions of the evolution of massive stars and their final fate, solid constraints on the physical parameters, and their changes along the evolution and in different environments, are required. Massive stars evolve through a variety of short transition phases, in which they can experience large mass-loss either in the form of dense winds or via sudden eruptions. The B[e] supergiants comprise one such group of massive transition objects. They are characterized by dense, dusty disks of yet unknown origin. In the Milky Way, identification and classification of B[e] supergiants is usually hampered by their uncertain distances, hence luminosities, and by the confusion of low-luminosity candidates with massive pre-main sequence objects. The extragalactic objects are often mistaken as quiescent or candidate luminous blue variables, with whom B[e] supergiants share a number of spectroscopic characteristics. In this review, proper criteria are provided, based on which B[e] supergiants can be unambiguously classified and separated from other high luminosity post-main sequence stars and pre-main sequence stars. Using these criteria, the B[e] supergiant samples in diverse galaxies are critically inspected, to achieve a reliable census of the current population.

Keywords: stars: massive; stars: emission line, Be; supergiants; stars: winds, outflows; circumstellar matter

1. Introduction

Massive stars play a major role in the evolution of their host galaxies. Via stellar winds, they strongly enrich the interstellar medium (ISM) with chemically processed material and deposit large amounts of momentum and energy into their surroundings during their entire lifetime, from the main-sequence up to their final fate as spectacular supernova explosions (e.g., [1–3]). The released energy provides the ionizing radiation, substantially supplies the global energy budget of the host galaxy and significantly contributes to shaping the local ISM, whereas the released material condenses into molecules and dust, providing the cradles for the next generation of stars and planets (e.g., [4,5]).

Despite their great importance, stellar evolution theory is most uncertain for massive stars due to the often still poor understanding of some physical processes in the stellar interiors (e.g., core convective overshooting, chemical diffusion, internal differential rotation law and angular momentum transport), the excitation and propagation of pulsation instabilities within their atmospheres, the amount of mass loss via stellar (often asymmetric) winds and (irregular) mass ejections, and the role of binarity for certain phases.

From an observational point of view, the post-main sequence domain within the Hertzsprung–Russell (HR) diagram is populated with various types of extreme massive stars. These are found to be in transition phases, in which the stars shed huge amounts of material into their environments, typically via episodic, sometimes even eruptive events. These objects are luminous super- or hypergiants populating the upper part of the HR diagram and spreading from spectral type O to F or even later. The ejected material thereby accumulates in either nebulae, shells, or even disk-like structures.

The mass-loss of massive stars not only critically depends on the physical parameters, such as mass, effective temperature, and rotation speed, but also on the chemical composition of the star. The amount of mass that is lost, within each individual evolutionary phase, determines the fate of the object. It is thus not surprising that relative numbers of various types of massive stars can change drastically among galaxies with different metallicities (e.g., [6]). For reliable predictions of the evolutionary path of massive stars in any environment, solid constraints on the physical parameters, used in modern stellar evolution models, are indispensable. To obtain such constraints, the properties of the members within each class of objects need to be studied in great detail and within a variety of environments. This requires statistically significant samples of stars in each class of objects, suitable for a detailed analysis. The star-forming galaxies within the local Universe, in which the metallicities spread over a factor of about 25 between the most metal poor and the most metal rich representative, are the most ideal sites to tackle this challenge.

This review is devoted to the B[e] supergiants, which comprise one of the various classes of extreme massive stars in transition. The article is structured as follows. First, an overview on the general properties of these stars is given based mostly on the well-studied sample within the Magellanic Clouds (Section 2), followed by a review on how these objects are searched for in various environments (Section 3). A census of the currently known objects in the Local Group galaxies and slightly beyond is presented in Section 4, based on a critical inspection of the properties of the individual candidates. The discussion of the B[e]SG samples and our conclusions are finally summarized in Section 5.

2. B[e] Supergiants

The early-type supergiants include a class of emission-line objects, whose optical spectra display a peculiar character with strong Balmer emission along with narrow emission lines from permitted and forbidden transitions (e.g., [7–10]). The latter are indicative of a cool and slowly expanding medium. With the advent of ultraviolet (UV) observations taken with the International Ultraviolet Explorer (IUE), these stars were found to display very broad blueshifted resonance lines of highly ionized elements in this spectral range. These resonance lines originate from a hot and fast stellar line-driven wind which is very typical for supergiants in this temperature and luminosity range.

Another peculiar property of these stars was discovered in the near-infrared, in which these objects possess a pronounced excess emission pointing to hot circumstellar dust [11–16]. This dust was proposed to be most likely produced within the slow and cool component and to possibly populate a ring or disk-like region at far distances from the luminous central objects [15].

In the HR diagram, these objects are all found beyond the main-sequence and with luminosities spreading from about $\log L/L_\odot \sim 4$ to about $\log L/L_\odot \sim 6$, implying that they are all evolved, massive stars. This luminosity range was determined from the sample residing in the Magellanic Clouds (MCs), for which the luminosity determination is unquestionable, due to the low extinction towards the MCs and their well constraint distances. The classification of Galactic objects as supergiants bears much higher uncertainties due to their often poorly constrained distances, hence luminosities. We come back to this issue in Section 4.4.

The position of the MC sample in the HR diagram is shown in Figure 1 for the values of luminosity and effective temperature listed in Table 1. The stellar parameters (effective temperature T_{eff}, visual magnitude V, and color excess E(B-V)) of the sample have been taken from the references listed in the last column of Table 1. For the calculations of the luminosities, distance moduli of 18.5 and 18.9 mag, respectively, for the Large and Small Magellanic Clouds have been utilized (see the review paper by Humphreys, this volume) along with bolometric corrections from [17].

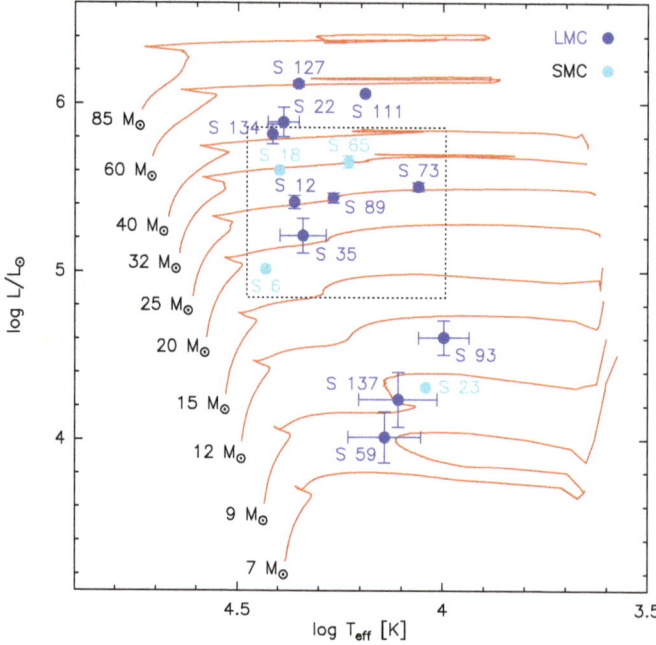

Figure 1. HR diagram showing the positions of the classical MC B[e]SG sample [18]. The stellar evolutionary tracks at SMC metallicity for stars rotating initially with 40% of their critical velocity are also included (from [19]). The dotted square contains objects that display CO band emission (except for S 89, see Section 2.3). For brevity and readability, the identifiers LHA 120 and LHA 115 for objects within the LMC and SMC, respectively, have been omitted.

Table 1. Names and parameters of the established B[e]SG sample within the Magellanic Clouds.

Object	Other Common Identifiers	log T_{eff}	log $L/L_\odot$	V	E(B–V)	Ref.
	Large Magellanic Cloud B[e]SGs					
LHA 120-S 12	SK −67 23	4.36	5.41 ± 0.04	12.6	0.2–0.25	[16]
LHA 120-S 22	HD 34664, SK −67 64	4.39 ± 0.04	5.89 ± 0.09	11.7	0.25–0.3	[16]
LHA 120-S 35	SK −66 97	4.34 ± 0.06	5.21 ± 0.10	12.5	0.07	[20]
LHA 120-S 59	...	4.14 ± 0.09	4.01 ± 0.15	14.4	0.05	[20]
LHA 120-S 73	RMC 66, HD 268835	4.06	5.51 ± 0.03	10.6	0.12–0.15	[13]
LHA 120-S 89	RMC 82, HD 269217	4.27	5.44 ± 0.03	12.0	0.20	[16]
LHA 120-S 93	SK −68 66	4.00 ± 0.06	4.61 ± 0.10	12.7	0.22	[20]
LHA 120-S 111 [a]	HD 269599	4.19	6.06	10.3	0.28	[21]
LHA 120-S 127	RMC 126, HD 37974	4.35	6.12 ± 0.02	10.9	0.25	[15]
LHA 120-S 134 [b,c]	HD 38489, SK −69 259	4.42	5.82 ± 0.06	12.0	0.2–0.25	[16]
LHA 120-S 137	...	4.11 ± 0.10	4.24 ± 0.16	14.0	0.17	[20]
	Small Magellanic Cloud B[e]SGs					
LHA 115-S 6 [b]	RMC 4, AzV 16	4.43	5.02	13.0	0.07	[22]
LHA 115-S 18 [b,c]	AzV 154	4.40	5.60	13.3	0.4	[23]
LHA 115-S 23	AzV 172	4.04	4.31	13.3	0.03–0.1	[24]
LHA 115-S 65	RMC 50	4.23	5.65 ± 0.04	11.6	0.15–0.2	[16]

Note: Former designations of some of the objects as Hen S # (see, e.g., [25]) were omitted here, as these are not SIMBAD identifiers. However, LHA 120-S and LHA 115-S, respectively, and the former Hen S numbers refer to the same objects. [a] The star is also listed as RMC 105 in SIMBAD, but this designation should be used for a neighboring, normal B-type star in this dense cluster (see [26]). [b] Confirmed or suspected binary. [c] X-ray source [27–29].

The presence of dust around an early-type (typically of spectral type B) supergiant, along with the often pure emission-line spectra with numerous forbidden lines predominantly of [Fe II] and [O I] finally resulted in the designation of these objects as B[e] supergiants (B[e]SGs)[1].

2.1. General Aspects of B[e]SG Stars' Disks

There is compelling evidence that B[e]SGs are surrounded by gaseous and dusty disks. The simultaneous presence of a hot and fast polar wind traced in the UV, and the cool and slow equatorial wind traced at optical wavelengths led to the assignment of a so-called hybrid or two-component wind model [15]. For this two-component wind, a density contrast between the equatorial and polar components of 100–1000 was proposed, meaning that the equatorial wind might be assigned the character of an outflowing disk [23].

The degree of non-sphericity of the envelopes and the latitude dependence of the wind density, respectively, are pursued by the measured net intrinsic polarization [30,31] and from spectropolarimetric observations [32,33]. The often high degree of intrinsic polarization support the idea of a combination of Thomson scattering by free electrons and Mie scattering by dust in a circumstellar disk [34].

If the disks of B[e]SGs are supposed to form from a high-density equatorial stellar outflow, there should be a transition zone between the atomic gas and the location of the dust, in which molecules can form in substantial amounts, because the high gas density can shield the material from the direct irradiation with dissociating UV photons coming from the hot luminous star. In fact, molecular emission, in particular of the first-overtone bands of carbon monoxide (CO), has been detected in the K-band spectra of a number of B[e]SGs in the Galaxy and the MCs (e.g., [21,35–44], see Table 2). To produce the characteristic observed emission spectra with several individual band heads, temperatures of the CO gas higher than $\sim$2000 K are required. These temperatures are in excess of the dust sublimation temperature, which is on the order of $\sim$1500 K, placing the CO emitting region closer to the star than the dust.

Additional hot molecular emission from silicon oxide (SiO) has been identified in four Galactic B[e]SGs [45], and a feature arising in the optical spectrum, which has been tentatively identified as emission from titanium oxide (TiO), was reported from six MC B[e]SGs [23,43,44,46]. However, to date, no systematic surveys for molecular emission has been performed, so that these numbers are not representative for the existence or absence of molecules in the environments of B[e]SGs. For instance, SiO emission has not been searched for yet in any of the MC B[e]SGs, and only those Galactic B[e]SGs with the most intense CO band emission have been observed in the wavelength range of the first-overtone band of SiO arising in the L-band. Hence, one might expect to find molecular emission from SiO in many more objects, but also emission from other yet undiscovered molecules that might form in the environments of B[e]SGs. What is interesting though is the fact that all MC stars displaying TiO emission also have CO emission, whereas the opposite does not hold. No detection of TiO from Galactic B[e]SGs has been reported so far.

Finally, the power of optical interferometry operating at near- and mid-infrared wavelengths should be mentioned when talking about the disks of B[e]SGs. Based on this technique, the disks of the closest and infrared brightest Galactic objects could be spatially resolved, providing precise measurements of the disk inclinations, disk sizes, and the distances of the emitting material (dust, CO gas, and ionized gas traced by the Br γ emission) from the central star (see [38,47–52]).

---

[1] Note that these objects have previously been abbreviated sgB[e] [25], to separate them from other stars showing the B[e] phenomenon. We prefer the designation B[e]SG, to be in line with the naming and abbreviation of other types of supergiants such as blue supergiant (BSG) and red supergiant (RSG).

Table 2. Presence of disk tracers in the optical and near-IR spectra of the Galactic and Magellanic Cloud B[e]SG samples.

Object	[Ca II]	[O I][a]	Ref.	CO	$^{12}C/^{13}C$	Ref.	TiO	SiO	Ref.
Large Magellanic Cloud B[e]SGs									
LHA 120-S 12	yes	no	[53]	yes	20 ± 2	[21,41]	yes	...	[23]
LHA 120-S 22	yes	yes	[53]	no	...	[21,41]	no	...	TW [b]
LHA 120-S 35	yes	yes	[44]	yes	10 ± 2	[41,44]	yes	...	[44]
LHA 120-S 59	no	yes	[54]	?	...	[41]	no	...	TW [b]
LHA 120-S 73	yes	yes	[43,53]	yes	9 ± 1	[21,41,43]	yes	...	[55]
LHA 120-S 89	no	no	TW [b]	no	...	[21,41]	no	...	TW [b]
LHA 120-S 93	yes	no	TW [b]	no	...	[41]	no	...	TW [b]
LHA 120-S 111	yes	yes	[53]	no	...	[21]	yes	...	[23]
LHA 120-S 127	yes	yes	[53,56]	no	...	[21,41]	no	...	TW [b]
LHA 120-S 134	yes	yes	[53]	yes	15 ± 2	[21,41]	yes	...	[23]
LHA 120-S 137	no	yes	TW [b]	no	...	[41]	no	...	TW [b]
Small Magellanic Cloud B[e]SGs									
LHA 115-S 6	?	yes	TW [b]	yes	12 ± 2	[36,41]	no	...	TW [b]
LHA 115-S 18	yes	yes	[53]	yes	20 ± 5	[37,41]	yes	...	[23,46]
LHA 115-S 23 [c]	?	no	TW [b]	no	...	TW [d]	no	...	TW [b]
LHA 115-S 65	yes	yes	[53,57]	yes [e]	20 ± 5	[40,41]	no	...	TW [b]
Galactic B[e]SGs/B[e]SG Candidates									
MWC 137	no	no	[55,58]	yes	25 ± 2	[39,41,59]	no	...	TW [b]
MWC 349	yes	...	[60]	yes	4 ± 1	[61,62]	...	no	[62]
GG Car	yes	no	[58]	yes	15 ± 5	[35,37,39,41,42,58]	no	...	TW [b]
Hen 3-298	yes	yes	[58,63]	yes	20 ± 5	[39,41,58,63]	no	...	TW [b]
CPD-52 9243	yes	no	[58]	yes	...	[35,39,58]	no	yes	TW [b], [45]
HD 327083	yes	no	[58]	yes	...	[39,58]	no	yes	TW [b], [45]
MWC 300	no	yes	TW [b]	no	...	[39,64]	no	...	TW [b]
AS 381	no	no	[65]	abs	?	[39,64,65]	...	...	...
CPD-57 2874	yes	no	[58]	yes	...	[35,39,58]	no	yes	TW [b], [45]
Hen 3-938	yes	yes	[54]	...	...	...	no	...	TW [b]
MWC 342	no	yes	[60,66]	...	...	...	...	...	...
Hen 3-303	no	no	TW [b]	no	...	[63]	no	...	TW [b]
CD-42 11721	no	yes	[67]	no	...	[39]	no	...	TW [b]
HD 87643	yes	no	[58]	yes [e]	...	[39,58]	no	...	TW [b]
HD 62623 [c]	yes	yes	[58]	yes	...	[39,58]	no	yes	TW [b], [45]

Note: TW, This work; abs, in absorption; ?, uncertain detection/no value available; ..., no information. [a] Refers to the presence of [O I]λ5577. All sample stars show emission of [O I]$\lambda\lambda$6300,6364. [b] Based on (unpublished) high-resolution optical spectra taken between 2005 and 2017 with FEROS at the MPG 2.2 m telescope. [c] A[e]SG due to early-A spectral type assignment [68,69]. [d] No indication of CO band features seen in a K-band spectrum (unpublished) taken on 20 October 2013 with OSIRIS at the Southern Astrophysical Research (SOAR) Telescope. [e] No CO emission was detected during the observations taken between 1987 and 1989 [35,36].

2.2. Disk Dynamics and Structure

Determination of the kinematics within dense circumstellar environments requires the use of reliable tracers. High-resolution near-infrared spectroscopic observations have revealed that the band heads of the CO emission from B[e]SGs typically display a characteristic shape, consisting of a blue-shifted shoulder and a red-shifted maximum. For the generation of such a band head profile, the individual CO rotation-vibration lines, superimposing within the region of the band head, must display double-peaked profiles (see Figure 2). Such line profiles can originate either from a circumstellar ring of gas expanding with constant velocity (constant outflow), or from rotational motion of a ring of gas around the central object. To discriminate between the two scenarios, complementary tracers are needed.

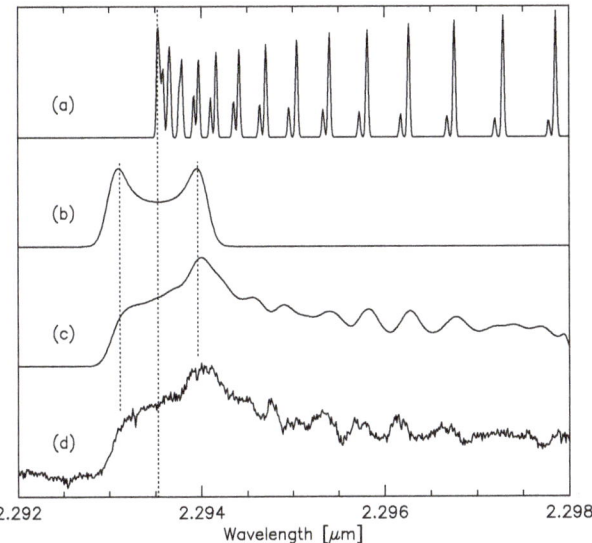

Figure 2. Sketch of the generation of the typical CO band head profile. (**a**) Spectrum around the (2-0) band head of the CO first-overtone bands for a hot gas with velocity dispersion of a few km s^{-1}. (**b**) Profile of a single line from a rotating gas ring with a velocity, projected to the line of sight, of 66 km s^{-1} as seen with a spectral resolution of 6 km s^{-1}. (**c**) Total synthetic CO band head spectrum resulting from the convolution of the band transitions in (**a**) with the profile of the ring in (**b**). (**d**) CO band head observations of the Galactic B[e]SG CPD-57 2874 [58].

The SiO band emission seen in four Galactic B[e]SGs displays a similar shape of the band heads. Detailed modeling revealed that in each object the SiO bands required a slightly lower value of the velocity [45] than the CO bands. The SiO molecule is less stable than CO, meaning that it can form and persist only at lower temperatures. This fact naturally places the region where SiO molecules are expected to form, and hence the SiO band emitting region, at slower orbital velocities and greater distances from the central object.

As CO is the most stable molecule, its formation and emission region marks the inner edge of the molecular disk. Closer to the star, tracers for the kinematics need to be found from line emission of the atomic gas. Here, the lines from forbidden transitions are most suitable, because their emission is optically thin, so that their profile shapes contain the full velocity information of their formation region [56,57]. Of particular interest are hereby the [O I] lines, because they are one of the defining characteristics of the B[e] phenomenon and hence observed in all B[e]SGs. The ionization potential of O I is about the same as the one for H I, which means that within the [O I] line forming regions, hydrogen should be basically neutral as well, restricting the formation region of the [O I] line emission to the neutral regions within the circumstellar disk. While recombination in the equatorial region close to the star might be achieved, e.g., with the model of a latitude dependent wind [70–72], the requirement of a hydrogen neutral environment severely limits the number of free electrons that will be available to collisionally excite the levels within O I from which the forbidden transitions emerge. Consequently, the [O I] lines arise in regions with high total density, but low electron density.

The profiles of the [O I] lines often display double-peaks, in line with their formation in the disk. Typically, the [O I] λ5577 line, which arises from a higher level than the $\lambda\lambda$6300,6364 lines, is broader, indicating spatially distinct formation regions of the emissions, with the [O I] λ5577 line being formed at higher velocities and higher densities and hence closer to the star than the other two lines.

With the identification of the lines of [Ca II] $\lambda\lambda$7291,7324 in the spectra of numerous B[e]SGs, a further highly valuable tracer for the disk kinematics has been found. These lines typically display double-peaked profiles as well, with velocities comparable to or even greater than the one traced by the [O I] λ5577 line [53,57,58]. This implies that they form in the same region, or at least very close to each other, which is in agreement with their comparably high critical density. Since the [O I] λ5577 line is not always detectable, the [Ca II] lines thus provide a suitable, complementary benchmark for the dynamics within the disks of B[e]SGs.

In summary, the optical and near-infrared spectra provide emission features from several species, which are suitable to pin down the kinematics within the disks of B[e]SGs at various distances from the star, and Table 2 includes the information on the detection of the individual tracers in the MC sample. Based on the physical constraints outlined above, the logical order of the appearance of the divers tracers from inside out would be: [Ca II], [O I], CO bands and SiO bands. The velocity information carried by these species thereby implies a decrease with increasing distance from the star. While an equatorial, outwards decelerating outflow might be able to explain some of the observed line profiles [56], the velocity patterns seem to be in better agreement with (quasi-)Keplerian rotation. In this respect, it is interesting to note that Keplerian rotation has been made directly discernible by means of spectro-interferometric observations. The rotational motion of the CO gas has been derived based on the differential phase spectrum [38,52], and the rotational motion of the ionized gas based on the spatially resolved Brγ emission [49].

While, in general, the rotational motion of the material within the circumstellar disks of B[e]SGs seems now to be well established, possibly in connection with a (very) slow outflow component [57], recent investigations of the spatial distribution of the circumstellar gas revealed that it is more likely accumulated in multiple rings, partial rings, and possible spiral arm-like structures rather than in a smooth disk [43,44,58]. These rings might result from multiple mass ejection phases caused by (pulsational) instabilities acting in the outer layers of these luminous objects (e.g., [73–75]), or from binary interaction in close systems, as seems to be the case for some of the Galactic objects [58], in which the rings are circumbinary rather than circumstellar. Other disk-forming mechanisms that have been proposed in the literature over the years include equatorial mass-loss from a critically rotating star [76,77], the rotationally induced bi-stability mechanism [78], the slow-wind solution [79], and the combination of the latter two [80]. For an overview including a detailed description of the various models and their limitations, see [81].

The circumstellar material of many MC objects appears durable. This is evidenced by their emission features and their infrared photometry that both display no considerable variability over several decades, in combination with chemically processed dust displaying emission from crystalline silicates [82]. It is tempting to imagine that in such an environment even minor bodies might have formed from the long-lived disk material, creating gaps within the disk in radial direction and hence leading to the formation of the presumed ring structures [43]. These minor bodies or possible planets can also stabilize the neighboring rings, in analogy to the shepherd moons in planetary systems. However, thus far, there is insufficient observational evidence that might support the validity of such a scenario.

2.3. Current Evolutionary State of B[e]SGs

The formation mechanism of the observed gaseous and dusty rings or disk-like structures around B[e]SGs is certainly one of the most important yet unsolved issues. Equally important questions arise: What is the evolutionary phase of B[e]SGs? What is their evolutionary connection to other evolved massive stars? Does such a connection exist at all? While the question on the relation between B[e]SGs and other evolved objects is beyond the scope of this review, we briefly elucidate the current knowledge about the evolutionary status of B[e]SGs. Considering the MC sample, it is obvious from Figure 1 that all objects have evolved off the main sequence. Whether this occurred only recently, or whether B[e]SGs might be on a blue loop or blueward evolution after having passed through the turning point on the cool edge of their track, is still an open issue, in particular, since we lack clear methods for age determinations of these emission-line objects, which only very rarely are detected in clusters[2].

In this regard, the detection of clear signs of the ^{13}C isotope in the form of ^{13}CO band emission from a number of MC B[e]SGs ([41,86], see Table 2) was one major step forward. The appearance of these bands has been predicted based on theoretical model computations for a variation of the carbon isotope ratio ^{12}C/^{13}C [87]. As surface abundance calculations have shown, this ratio will drop during the evolution of massive stars from an initial, interstellar value of ~90 down to values <5, depending on the initial mass of the star and its initial rotation speed. The surface material enriched in ^{13}C is transported via winds to the environments, where it will cool and condense into ^{13}CO molecules, whose emission can be observed in the K-band, together with the emission from the main isotope, ^{12}CO (see Figure 3). Hence, the detected amount of ^{13}CO is a measure for the stellar surface enrichment in ^{13}C at the time the material, which is currently traced in the molecular emission, has been released from the stellar surface.

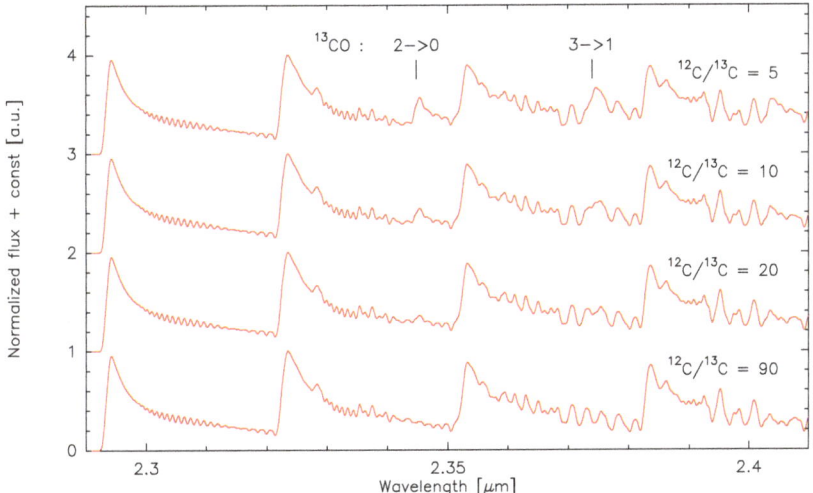

Figure 3. Synthetic spectra of the combined emission from ^{12}CO and ^{13}CO for different values of the ^{12}C/^{13}C ratio. The computations have been performed for the following physical parameters: a ^{12}CO column density of 2×10^{21} cm^{-2}, a gas temperature of 3000 K, a line-of-sight rotational velocity of 66 km s^{-1}, and a spectral resolution of 50 km s^{-1}.

From the MC sample of 15 objects, seven have been found to display CO band emission, and all of them display clear indication of enrichment in ^{13}C (see Table 2). Interestingly, all these objects with CO emission cluster in the same region of the HR diagram, as indicated by the dotted black square in Figure 1, i.e., in the luminosity range $\log L/L_\odot = 5.0$–5.8. None of the three most luminous stars (S 22, S 111, and S 127) or of the four low-luminosity objects (S 23, S 59, S 93, and S 137) displays clear signs of CO emission. One outlier in the luminosity domain occupied by the CO emitting B[e]SGs is the star S 89, which also has no detectable CO band emission [41].

The absence of measurable CO band emission might have different reasons. Either the intensity of the emission is too low to be detectable against the strong near-IR continuum[3], or the density of

---

[2] Currently, only four B[e]SGs are reported to be cluster members: the two LMC objects LHA 120-S 111 in the compact cluster NGC 1994 [26] and LHA 120-S 35 in SL482 [44], and the two Galactic sources MWC 137 in SH 2-266 [83] and Wd1-9 in Westerlund 1 [84,85].
[3] This spectral region suffers from strong telluric contamination, which is not always easy to remove, so that especially weak CO emission features might be hidden within telluric remnants.

the molecular gas might be very high, resulting in optically thick emission, which no longer has a characteristic band head structure. Another possibility would be that the CO emission from these stars might have variable CO band emission and they have thus far always been observed in phases of no emission. In this context, it is interesting to refer to the SMC object LHA 115-S 65, in which CO band emission suddenly occurred, while observations taken about nine months earlier did not detect any molecular features [40]. Alternatively, since we now know that the material is most probably concentrated in rings, the conditions within the circumstellar environment in terms of density and temperature might not be favorable for the excitation of the first-overtone bands. For those stars, observations in the spectral region of the fundamental bands might therefore be a possibility to search for cooler CO gas.

The measured ^{12}C/^{13}C isotope ratios of the MC B[e]SGs are all very similar, spreading from 9 to 20. This might point towards a similar formation history of the circumstellar material, i.e., a similar phase in the evolution (considering they are single stars or at least unaffected by a possible companion) when the enriched material was ejected. Considering that stars are typically born with some intrinsic rotation velocity, rotational mixing in combination with enhanced mass-loss may drive the enrichment of the stellar surface with ^{13}C already in early stages of the evolution of massive stars. This can be seen in Figure 4, where the evolutionary tracks of a 32 $M_\odot$ star with solar metallicity [88] and for a variety of initial rotation velocities are shown. The covered rotation speeds spread from $v/v_{\text{crit}} = 0$ to $v/v_{\text{crit}} = 0.4$, which correspond to values of $\Omega/\Omega_{\text{crit}}$ from 0 to 0.568. The interpolation of the tracks has been performed with the SCYCLIST tool[4] provided by the Geneva group.

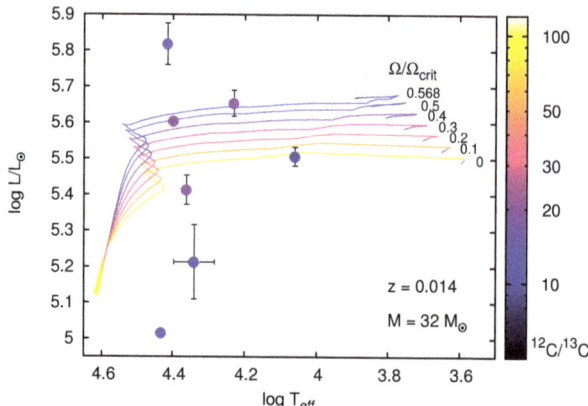

Figure 4. Evolution of the ^{12}C/^{13}C isotope ratio along the solar metallicity tracks of a star with initial mass of 32 $M_\odot$ and initial rotation speeds v/v_{crit} ranging from 0 to 0.4 (corresponding to $\Omega/\Omega_{\text{crit}} = 0.0; 0.1; 0.2; 0.3; 0.4; 0.5; 0.568$). The individual tracks have been obtained from the interpolation tool SCYCLIST provided by the Geneva group. For clarity of the plot, we truncated the evolutionary tracks within the red supergiant regions. Included are the positions of the MC B[e]SG sample from Table 2 with known ^{12}C/^{13}C ratio, following the same color coding as for the tracks. The Galactic objects are excluded due to their highly uncertain luminosities. Depending on the initial rotation speed of the star, the observed ratio can be reached either in the pre-RSG (moderate rotator) or post-RSG (slow rotator) phase.

The color coding along the tracks refers to the values of the ^{12}C/^{13}C isotope ratio on the stellar surface. Figure 4 also includes the positions of the seven MC B[e]SGs with known values of the

---

[4] https://www.unige.ch/sciences/astro/evolution/en/database/syclist/.

^{12}CO/^{13}CO isotope ratios. Their colors correspond to the same color coding as the evolutionary tracks. Obviously, the observed ratios for the sample stars might be reached either along or after the main-sequence evolution for stars rotating initially with rates $\Omega/\Omega_{crit} \geq 0.3$. However, they might also be reached during or after the red supergiant stage for stars rotating initially with rates smaller than $\Omega/\Omega_{crit} \leq 0.3$. As we do not know the initial rotation speeds of the progenitor stars of these B[e]SGs, the measured values of the ^{12}CO/^{13}CO isotope ratio alone cannot solve the issue with the current evolutionary state of the objects.

If B[e]SGs represent a specific phase in the evolution of massive stars, then these objects should also exist in other environments with high content of massive stars. Searching for representatives of B[e]SGs in other galaxies and studying their properties and number statistics at various metallicities might help to unveil their disk/ring formation mechanism, to pin down their evolutionary phase (pre- versus post-RSG), and to set constraints on the evolution of massive stars in general.

3. Identification of B[e] Supergiants in the Local Universe

Identifying and classifying B[e]SGs is a tedious job, whether in the Milky Way, where we face large amounts of foreground extinction and the issue with the often unknown distances, or in other galaxies, where we have strong contamination with foreground sources, often crowded regions, and the faintness of the objects. While the first B[e]SGs have been found rather accidentally, nowadays dedicated surveys can make use of the established classification criteria. As mentioned in Section 2, there are basically four characteristics a star should fulfill to be classified as B[e]SGs [18,25]. It should display:

- A spectral-type B (with extensions to late-O and early-A types), as evidenced from the hot underlying continuum or from photospheric absorption features[5], and a luminosity higher than $\log L/L_\odot \sim 4$.
- Very intense Balmer lines dominating the optical spectrum with an equivalent width of Hα reaching up to 1000 Å.
- A spectrum indicating the hybrid character of the environment, consisting of a hot, fast line-driven wind coexisting with a cool, slow component from which the forbidden emission lines from neutral and low-ionized metals originate ([O I], [Fe II]).
- Strong near- and mid-infrared excess emission, indicative of hot circumstellar dust.

Despite these defining criteria, the identification of extragalactic B[e]SGs is not as straightforward, because suitable candidates need first to be found based on other means. These candidates can then be further investigated to search for these, mostly spectroscopic, characteristics.

A suitable approach is to search for B[e]SG candidates among the luminous, blue objects identified in imaging surveys that have been performed for several galaxies over the past ~30 years. For instance, the early photographic and photometric surveys of M31 [89,90] and M33 [91] revealed (amongst many other objects) the most luminous hot stars and the brightest blue supergiants, which could then be studied spectroscopically to obtain indications for their possible nature (e.g., [92]). A milestone for the identification of evolved massive stars was certainly the Local Group Galaxies Survey (LGGS) project [93,94], which resulted in the discovery of numerous putative emission-line stars in M31, M33, and seven more dwarf galaxies. Spectroscopic follow-ups were used to sort out H II regions, and to match the remaining objects with the various known categories of evolved massive stars. A major result of this survey was the identification of numerous objects that were dubbed as luminous blue variable (LBV) candidates [6,95], based on the appearance of their blue emission-line or P Cygni-type spectra that resemble confirmed LBVs in quiescence. Since the blue optical spectra of LBVs in their

---

[5] We would like to stress that the high-luminosity B[e]SGs barely display photospheric absorption lines, whereas the low-luminosity B[e]SGs typically do.

quiescence state display a number of common characteristics with B[e]SGs [95–97], it was expected that this sample contains a number of B[e]SGs.

Surprisingly, when analyzed in more detail [97–99], this bunch of newly identified LBV candidates in M31 and M33 turned out to be a mixed bag containing not only B[e]SGs candidates, but also so-called Fe II emission stars (with neither [O I] nor [Fe II] emission and lacking warm dust), and warm hypergiants (with lots of dust, possible [O I] and/or [Fe II] but of spectral type A–F), with only a few objects left to be considered as LBV candidates (with no [O I] emission and lacking hot dust).

As both LBVs and B[e]SGs are luminous blue supergiants, they share the same optical colors. Hence, one step to distinguish these two groups of objects is to inspect their location in infrared color–color diagrams (e.g., [16,20,41,97,99]). The hot ($\sim$1000 K) circumstellar dust of B[e]SGs results in significantly increased near-IR emission. On the other hand, LBVs can be associated to cold, dusty environments such as the circumstellar shells recently discovered around many LBVs and LBV candidates with the Spitzer Space Telescope at 24 µm (e.g., [100,101]). The separation of B[e]SGs from quiescent LBVs, based on their diverse IR properties, is demonstrated in Figure 5 for the known samples of MC objects, limiting to the confirmed and generally accepted LBVs in the LMC [102] and including one confirmed object from the SMC (R40, [103]). Shown are two different color–color diagrams (J–H versus H–K and W1–W2 versus W2–W4). The IR colors of the objects are listed in Table 3. They result from the JHK-band magnitudes obtained from the 2MASS point source catalog[6] [104], and from the mid-IR magnitudes (W1, W2, W4) collected with the Wide-field Infrared Survey Explorer[7] (WISE [105]). From the many possibilities of near- and mid-IR color–color diagrams, these two show the clearest separation between the two groups of objects.

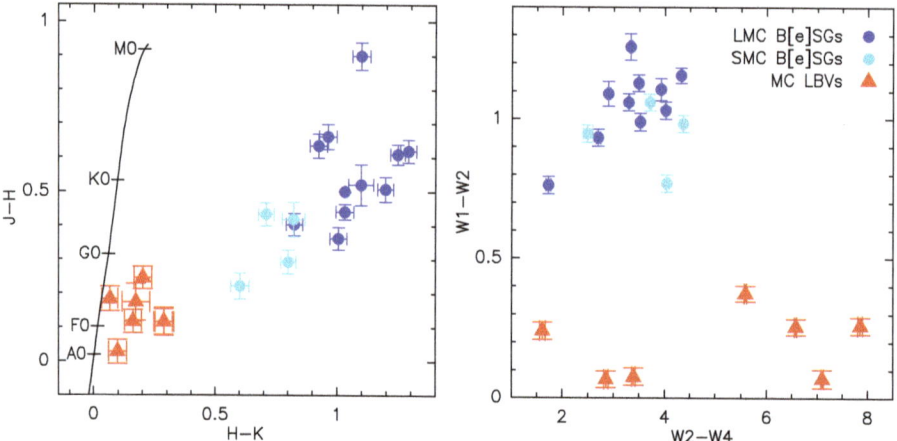

Figure 5. Demonstration of the separation of the B[e]SGs from the quiescent LBVs within the near-IR (J–H versus H–K diagram (**left**)) and the WISE diagram (W1–W2 versus W2–W4 (**right**)). Shown are the positions of the classical MC B[e]SG sample and of the MC LBV sample. IR colors of the objects are provided in Table 3. The solid line represents the positions of regular supergiants with empirical colors taken from [106] for solar metallicity stars.

Another clear distinctive feature between B[e]SGs and LBVs is the S Dor cycle of the latter, while B[e]SGs are typically not undergoing this type of variability. However, to identify such S Dor excursions of the stars within the HR diagram is a time-consuming (though important) task, because it requires regular monitoring of the whole sample.

---

[6] http://cdsarc.u-strasbg.fr/viz-bin/cat/II/246.
[7] http://cdsarc.u-strasbg.fr/viz-bin/cat/II/311.

In addition, dedicated spectroscopic observations are required to search for the characteristic forbidden emission lines of [O I] and possible [Ca II] in the red portion of the optical spectra, and to search for possible molecular emission of CO in their near-IR spectra, because LBVs typically do not show these sets of forbidden lines and CO band emission[8] [21,35,41].

Table 3. IR colors of Magellanic Cloud B[e]SGs and LBVs.

Object	J–H	H–K	W1–W2	W2–W4	Class
Large Magellanic Cloud					
LHA 120-S 12	0.634 ± 0.035	0.924 ± 0.035	0.761 ± 0.031	1.723 ± 0.051	B[e]SG
LHA 120-S 22	0.619 ± 0.033	1.289 ± 0.033	1.258 ± 0.047	3.328 ± 0.032	B[e]SG
LHA 120-S 35	0.520 ± 0.060	1.096 ± 0.051	1.156 ± 0.027	4.323 ± 0.021	B[e]SG
LHA 120-S 59	0.661 ± 0.037	0.962 ± 0.035	0.988 ± 0.031	3.514 ± 0.033	B[e]SG
LHA 120-S 73	0.403 ± 0.033	0.825 ± 0.033	1.030 ± 0.031	4.015 ± 0.027	B[e]SG
LHA 120-S 89	0.440 ± 0.022	1.030 ± 0.036	1.059 ± 0.031	3.289 ± 0.026	B[e]SG
LHA 120-S 93	0.507 ± 0.036	1.195 ± 0.033	1.128 ± 0.030	3.486 ± 0.025	B[e]SG
LHA 120-S 111 [a]	0.500 ± 0.000	1.030 ± 0.000	...	...	B[e]SG
LHA 120-S 127	0.362 ± 0.033	1.004 ± 0.033	1.106 ± 0.039	3.922 ± 0.028	B[e]SG
LHA 120-S 134	0.609 ± 0.030	1.245 ± 0.030	1.090 ± 0.045	2.892 ± 0.035	B[e]SG
LHA 120-S 137	0.899 ± 0.040	1.097 ± 0.037	0.931 ± 0.030	2.691 ± 0.045	B[e]SG
LHA 120-S 96 (=S Dor)	0.173 ± 0.054	0.172 ± 0.057	0.374 ± 0.028	5.586 ± 0.022	LBV
LHA 120-S 116 (=R110)	0.116 ± 0.034	0.161 ± 0.036	0.076 ± 0.032	3.391 ± 0.034	LBV
LHA 120-S 128 (=R127)	0.115 ± 0.037	0.286 ± 0.036	0.255 ± 0.029	6.573 ± 0.021	LBV
LHA 120-S 155 (=R71)	0.027 ± 0.035	0.097 ± 0.038	0.069 ± 0.033	7.095 ± 0.024	LBV
CPD-69 463 (=R143)	0.244 ± 0.033	0.201 ± 0.035	0.259 ± 0.030	7.851 ± 0.049	LBV
LHA 120-S 83 (=Sk −69 142a)	0.115 ± 0.041	0.287 ± 0.041	0.239 ± 0.031	1.609 ± 0.055	LBV
Small Magellanic Cloud					
LHA 115-S 6	0.433 ± 0.034	0.709 ± 0.0319	0.769 ± 0.031	4.037 ± 0.035	B[e]SG
LHA 115-S 18	0.418 ± 0.050	0.822 ± 0.0460	1.059 ± 0.030	3.705 ± 0.031	B[e]SG
LHA 115-S 23	0.221 ± 0.038	0.601 ± 0.0382	0.946 ± 0.031	2.484 ± 0.096	B[e]SG
LHA 115-S 65	0.292 ± 0.035	0.800 ± 0.0311	0.982 ± 0.030	4.363 ± 0.027	B[e]SG
LHA 115-S 52 (=R40)	0.182 ± 0.036	0.065 ± 0.0330	0.066 ± 0.031	2.851 ± 0.043	LBV

Note: IR photometry for all objects is taken from the 2MASS point source catalog (J, H, and K [104]), except for the stars LHA 120-S 111 [21] and LHA 120-S 89 [108], and from the WISE All-Sky Data Release (W1, W2, and W4 [105]). [a] Despite of the lack of WISE colors, the presence of warm dust is proven by its IR excess seen in the *Spitzer* data, and its IR spectrum that looks like a twin of the one of LHA 120-S 73 (see [82]).

4. A Census of B[e]Sgs

With clearly defined classification characteristics and the proper observational tools at hand, the massive star population within the local Universe can be scanned for suitable B[e]SG candidates. At the moment of writing this review, this is still an ongoing project that requires patience and sufficient telescope time at both optical and infrared facilities. Nevertheless, many new, particularly extragalactic B[e]SG star discoveries were reported in the literature within the past 20 years. The aim of this section is, therefore, to take a closer and critical look at the suggested B[e]SG candidates in order to sort out possible misclassified objects, to check what type of observations are still missing for unambiguous classification of the candidates, and to compile updated lists of confirmed B[e]SGs for the galaxies with a reported B[e]SG population. The starting points for this investigation are the Magellanic Clouds (Section 4.1), moving further out into the Local Group (Section 4.2) and beyond (Section 4.3), before we finally return to the Milky Way (Section 4.4).

---

[8] One exception to this rule is the LBV star HR Car, which occasionally showed CO band emission related to phases when the star was dimmer [107].

The samples in each galaxy are presented in tables, which follow the same structure. The objects are listed under a homogeneous SIMBAD identifier (if possible) in Column 1, reference(s) for the B[e]SG classification of the stars follow in Column 2. Where available, E(B–V) values and their references are provide in Columns 3 and 4, and the four colors (J–H, H–K, W1–W2, and W2–W4) are given along with their errors in the last four columns. The tables are furthermore organized such that in the top part appear the confirmed B[e]SGs. These are stars that fulfill all classification criteria. In the middle part of each table stars with uncertain or controversial classification are listed with their names in parentheses. These are objects that lack one or more of the classification criteria due to incomplete observational datasets. Objects for which different research teams find controversial results such that clarification is needed are also included here. In the bottom of each table, erroneously classified objects are gathered with their names in italic and within parentheses. These are stars for which observational evidence (e.g., a specific color) excludes them from belonging to the class of B[e]SGs.

4.1. Magellanic Clouds

As mentioned in Section 2, the "classical" sample of B[e]SGs resides within the Magellanic Clouds. The first such object, for which the hybrid character was reported, was the LMC star RMC 126 (LHA 120-S 127 [15]), and soon after followed the identification of ten more B[e]SGs in the LMC and four in the SMC ([16,20,22–24,109], see Table 1) based on a dedicated search for similar objects. Since then, a few more stars have been suggested as B[e]SGs candidates. These are presented and discussed in the following. The classical sample of B[e]SGs in the MCs, which fulfill all classification criteria, is listed in Table 1 above, with their IR colors provided in Table 3.

4.1.1. Large Magellanic Cloud

New B[e]SG candidates have been found either by dedicated searches, e.g., from cross-matching catalogs of emission-line stars with near-IR catalogs (e.g., [110]), or more serendipitously as a by-product of deep spectroscopic surveys of specific regions, such as the VLT-FLAMES Tarantula Survey (VFTS [111–113]) that was devoted to the 30 Doradus starburst region. In addition, surveys for other purposes, such as the search for post-asymptotic giant branch stars [114], resulted in new B[e]SG candidates[9]. Seven new B[e]SG objects have been proposed in total, of which only two fulfill the criteria for B[e]SGs, two are considered candidates, and three appear to be misclassified. All seven stars are listed in Table 4. Their locations in the two color–color diagrams are shown in Figure 6 in comparison with the confirmed B[e]SGs and LBVs from Table 3.

A sample of confirmed MC late-type stars and supergiants [115] is included in these color–color diagrams. These serve as reference for objects that might have a late-type companion. The outliers of the late-type stars, especially in the WISE diagram, are objects with high mass-loss.

A sample of Galactic Herbig AeBe (HAeBe) stars is also shown [116]. Only stars with known extinction values and solid magnitudes in all four bands were selected (rejecting objects with reported contamination). As some of these pre-main sequence objects suffer from very high extinction ($A_V > 5$ mag), their colors were corrected using an R_V-dependent extinction law [117] with $R_V = 5.0$, which has been found to be reasonable for HAeBe stars [116]. These pre-main sequence stars were included to check for possible misclassification of objects that have a proposed luminosity ranging around the lower limit for B[e]SGs, because this luminosity range is shared by the most massive HAeBes. In the near-IR diagram, the HAeBes populate a stripe that appears to be parallel and seems to connect seamlessly to the region occupied by the classical B[e]SGs. In the WISE diagram, the HAeBes also seem to populate a stripe adjacent to the B[e]SG domain.

---

[9] I would like to point out that from the proposed 12 newly discovered B[e] stars only one was found to be a supergiant (see Table 4). The others have been carefully inspected in collaboration with Devika Kamath, and they did not fulfill the requirements. These results are yet unpublished.

In the following, the reasons for classification of the new LMC objects as either confirmed or candidate, or for rejecting them as B[e]SGs are briefly presented.

Table 4. Confirmed and candidate B[e]SGs in the LMC. Misclassified objects are listed in the bottom part of the table.

Object	Ref.	E(B–V)	Ref.	J–H	H–K	W1–W2	W2–W4
Confirmed B[e]SGs							
LHA 120-S 165	[114]	...	...	0.735 ± 0.041	0.98 ± 0.040	0.886 ± 0.029	3.385 ± 0.037
ARDB 54	[54,118]	0.11	[54]	0.340 ± 0.057	0.810 ± 0.04	0.402 ± 0.023	2.057 ± 0.054
Uncertain or controversial classification							
(VFTS 1003)	[111]	...	...	0.540 ± 0.094	0.87 ± 0.0539	...	...
(VFTS 822)	[112,113]	0.56	[113]	0.689 ± 0.039	1.261 ± 0.035	0.952 ± 0.033	6.09 ± 0.046
Erroneous classification							
(VFTS 698)	[119]	0.6	[119]	0.436 ± 0.030	0.444 ± 0.034	0.585 ± 0.030	9.81 ± 0.023
([L72] LH 85-10)	[96]	...	...	0.108 ± 0.036	0.211 ± 0.048	...	...
(NOMAD1 0181-0125572)	[118]	...	...	...[a]	...[a]	...	...

Note: IR photometry for all objects is taken from the 2MASS point source catalog (J, H, and K [104]) and from the WISE All-Sky Data Release (W1, W2, and W4 [105]). [a] The JHK magnitudes listed in SIMBAD were mistakenly taken from the paper of Levato et al. [118], but these belong to the star LHA 120-S 165.

LHA 120-S 165

This object was first listed as a candidate young stellar object [120], but was later classified as possible B[e]SG, based on its optical spectrum displaying all characteristic emission features (SSTISAGEMC J052747.62-714852.8 [114]). Its optical and infrared brightness together with its position in both color–color diagrams support this classification.

ARDB 54

Recent analysis of ARDB 54 revealed that it belongs to the (thus far only few) A[e]SGs, being the first of its kind in the LMC [54]. Its position in the near-IR diagram supports this classification, although the object is displaced from the region of classical B[e]SGs in the WISE diagram, where it is located closer to the LBV region. Its luminosity of $\log L/L_\odot \simeq 4.4$ [54] is too low to be considered as LBV candidate, and a bit too high to be considered as HAeBe star. From the latter, ARDB 54 is also offset in both color–color diagrams, so that a classification as A[e]SG seems to be the most reasonable, despite its exposed location in the WISE diagram, whose cause should be examined more closely.

(VFTS 1003)

This star was found from the VLT-FLAMES Tarantula Survey [111]. In high angular-resolution near-IR images, it appears as a single, isolated object [121]. The star's blue emission-line spectrum resembles closely the one of the Galactic B[e]SG GG Car. VFTS 1003 was suggested to be either a Herbig B[e] star or a B[e] supergiant [111]. From its position in the near-IR diagram, a supergiant classification seems more likely. However, the lack of WISE photometry allows assigning VFTS 1003 only a candidate status. In addition, for a definite B[e] classification, red optical spectra are needed, because the optical spectra reported in the literature do not cover the region of the [O I] lines.

(VFTS 822)

This late B-type star [113] is another object that was identified in the VLT-FLAMES Tarantula Survey [112] based on its blue emission-line spectrum. As with VFTS 1003, a red spectrum is needed to check for the presence of [O I] emission. Similar to that star, VFTS 822 has been proposed as possible

Herbig B[e] pre-main sequence candidate. It displays a UV excess typical for pre-main sequence stars, but its luminosity of $\log L/L_\odot \sim 4$ is not conclusive and leaves as well room for an interpretation as evolved star. In the near-IR diagram, VFTS 822 appears within the B[e]SG domain with a clear separation from the region hosting HAeBe stars, but from the WISE diagram the star is located at the border region with the HAeBes.

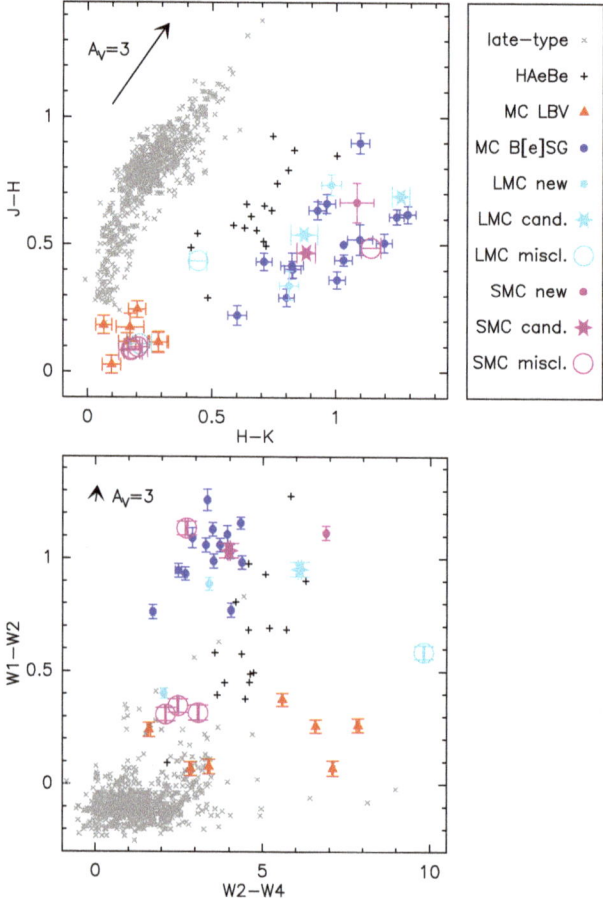

Figure 6. Location of the new LMC (light blue) and SMC (purple) samples with respect to those MC B[e]SGs that meet all the required classification criteria (dark blue) and LBVs (red triangles) in the near-IR (**top**) and the WISE diagram (**bottom**). Filled circles are used for confirmed B[e]SGs, filled stars for candidates, and empty circles for misclassified objects. A sample of late-type stars and supergiants in the MCs (small gray crosses, from [115]) and a sample of (dereddened, see text) Galactic HAeBe stars (black plus signs, from [116]) are also included. The arrow in each panel indicates the direction of the reddening, and their length complies with a value of $A_V = 3$. As the color excess is not known for all MC objects, no extinction correction has been applied. However, the MC stars have in general relatively small color excess values (see Tables 4 and 5), which would shift them only marginally in the diagrams.

(*VFTS 698*)

This is the third object identified within in the VLT-FLAMES Tarantula Survey and suggested as B[e]SG [119]. It also lacks a red optical spectrum to ascertain the presence of [O I] emission in order to be considered a B[e] star. Its spectral variability and photometric light curve unveiled that the object is most likely an interacting binary consisting possibly of an early B-type star and a veiled more massive companion. Its IR colors place it in both diagrams into the region occupied by HAeBe stars, but the luminosity estimates for both stars with $\log L/L_\odot \sim 5$ and 5.3 are too high for a pre-main sequence classification. Correcting for the extinction towards VFTS 698 would shift the object closer to the LBV domain. In the WISE diagram, the object appears particularly far off from the location of the classical B[e]SGs. Hence, this object needs further investigations for a proper classification.

([L72] LH 85-10)

Searching for LBV candidates in OB associations in the LMC, the object number 10, within the association LH 85, was found to be a luminous emission-line star. Due to the similarity of its blue spectrum with other B[e]SGs, it has been suggested as a new member of that group [96]. To date, no information about the presence of [O I] emission from that object is available. Moreover, its spectral energy distribution does not display the characteristic near-IR excess emission of B[e]SGs questioning the validity of this classification [108]. The star's position in the near-IR diagram coincides with the region populated by LBVs, supporting the erroneous classification. No WISE photometry is available for this object.

(NOMAD1 0181-0125572)

This star appears neither in the 2MASS nor in the WISE catalog, although it was reported to have 2MASS photometry [118]. From the coordinates of the source labeled NOMAD1 0181-0125572 by Miroshnichenko et al. [110] it is obvious that it was confused with the object LHA 120-S 165. The star with SIMBAD identifier NOMAD1 0181-0125572 is not a B[e]SG.

4.1.2. Small Magellanic Cloud

In the SMC, a total of six new B[e]SGs has been reported. Five of them were discovered as by-products from the spectroscopic survey of the hot, luminous stars in the SMC (2dFS [122]), from the Runaways and Isolated O Star Spectroscopic Survey of the SMC (RIOTS4 [123]), and from the SMC photometric catalog [124]. One more object was serendipitously detected [125]. From a critical inspection of the properties of these six stars, only one object appears to fulfill all required criteria, one object appears as very promising candidate that lacks only some complementary information for unambiguous classification, whereas the remaining four stars are considered as misclassified. All objects are presented in the following, their IR colors are provided in Table 5, and their positions within the two color–color diagrams are included in Figure 6.

[MA93] 1116 = NGC 346:KWBBe 200

The star is one of the few objects residing in a cluster, the SMC cluster NGC 346. Due to its Hα emission, it was first classified as a compact H II region [126], then as a Be star based on its optical photometry [127], and in the following as B[e]SG based on its optical appearance and near-IR excess emission [125]. In the mid-IR, [MA93] 1116 displays silicate emission and strong PAH bands, features often seen in HAeBe stars (e.g., [128,129]). Additional cold dust is surrounding the object as is implied by its detection at 24 μm with Spitzer/MIPS. These characteristics led to the suggestion that [MA93] 1116 might be an evolved young stellar object [130] or a HAeBe star [131], despite the lack of clear evidence for infall, e.g., in form of inverse P Cygni profiles. If true, its luminosity of $\log L/L_\odot \sim 4.4$ [54,125] would place it at the upper limit for Herbig objects, while it would get in lane with the group of less-luminous B[e]SGs in the evolved scenario. The position of [MA93] 1116 within the WISE diagram leaves room for a possible classification as HAeBe star where it falls on the edge of the region populated by HAeBe stars, whereas it appears clearly off the HAeBe region in the near-IR color–color diagram. It is hence considered as B[e]SG, and the fifth object within a cluster (the first one in the SMC).

Table 5. Confirmed and candidate B[e]SGs in the SMC. Misclassified objects are listed in the bottom part of the table.

Object	Ref.	E(B–V)	Ref.	J–H	H–K	W1–W2	W2–W4
			Confirmed B[e]SGs				
[MA93] 1116	[54,125]	0.42	[54]	0.665 ± 0.077	1.084 ± 0.0689	1.112 ± 0.03	6.883 ± 0.034
			Uncertain or controversial classification				
(LHA 115-S 38)	[124]	0.13	[132]	0.468 ± 0.038	0.879 ± 0.038	1.034 ± 0.031	3.978 ± 0.035
			Erroneous classification				
(LHA 115-N 82)	[122,133]	0.12	[133]	0.488 ± 0.040	1.141 ± 0.040	1.133 ± 0.031	2.723 ± 0.038
(LHA 115-S 29)	[123]	0.05	[123]	0.089 ± 0.030	0.180 ± 0.032	0.346 ± 0.030	2.479 ± 0.020
(LHA 115-S 46)	[123]	0.05	[123]	0.096 ± 0.037	0.204 ± 0.040	0.308 ± 0.030	2.108 ± 0.020
(LHA 115-S 62)	[123]	0.11	[123]	0.084 ± 0.039	0.175 ± 0.048	0.315 ± 0.034	3.076 ± 0.024

Note: IR photometry for all objects is taken from the 2MASS point source catalog (J, H, and K [104]) and from the WISE All-Sky Data Release (W1, W2, and W4 [105]).

(LHA 115-S 38 = 2dFS 1804)

This object was found from the spectroscopic survey of the hot, luminous stars in the SMC [122]. The blue spectrum displays numerous emission lines, in particular of [Fe II], but no information about [O I] emission is available. The spectral energy distribution indicates a near-IR excess [124]. With a luminosity estimate of $\log L/L_\odot \simeq 4.1$ the star was also considered as a post-AGB object [132], but its positions within the two color–color diagrams places the star clearly within the B[e]SG domain. As such, it is a strong candidate for another low-luminosity B[e]SG.

(LHA 115-N 82 = 2dFS 2837 = LIN 495)

This star displays all spectroscopic B[e]SG characteristics [122,133] and an IR excess emission [124]. The optical spectrum was reported to be composite [122], and the detected photospheric lines display radial velocity variations, whereas the emission from the circumstellar matter appears to be stable [54]. The IR colors of LHA 115-N82 locate it clearly within the B[e]SG domains. Curiously, the V and I band light curves display a long-term brightening, which resembles LBV outbursts and is not common in B[e]SGs. On the other hand, it displays both [O I] and [Ca II] emission, which are typically not seen in LBVs. Since the star has too low luminosity ($\log L/L_\odot \simeq 3.8$) to be an LBV (or even a B[e]SG), it was recently assigned a classification as "LBV imposter" [54]. This object clearly requires further investigations to pin down its status.

(LHA 115-S 29 = RMC 15), (LHA 115-S 46 = RMC 38), (LHA 115-S 62 = RMC 48)

The blue optical spectra of these three objects show emission-line features similar to B[e]SGs, and the luminosities derived for these objects, ranging from $\log L/L_\odot \simeq 4.4$ to 4.8, assigns them a supergiant status [123]. However, it is currently not known whether these objects display [O I] line emission, one of the defining characters of the B[e] phenomenon. Due to the lack of a pronounced near-IR excess emission, these three stars have been proposed to be dust-poor B[e]SGs [123]. However, the presence of warm dust is another main classification criteria for a star to be considered as B[e]SG. It is not surprising that all three objects fall clearly outside the B[e]SG domains in the color–color diagrams. Instead, their positions coincide with the regions populated by LBVs, which implies that these three stars have dense winds and circumstellar ionized gas, which is exemplified by their emission-line spectra. These objects require further investigations to unveil their true nature.

4.2. Local Group Galaxies beyond the Magellanic Clouds

Moving further away, beyond the Magellanic Clouds, we may expect to find B[e]SG stars and candidates in those galaxies, in which star-formation is ongoing. For these galaxies, surveys such as the

LGGS project mentioned in Section 3, but also earlier surveys (e.g., [134,135]) provide indispensable information on the population of luminous, evolved massive stars, and provide the base for systematic investigations for unambiguous classification of these objects.

To date, systematic spectroscopic studies have been performed in the two large spiral galaxies of the Local Group, M31 and M33, in which a number of B[e]SGs were found amongst the putative LBV candidates [97,99,136–138]. The suggested B[e]SG populations in each of these galaxies is presented in the following subsections. The samples are provided in Tables 6 and 7, in which all objects are listed under their Local Group Galaxy Survey (LGGS) identifiers (Column 1). The properties, based on which the decision to categorize an individual star as either a candidate or a misclassified object has been made, are briefly depicted. The locations of the confirmed, candidate, and possibly misclassified B[e]SGs in the two color–color diagrams are shown in Figure 7.

Table 6. Confirmed and candidate B[e]SGs in M31. Misclassified objects are listed at the bottom of the table.

LGGS	Ref.	J–H	H–K	W1–W2	W2–W4
		Confirmed B[e]SGs			
J004320.97+414039.6	[99,139]	0.358	1.522 ± 0.143	0.983 ± 0.051	5.284 ± 0.084
J004415.00+420156.2 [a]	[99,139]	0.741 ± 0.253	1.510 ± 0.210	0.961 ± 0.033	3.645 ± 0.230
J004417.10+411928.0 [b,c]	[99,136]	0.387 ± 0.148	0.853 ± 0.149	0.882 ± 0.068	4.455 ± 0.247
J004522.58+415034.8 [c,d]	[136]	0.684 ± 0.223	0.877 ± 0.273	1.091 ± 0.039	4.303 ± 0.187
		Uncertain or controversial classifications			
(J004220.31+405123.2)	[99]	...	...	0.880 ± 0.062	4.495 ± 0.298
(J004221.78+410013.4)	[99]	...	...	0.927 ± 0.072	5.253 ± 0.107
(J004229.87+410551.8) [e]	[99,139]	...	...	0.870 ± 0.045	4.484 ± 0.082
(J004411.36+413257.2)	[139]	0.187 ± 0.123	0.914 ± 0.165	...	...
(J004442.28+415823.1)	[99,139]	...	...	0.781 ± 0.076	4.994 ± 0.406
(J004444.52+412804.0) [c,d]	[137]	0.576 ± 0.119	0.858 ± 0.114	0.883 ± 0.039	5.247 ± 0.046
(J004621.08+421308.2) [d]	[139]	0.374 ± 0.171	1.283 ± 0.128	1.091 ± 0.039	4.303 ± 0.188
		Erroneous classifications			
(J004043.10+410846.0)	[99,139]	1.616	0.523 ± 0.162	0.974 ± 0.045	4.030 ± 0.250
(J004057.03+405238.6)	[99]	...	...	...	...

Note: IR photometry is taken from the 2MASS point source catalog (J, H, K [104]) and from the WISE All-Sky Data Release (W1, W2, W4 [105]). Colors resulting from uncertain photometric values are written in italic. [a] Possible contamination in the photometric bands J, W1 and W2 due to crowding. [b] Possible contamination in the photometric bands J and H due to crowding. [c] Has CO band emission [136,137]. [d] Has also been classified as warm hypergiant [140]. [e] Possible contamination in the photometric bands W1 and W2 due to crowding.

4.2.1. M31

A total of 13 B[e]SGs has been proposed in M31 to date. Twelve of them resulted from optical spectroscopic observations of LBV candidates that display the typical B[e]SG characteristics [99,139]. Infrared spectroscopic observations of a small LBV candidate sample revealed so far that three display CO band emission [136,137], making their classification as B[e]SG very likely, especially since two of the CO band emission objects were also found from optical spectroscopy to be a possible B[e]SG. However, a closer look at this total sample of 13 objects, including the available information from near- and mid-IR photometric observations, reveals that only four stars can be considered as confirmed B[e]SGs. Seven objects require further clarifications, and two appear to be misclassified (see Table 6).

Table 7. Confirmed and candidate B[e]SGs in M33. Misclassified objects are listed at the bottom of the table.

LGGS	Ref.	J–H	H–K	W1–W2	W2–W4
		Confirmed B[e]SGs			
J013333.22+303343.4 [a]	[97,99,138]	1.144 ± 0.234	1.128 ± 0.154	0.900 ± 0.063	5.269 ± 0.098
J013350.12+304126.6	[97,99]	0.653 ± 0.089	0.683 ± 0.091	0.764 ± 0.041	4.132 ± 0.186
		Uncertain or controversial classifications			
(J013324.62+302328.4)	[97,99]	...	...	1.050 ± 0.047	4.235 ± 0.414
(J013342.78+303256.3)	[99]	...	...	...	...
(J013349.28+305250.2)	[99]	...	...	0.746 ± 0.054	*5.394 ± 0.040*
(J013426.11+303424.7)	[97,99]	...	...	1.208 ± 0.066	4.845 ± 0.278
(J013459.47+303701.9)	[97,99]	...	...	0.898 ± 0.053	*4.996 ± 0.038*
(J013500.30+304150.9)	[97,99]	...	...	0.939 ± 0.040	*4.142 ± 0.029*
		Erroneous classifications			
(J013242.26+302114.1)	[97,99]	0.861 ± 0.059	0.376 ± 0.064	0.050 ± 0.040	*4.543 ± 0.031*
(J013406.63+304147.8) [b]	[97,138,141]	0.217 ± 0.115	0.249 ± 0.149	...	...

Note: IR photometry is taken from the 2MASS point source catalog (J, H, and K [104]) and from the WISE All-Sky Data Release (W1, W2, and W4 [105]). Colors resulting from uncertain photometric values are written in italic. [a] Has CO band absorption [138]. [b] Has CO band emission [138].

J004320.97+414039.6, J004415.00+420156.2, J004417.10+411928.0

These three objects fulfill all classification criteria of the confirmed B[e]SGs. Their positions in the near-IR and WISE diagrams coincide with the domain populated by the B[e]SGs, except for J004320.97+414039.6, which appears a bit off due to the uncertainties in its J and H magnitudes, which are only upper limits.

J004522.58+415034.8

The star was assigned a warm hypergiant status due to detected photospheric features pointing towards an A2Ia spectral-type [140]. As it otherwise fulfills all criteria of a B[e]SG (even displaying CO band emission [136]), it is included into the list of confirmed B[e]SGs, but keeping in mind that with its A spectral type it is actually a representative of the (though few) A[e]SGs. Recent modeling of the star's spectral energy distribution resulted in an effective temperature estimate of 11,000 K [142], which would point towards a late-B spectral type.

(J004220.31+405123.2), (J004221.78+410013.4), (J004229.87+410551.8), (J004442.28+415823.1)

These stars display the optical characteristics of B[e]SGs [99], and their WISE colors place them within the B[e]SG domain. However, all four objects lack JHK band photometry, so that they cannot be located in the J–H versus H–K diagram. The presence of He I emission in J004221.78+410013.4 [99] contradicts the proposed effective temperature of 7200 K [142] and requires clarification.

(J004411.36+413257.2)

The star is reported to display He I, Fe II, and [Fe II] emission and was suggested as B[e]SG star [139]. Whether it displays [O I] emission was not discussed by these authors and is hence yet unclear, although the spectrum covers that spectral region. Its J–H color is smaller than for typical B[e]SGs (but the H magnitude is uncertain), placing the star at the lower boundary for the B[e]SGs in the color–color diagram. No WISE photometry exists. The closest IR source is more than 6 arcsec away. The object is not reported in other published spectroscopic investigations to display all B[e]SG characteristics. Instead, it was listed as Fe II emission line star [98] and more recently as LBV candidate [142].

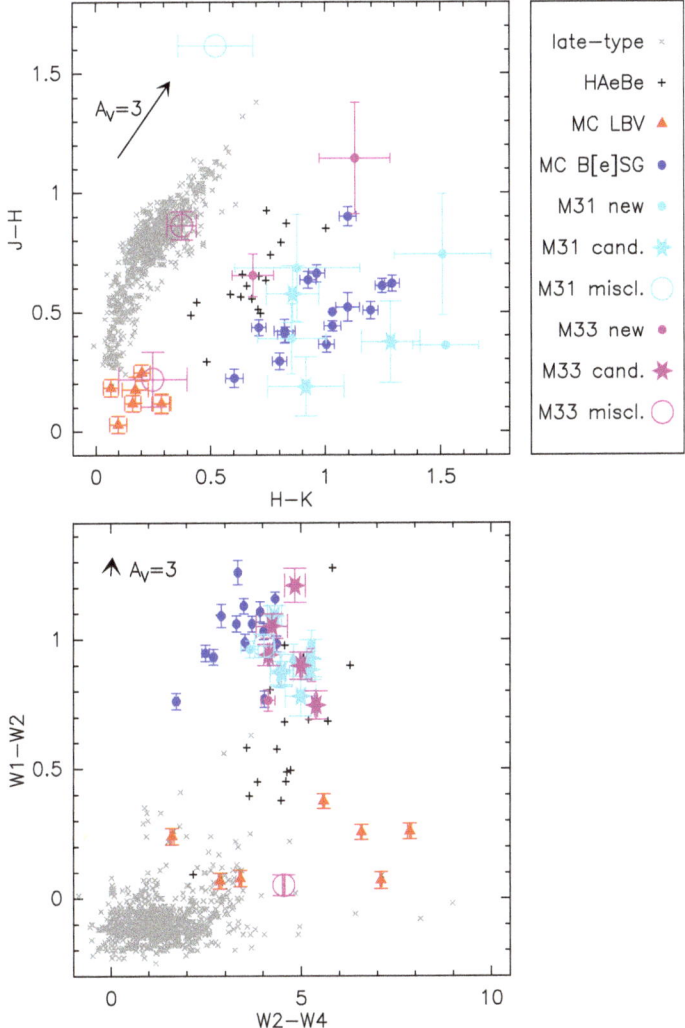

Figure 7. As Figure 6 but for the location of the M31 (light blue) and M33 (purple) samples in the near-IR (**top**) and the WISE diagram (**bottom**) based on their observed colors. Typical values for the objects' reddening are $A_V \leq 1.5$ mag [142].

(J004444.52+412804.0)

This object was first proposed to be a P Cyg-type LBV candidate [95]. With the detection of [Ca II] emission and circumstellar dust [98], and with the determination of an effective temperature of 6600 K [142], it was classified as a warm hypergiant. On the other hand, a B[e]SG classification with a central star of 15,000–20,000 K has been suggested, based on the presence of He I lines displaying P Cyg profiles and the detection of intense CO band emission [137]. This controversial classification, based on spectra taken at similar epochs, requires clarification. Infrared photometry places J004444.52+412804.0 within the region occupied by B[e]SGs in the two color–color diagrams.

(J004621.08+421308.2)

A status of a late A-type warm hypergiant has been assigned to this object based on the lack of detected He I emission [143]. In contrast to this, He I emission lines were reported by Sholukhova et al. [139], and an effective temperature of 10,000 K has been estimated from the modeling of the stars' spectral energy distribution [142]. The position of the object in the near-IR and WISE diagrams coincides with the B[e]SG domain.

(J004043.10+410846.0)

Its optical spectrum displays all typical characteristics of a B[e]SG [99], and its WISE colors confirm the presence of warm dust, but its (though uncertain) unusually high J–H value of 1.6 at a low value of H–K fits neither to B[e]SGs nor to LBVs.

(J004057.03+405238.6)

No IR photometry is available, so that currently no statement can be made about the presence of (hot) dust. The absence of He I lines in the optical spectra along with an effective temperature of 7700 K [142] speak against a B[e]SG classification.

4.2.2. M33

In M33, the sample of putative LBV candidates [6,95] served also as a starting point for a more detailed investigation of the individual objects. Optical [97,99,141], and infrared [138] spectroscopic studies revealed so far a total of ten possible B[e]SGs. They are listed in Table 7. Combined with the available information about their infrared colors, only two objects can currently be considered as confirmed B[e]SGs, six need further clarification, and two objects appear to be misclassified.

J013333.22+303343.4, J013350.12+304126.6

Both objects display all classification characteristics of B[e]SGs. Their locations in the WISE diagram coincide with the B[e]SG domain. In the near-IR diagram they appear slightly off the B[e]SG region, shifted towards the HAeBe domain, but their high-luminosities clearly classify them as supergiants. The CO band absorption detected from J013333.22+303343.4 cannot be explained with a cool companion, but might originate from a pole-on seen jet [138]. Whether also CO band emission is present in this object, but veiled by the intense absorption component, is currently unknown.

(J013324.62+302328.4), (J013342.78+303256.3), (J013349.28+305250.2), (J013426.11+303424.7), (J013459.47+303701.9), (J013500.30+304150.9)

All six objects lack near-IR photometry, which renders it difficult to unambiguously classify them as B[e]SGs, although their mid-IR colors and their optical spectroscopic appearances support such a classification. The star J013342.78+303256.3 has even no mid-IR photometry, so that it is unclear whether it is surrounded by (warm) dust at all. There is an IR source at a distance of $\sim$4.5 arcsec which is considered as too far off to be identified with J013342.78+303256.3. The sources J013324.62+302328.4 and J013500.30+304150.9 were reported as intrinsically faint objects with little extinction, and were suggested as counterparts of the low-luminosity B[e]SG in the MCs [97]. However, according to the luminosity estimates for these stars, for which values of $\log L/L_\odot \simeq 4.8$–5.0 were obtained [142], these stars fall rather onto the lower boundary of the high-luminosity B[e]SGs sample within the MCs (see Figure 1).

(J013242.26+302114.1)

The optical spectra of this object display all typical B[e]SG characteristics, but the star's IR colors displace it clearly from the sample of confirmed B[e]SGs. It has excess emission at 8 µm due to PAHs [99]. The JHK magnitudes suggest that J013242.26+302114.1 falls into the region of late-type stars and stars with a late-type companion, respectively, [20,144]. This classification is supported by the WISE colors, although regular late-type stars typically have slightly negative W1–W2 colors. Considering only the WISE photometry, J013242.26+302114.1 would fit to the region populated by LBVs. This objects clearly needs to be investigated in more detail for a proper classification.

(J013406.63+304147.8)

This star is also known as [HS80] B416. Despite the detected CO band emission from this object [138], a classification as B[e]SG seems not appropriate (as previously mentioned by Humphreys et al. [98,99]), because its near-IR colors show clearly the lack of warm dust and would place the object to the region populated by LBVs. In addition, no [OI] emission is seen from J013406.63+304147.8 [98]. However, thus far, no photometric variability, typical for LBVs, is recorded [97], and the star can only be considered as a possible LBV candidate, as was previously suggested [99]. The star is found to be surrounded by an expanding ring-like nebula [141] and shows excess emission at 8 µm due to PAHs [138]. The spectral energy distribution leaves space for a companion [98,138], and its spectral lines display radial velocity variations with a period of 16.13 d, leading to the suggestion that the system might be an interacting binary system causing mass loss in the equatorial plane [145].

4.3. Beyond the Local Group

Not much is known yet about the B[e]SG population beyond the Local Group. Most promising for dedicated searches are the nearby large spiral galaxies M81, M101, and NGC 2403, for which pioneering ground-based surveys have been conducted already in the 1980s and 1990s [146–149] and which are nowadays extended to fainter objects thanks to the capabilities of the *Hubble Space Telescope* (e.g., [150]). Spectroscopic follow-up investigations revealed a number of variable luminous objects [151–155], but only in one of these galaxies, M81, three B[e]SG candidates have been found so far [156]. They all display the typical B[e] features in their optical spectra, but without complementary information about the presence of (warm) circumstellar dust, their classification remains preliminary. These objects will certainly not remain the only of their kind, because the search for more candidates has just begun.

The B[e]SG candidates in M81 are listed in Table 8. For lack of proper SIMBAD identifiers for these objects, the table contains the star ID [156] along with the coordinates and, where available, the V band magnitudes.

Table 8. Candidate B[e]SGs in M81.

Star ID	RA(J2000)	Dec(J2000)	V	Ref.
(10584-8 4)	9:54:50.03	+69:06:55.47	...	[156]
(10584-4-1)	9:54:54.05	+69:10:23.00	19.68	[156]
(10584-9-1)	9:55:18.97	+69:08:27.54	19.10	[156]

In the past few years, additional surveys have been performed, using large ground-based facilities (e.g., [157]) and space telescopes (e.g., [158,159]). These surveys were aimed at revealing the luminous and variable massive star populations of other galaxies even further away. They provide an excellent basis for follow-up spectroscopic studies to classify their massive star content, so that many more LBV candidates and B[e]SGs may be found in the (near) future. In addition, the Local Group galaxies still need to be explored in more detail, in particular those galaxies that were already found to possess (even though in very small numbers) LBV candidates [95] that are awaiting their proper classifications. The upcoming era of the Extreme Large Telescope (ELT) promises to become particularly fruitful. The next generation of high-sensitivity instruments combined with the large collecting area of the telescope will facilitate ground-based spectroscopic observations of faint objects with very high spatial resolution.

4.4. Galactic Objects

Finally, we return to our own Galaxy. Searching for B[e]SG stars in the Milky Way is a difficult task. Though many B[e] stars are known, the assignment of a supergiant status is significantly hampered due to highly uncertain distances, hence luminosities. The situation will hopefully change with the final data release from the GAIA mission, from which one hopes for accurate parallax measurements. However, for now, the luminosities of the objects are subject to large uncertainties, so that only objects with a reported luminosity of at least $\log L/L_\odot \geq 5$ are considered as serious B[e]SG candidates.

The lower luminosity boundary of $\log L/L_\odot \sim 4.0$ for an evolved star to be assigned a supergiant status is a further hindrance in the classification of objects as B[e]SGs, because this luminosity domain is shared with the massive pre-main sequence (HAeBe) stars. The latter have emission-line spectra with numerous forbidden emission lines from [Fe II] and [O I] similar to the B[e]SGs, and the stars are surrounded by significant amounts of circumstellar dust within their massive accretion disks causing considerable IR excess emission, just as the B[e]SGs. Hence, it is not surprising that confusion exists about the proper classification for a number of objects within this luminosity domain of $4.0 < \log L/L_\odot < 4.5$, and that Galactic B[e]SG candidates also appear as candidates in catalogs of HAeBe stars (see, e.g., [160]). In the absence of clear indications for infall of material, which is a typical characteristic of pre-main sequence stars, alternative discriminators for the classification of such objects are needed.

A reasonable approach to this is to search for ^{13}CO emission from the circumstellar environments of the uncertain candidates. Many HAeBe stars have been reported to display CO band emission from their massive accretion disks (e.g., [161–165]). As these disks form from material provided by the interstellar medium in which the ^{12}C/^{13}C isotope abundance ratio has typically a value of about 90 [88], these pre-main sequence disks can clearly be distinguished from the disks around evolved massive stars, which should be enriched in ^{13}C and hence give rise to clearly measurable emission in ^{13}CO. K-band spectra of these objects, covering the first-overtone bands of both ^{12}CO and ^{13}CO (see Section 2.3 and Figure 3), are thus key for a proper discrimination between a young (pre-main sequence) and an evolved status.

From the currently proposed 15 Galactic B[e]SGs listed in Table 2, nine have been reported to display CO band emission. However, thus far, only four of them have been observed in the region around the ^{13}CO bands. The spectra of all four stars have been found to display clear signatures of ^{13}CO emission, and model results revealed that the environments of all four objects are clearly enriched in ^{13}C. Two of these objects were already known to be supergiants based on their confirmed high luminosities: the stars GG Car [41,42] and Hen 2-398 [41]. For the other two, which so far have also been considered as HAeBe candidates (see Table 9), the detection of chemically processed material can hence be regarded as the ultimate proof of their evolved, supergiant nature. These are the objects MWC 137 [59] and MWC 349 [62]. These results are very promising and encouraging, and they demonstrate that the ^{13}CO molecular emission provides a solid tool to unambiguously classify a star as either a pre-main sequence or an evolved object. Clearly, more observational effort needs to be undertaken to search also for the signatures of ^{13}CO in the spectra of the remaining objects.

When collecting the IR magnitudes of the Galactic sample, it turned out that only the near-IR measurements are reliable, whereas the WISE measurements for all objects have been flagged as being contaminated by neighboring objects. The latter are hence useless for classification purposes, and one can currently only rely on the JHK-band magnitudes. The list of objects, their observed colors, and literature values of their color excess are listed in Table 9. The relatively high values of the color excess requires correction for extinction before placing the objects to the near-IR diagram. Corrections have been performed with the galactic extinction curve using an R_V value of 3.1 [117]. The extinction corrected colors are included in Table 9, and the positions of the objects are shown in Figure 8 separately for the confirmed (Figure 8, top) and candidate objects (Figure 8, bottom).

Table 9. Confirmed and candidate B[e]SGs in the Milky Way.

Object	Ref.	$E(B-V)$	Ref.	(J–H)	(H–K)	$(J-H)_0$	$(H-K)_0$
				Confirmed B[e]SGs			
MWC 137 [a]	[59,166]	1.22	[166]	0.922 ± 0.040	1.217 ± 0.035	0.572 ± 0.040	0.930 ± 0.035
MWC 349 [a]	[62,167]	~3.2	[168]	1.472 ± 0.038	1.603 ± 0.369	0.554 ± 0.038	0.851 ± 0.369
GG Car [b]	[41,42,169]	0.51	[169]	0.818 ± 0.049	0.964 ± 0.049	0.672 ± 0.049	0.844 ± 0.049
Hen 3-298	[41,63]	1.7	[63]	1.009 ± 0.065	1.139 ± 0.060	0.522 ± 0.065	0.739 ± 0.060
CPD-52 9243	[50]	1.7	[35]	0.919 ± 0.029	0.948 ± 0.026	0.432 ± 0.029	0.548 ± 0.026
HD 327083 [b]	[170]	1.8	[170]	1.000 ± 0.042	1.309 ± 0.188	0.484 ± 0.042	0.886 ± 0.188
MWC 300 [a,b]	[171,172]	1.2	[172]	1.150 ± 0.056	1.951 ± 0.055	0.806 ± 0.056	1.669 ± 0.055
AS 381 [b]	[65]	2.3	[65]	1.285 ± 0.029	1.117 ± 0.028	0.625 ± 0.029	0.576 ± 0.028
CPD-57 2874	[47]	1.75	[47]	0.803 ± 0.062	0.678 ± 0.304	0.301 ± 0.062	0.267 ± 0.304
				Uncertain or controversial classifications			
(Hen 3-938)	[54]	1.64	[54]	1.593 ± 0.035	1.487 ± 0.034	1.124 ± 0.035	1.101 ± 0.034
(MWC 342) [a,b]	[173]	1.5	[173]	1.179 ± 0.027	1.121 ± 0.025	0.749 ± 0.027	0.768 ± 0.025
(Hen 3-303)	[63]	1.7	[63]	1.366 ± 0.063	1.443 ± 0.063	0.879 ± 0.063	1.042 ± 0.063
(CD-42 11721) [a]	[67]	1.4–1.6	[50,67]	1.317 ± 0.053	1.398 ± 0.050	0.901 ± 0.053	1.057 ± 0.050
(HD 87643) [a,b]	[35,174]	1.0	[35]	1.461 ± 0.271	1.300 ± 0.338	1.174 ± 0.271	1.065 ± 0.338
(HD 62623) [b]	[69]	0.17	[175]	0.395 ± 0.380	0.693 ± 0.368	0.346 ± 0.380	0.653 ± 0.368

Note: IR photometry is taken from the 2MASS point source catalog (J, H, and K [104]). [a] Star appears also in HAeBe catalogs (see, e.g., [160]). [b] Confirmed or suspected binary.

The separation of confirmed from candidate B[e]SGs is based on two characteristics: (i) stars with detected enrichment in ^{13}CO of their circumstellar environments are considered as confirmed; and (ii) stars with reported (by more than one research team) luminosity values of $\log L/L_\odot \geq 5.0$. Objects with lower luminosities $4.0 < \log L/L_\odot < 5.0$ are assigned a candidate status. Based on these criteria, the Galactic sample splits into nine confirmed B[e]SGs and six candidates (see Table 9).

MWC 137, MWC 349, GG Car, Hen 3-298

These four objects are considered as confirmed B[e]SGs based on the detected enrichment of their circumstellar disk material with ^{13}CO (see Table 2). They all fall into the region of the confirmed B[e]SGs in the near-IR diagram, regardless of the large error bar for MWC 349.

CPD-52 9243, HD 327083, MWC 300

All three stars fulfill the high luminosity criterion. The near-IR colors of CPD-52 9243 and HD 327083 place these two objects within the B[e]SG domain. MWC 300 appears regularly in studies of HAeBe stars. Its near-IR colors locate this star close to the B[e]SGs but far away from the HAeBe region, making a pre-main sequence nature of this object rather unlikely. Its relatively high color values might be influenced by a possible companion [51,172].

AS 381, CPD-57 2874

These two objects are also known to have high luminosities, but they reside slightly outside the classical B[e]SG domain in the near-IR diagram. AS 381 is a reported binary [64,65] consisting of a luminous B[e]SG and a K-type companion, which seems to (significantly) contribute to the total near-IR flux, hence altering the colors of the B[e]SG. The near-IR colors of CPD-57 2874 place it closer to the LBVs rather than to the B[e]SGs, although its H–K color is subject to large uncertainty. Its pronounced emission in CO, and in the [O I] and [Ca II] forbidden lines [58] speak against an LBV classification.

(Hen 3-938), (MWC 342), (Hen 3-303), (CD-42 11721)

The rather low luminosities of these four objects and the closeness or even coincidence of their location with the HAeBe domain clearly requires further studies for an unambiguous classification. While Hen 3-938 and MWC 342 have to our knowledge not yet been spectroscopically observed in the K-band, the other two stars showed no evidence for CO band emission [39]. This renders their classification more difficult.

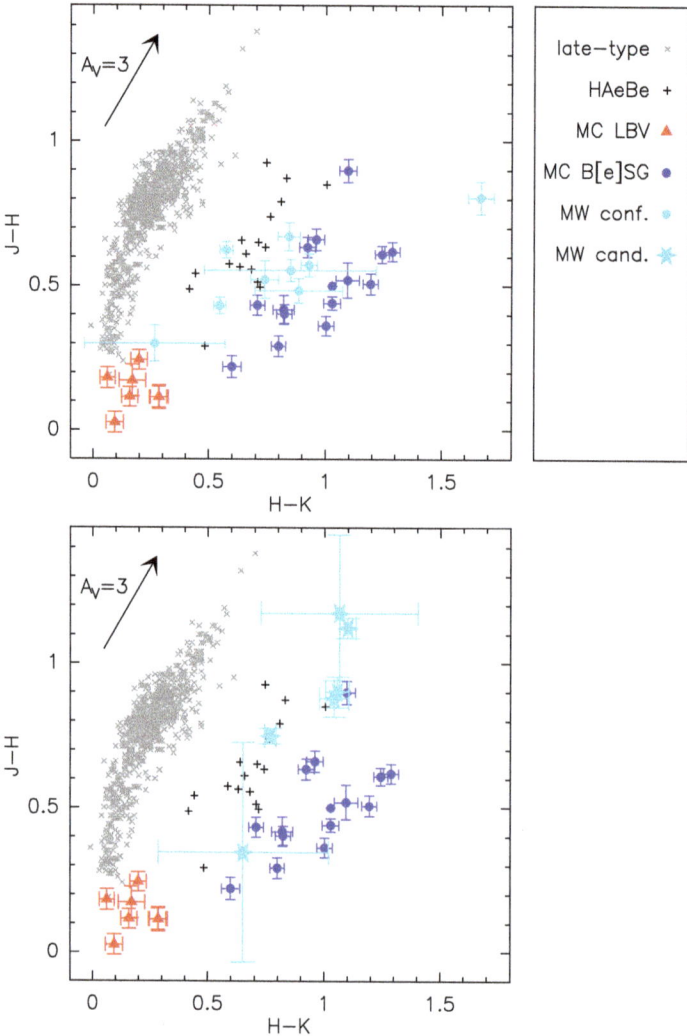

Figure 8. Near-IR diagrams as in Figure 6, showing the locations of the Galactic confirmed B[e]SGs (**top**) and B[e]SG candidates (**bottom**). The colors of the Galactic objects have been corrected for interstellar extinction (Table 9).

(HD 87643)

The near-IR colors of HD 87643 place it to the high end of the HAeBe regime. However, these colors have large errors and require refinement. Studies based on long-baseline interferometry revealed that this object consists of two B-type stars, and each component might be surrounded by a dusty disk [174]. Whether this object is a physical binary is currently not known.

(HD 62623)

It appeared in the literature as the first A[e]SG [49,176] and is known to be surrounded by a detached gas and dust disk, from which also CO band emission has been detected [39,58]. It has been speculated that the gap between the star and the inner rim of the gas disk might have been cleared by a companion. However, no clear evidence of such a companion has been found yet. Its low luminosity and large errors in the near-IR photometric measurements make it challenging to unambiguously assign the star a B[e]SG status, although its position within the near-IR seems to coincide more with the B[e]SG domain rather than with any other classification. This object is an ideal candidate to clarify its nature based on dedicated K-band observations to analyze the ^{13}CO content within its circumstellar disk.

5. Discussion and Conclusions

B[e]SGs form a special class of evolved massive stars and are thought to represent a short-lived transition phase either in their red-ward, post-main sequence evolution, or in their blue-ward post-RSG evolution. The total number of currently known B[e]SGs is low, supporting the idea of a short transition phase. However, in which direction the stars evolve, and whether all B[e]SGs evolve in the same direction, is still an open issue. For the objects in M31 and M33, it has been argued that the B[e]SGs are more isolated than LBVs and hardly found in stellar associations, so that a post-RSG (or post-yellow supergiant) evolution was proposed to be more likely [142]. This assessment was based on the available numbers of putative B[e]SGs in these two galaxies at that time. However, after revision of the two samples (Tables 6 and 7), two of the four confirmed B[e]SGs in M31 and both confirmed B[e]SGs in M33 are associated with stellar groups, questioning the conclusion that B[e]SGs are isolated and thus post-RSGs.

The best age indicator for B[e]SGs we have to date is their surface abundance enrichment in ^{13}C, as discussed in Section 2.3. If B[e]SGs were post-RSGs, their progenitors would all have started with a (very) low rotation speed. While such a scenario cannot be excluded, it may not be very likely considering that stars are born on average with a rotation rate of about 40% of their critical velocity [88].

Interaction within a (close) binary system, possibly even up to a binary merger, seems to be an alternative and popular scenario (e.g., [51,177]), but the number of currently confirmed binaries amongst B[e]SGs is still rather low to give preference to the binary channel as the sole possible way for the formation of B[e]SGs. Likewise, some B[e]SGs have been suggested as suitable supernova candidates. For example, the Galactic object MWC 137 appears similar to Sher 25 [178] and SBW1 [179], which both look like the progenitor of SN1987A. In addition, the SMC object LHA 115-S 18 has been proposed to be a viable SN1987A progenitor [27]. In this respect, it is vital to resolve B[e]SG populations and to study their properties.

In this review, a census of the currently known B[e]SG population in the Milky Way and in nearby star-forming galaxies within and beyond the Local Group is presented. The proposed candidates have been undertaken a critical examination, sorted into confirmed B[e]SGs and B[e]SG candidates, and unsuitable objects have been flagged as "misclassified"[10].

During these investigations, a fundamental difference has been recognized between the identification issues for objects in our Galaxy compared to those in other galaxies. Extragalactic B[e]SGs bare the risk of being confused with LBVs in quiescence, which share very similar optical spectroscopic characteristics. To separate these two classes of objects, one can make use of clearly defined classification criteria based on certain sets of emission features identified in their optical and near-IR spectra, as outlined in Section 3. In addition, inspection of the location of possible B[e]SG candidates in

---

[10] We would like to caution that, with insufficient knowledge of stellar properties, individual objects may easily be misclassified, as it happened in recently published catalogs [6,102,180,181], in which erroneously a number of (even confirmed) B[e]SGs are listed as LBV candidates.

the IR color–color diagrams is highly advisable, because B[e]SGs and LBVs populate clearly separate domains. This fact might also be used as starting point for future investigations of extragalactic samples. However, for such future studies, infrared photometry with higher spatial resolution than what is currently provided by 2MASS and WISE is desirable to prevent from contamination with neighboring sources in densely populated regions. Moreover, precise distances, as soon provided by GAIA, will help to separate foreground stars in the directions to other galaxies that might have been misclassified as luminous extragalactic stars.

In the Milky Way, an additional complication occurs due to the often uncertain luminosity estimates, and the overlap of low-luminosity B[e]SGs with the most luminous massive pre-main sequence objects (HAeBe). Here, special care needs to be taken, especially since both classes of objects occupy adjacent regions in the near-IR color–color diagram with a probable overlap. Without additional distinctive features for such low-luminosity, borderline B[e]SG candidates, their real nature remains elusive. One complementary classification criteria that was discussed, is provided by the enrichment of the circumstellar material of evolved objects with processed material that has been released from the stellar surface, as opposed to the non-processed material with interstellar abundance patterns found around HAeBes. The most ideal element to search for is ^{13}C, which is bound in ^{13}CO molecules in the circumstellar disks. Measuring the ^{13}CO amount with respect to ^{12}CO provides immediate insight into the nature of the object. The current HAeBe samples might hide such low-luminosity B[e]SG candidates, which can only be identified as such by careful and honest analysis.

The result from this census, after strict application of the classification criteria, is that we count nine confirmed and six candidate B[e]SGs in the Galaxy. Moving out to the MCs, the numbers amount to 13 (+2) in the LMC and 5 (+1) in the SMC. The situation in other members of the Local Group is not much better, where the numbers drop to four (+7) in M31 and two (+6) in M33. Even further away, only three candidates have been reported from M81. The total number of B[e]SGs found in the various galaxies are too small for statistical analyses with respect to a metallicity dependence of their number, but the comparable quantities within the Milky Way and the LMC suggest only a mild dependence of the amount of B[e]SGs on (i.e., a possible drop with decreasing) metallicity.

Further, we report that the Galactic sample of 15 B[e]SGs and candidates contains currently seven confirmed or suspected binaries (see Table 9), which is less than half of the population. In the MCs, this number is even lower, because thus far only three stars have been reported to be possible binaries, of which one, the SMC star LHA 115-S 6 (= RMC 4) has been proposed to be the remnant of a binary merger within an initially triple system [182,183]. The other two objects, the LMC star LHA 120-S 134 and the SMC object LHA 115-S 18, have been identified as optical counterparts of X-ray sources [27,29]. Nothing is known about possible binarity in the B[e]SG samples from the other galaxies, although photometric variability was seen in at least two M31 objects: the confirmed B[e]SG star J004417.10+411928.0 [137,184] and the candidate J004444.52+412804.0 [185,186]. Whether this variability is a sign of binarity or just of semi-regular variability which might be interpreted with pulsation activity such as reported from the α Cygni variables, needs to be studied in more detail.

For the sake of completeness, a special class of objects, which have not been discussed yet, should be briefly mentioned as well. These are the high-mass X-ray binaries (HMXB) with possible B[e] supergiant (candidate) companion. Some objects with an assigned B[e]SG status have been found to be too luminous in X-rays for being considered single stars. These objects have been proposed to be binary systems, in which the high energy emission is caused by either accretion onto a compact object, or by shocks in a colliding wind binary with a second massive star. Members of this group are the Galactic objects Cl* Westerlund 1 W 9 (=Wd1-9), which is considered a colliding wind system [85]; CI Cam (=MWC 84), which might be interpreted as supernova imposter [187]; and the high-mass X-ray binary (HMXB) IGR J16318-4848, in which the compact object was proposed to be a neutron star [188]. Two ultra-luminous X-ray sources (ULXs), Holmberg II X-1 and NGC 300 ULX1, the latter being also named as supernova imposter SN2010da, have also been proposed to be HMXBs including a B[e]SG [189–191]. Whether these objects indeed host a B[e]SG clearly needs to be investigated in more detail. However, since the behavior of these sources is considerably different from the confirmed B[e]SGs, I hesitate to include them into the census.

The presented populations represent the current knowledge of B[e]SGs and B[e]SG candidates in the closest star-forming galaxies. With the ever growing sensitivities of instruments and telescope sizes, the future in B[e]SG star research is bright, because many more candidates will be identified in even more distant galaxies and with metallicities spreading over a large range. With statistically meaningful samples, it will finally be possible to unveil the nature and fate of these fascinating objects.

Funding: This project received funding from the Grant Agency of the Czech Republic (GAČR, grant number 17-02337S). The Astronomical Institute of the Czech Academy of Sciences, Ondřejov, is supported by the project RVO:67985815.

Acknowledgments: I wish to thank the editor, Roberta Humphreys, for the invitation to write this review. In addition, I am grateful to Michalis Kourniotis for inspiring discussions on searching for extragalactic B[e]SGs, and to Dieter Nickeler and Lydia Cidale for passionate discussions about B[e]SGs as well as for their proofreading of and suggestions on the draft versions. Moreover, I thank the anonymous referees for their careful reading and suggestions on the draft version. This research made use of the NASA Astrophysics Data System (ADS) and of the SIMBAD database, operated at CDS, Strasbourg, France. This publication makes use of data products from the 2MASS, which is a joint project of the University of Massachusetts and the Infrared Processing and Analysis Center/California Institute of Technology, funded by the National Aeronautics and Space Administration and the National Science Foundation. This publication makes use of data products from the Wide-field Infrared Survey Explorer, which is a joint project of the University of California, Los Angeles, and the Jet Propulsion Laboratory/California Institute of Technology, funded by the National Aeronautics and Space Administration.

Conflicts of Interest: The author declares no conflict of interest.

References

1. Groh, J.H.; Georgy, C.; Ekström, S. Progenitors of supernova Ibc: A single Wolf-Rayet star as the possible progenitor of the SN Ib iPTF13bvn. *Astron. Astrophys.* **2013**, *558*, L1. [CrossRef]
2. Groh, J.H.; Meynet, G.; Georgy, C.; Ekström, S. Fundamental properties of core-collapse supernova and GRB progenitors: Predicting the look of massive stars before death. *Astron. Astrophys.* **2013**, *558*, A131. [CrossRef]
3. Groh, J.H.; Meynet, G.; Ekström, S. Massive star evolution: Luminous blue variables as unexpected supernova progenitors. *Astron. Astrophys.* **2013**, *550*, L7. [CrossRef]
4. Pauldrach, A.W.A.; Vanbeveren, D.; Hoffmann, T.L. Radiation-driven winds of hot luminous stars XVI. Expanding atmospheres of massive and very massive stars and the evolution of dense stellar clusters. *Astron. Astrophys.* **2012**, *538*, A75. [CrossRef]
5. Hopkins, P.F.; Kereš, D.; Oñorbe, J.; Faucher-Giguère, C.-A.; Quataert, E.; Murray, N.; Bullock, J.S. Galaxies on FIRE (Feedback in Realistic Environments): Stellar feedback explains cosmologically inefficient star formation. *Mon. Not. R. Astron. Soc.* **2014**, *445*, 581–603. [CrossRef]
6. Massey, P.; Neugent, K.F.; Smart, B.M. A Spectroscopic Survey of Massive Stars in M31 and M33. *Astron. J.* **2016**, *152*, 62. [CrossRef]
7. Cannon, A.J. Peculiar Spectra in the Large Magellanic Cloud. *Harv. Coll. Obs. Bull.* **1924**, *801*, 1–3.
8. Henize, K.G. Catalogues of Hα-emission Stars and Nebulae in the Magellanic Clouds. *Astrophys. J. Suppl.* **1956**, *2*, 315. [CrossRef]
9. Smith, H.J. Spectra of Bright-Line Stars in the Large Magellanic Cloud. *Publ. ASP* **1957**, *69*, 137. [CrossRef]
10. Feast, M.W.; Thackeray, A.D.; Wesselink, A.J. The brightest stars in the Magellanic Clouds. *Mon. Not. R. Astron. Soc.* **1960**, *121*, 337. [CrossRef]
11. Allen, D.A.; Glass, I.S. Emission-line stars in the Large Magellanic Cloud: Spectroscopy and infrared photometry. *Astrophys. J.* **1976**, *210*, 666–669. [CrossRef]
12. Glass, I.S. A dusty star in the Small Magellanic Cloud. *Mon. Not. R. Astron. Soc.* **1977**, *178*, 9P. [CrossRef]
13. Stahl, O.; Wolf, B.; Zickgraf, F.J.; Bastian, U.; de Groot, M.H.J.; Leitherer, C. R66 (Aeq), an LMC B Supergiant with a Massive, Cool and Dusty Wind. *Astron. Astrophys.* **1983**, *120*, 287–296.
14. Stahl, O.; Leitherer, C.; Wolf, B.; Zickgraf, F.-J. Three new hot stars with dust shells in the Magellanic clouds. *Astron. Astrophys.* **1984**, *131*, L5–L6.
15. Zickgraf, F.-J.; Wolf, B.; Stahl, O.; Leitherer, C.; Klare, G. The hybrid spectrum of the LMC hypergiant R 126. *Astron. Astrophys.* **1985**, *143*, 421–430.

16. Zickgraf, F.J.; Wolf, B.; Stahl, O.; Leitherer, C.; Appenzeller, I. B(e)-supergiants of the Magellanic Clouds. *Astron. Astrophys.* **1986**, *163*, 119–134.
17. Gray, D.F. *The Observation and Analysis of Stellar Photospheres*; Cambridge University Press: Cambridge, UK, 2005.
18. Zickgraf, F.-J. B[e] Supergiants in the Magellanic Clouds. In *Stars with the B[e] Phenomenon*; ASP Conference Series; Astronomical Society of the Pacific: San Francisco, CA, USA, 2006; Volume 355, p. 135.
19. Georgy, C.; Ekström, S.; Eggenberger, P.; Meynet, G.; Haemmerlé, L.; Maeder, A.; Granada, A.; Groh, J.H.; Hirschi, R.; Mowlavi, N.; et al. Grids of stellar models with rotation. III. Models from 0.8 to 120 M$_\odot$ at a metallicity Z = 0.002. *Astron. Astrophys.* **2013**, *558*, A103. [CrossRef]
20. Gummersbach, C.A.; Zickgraf, F.-J.; Wolf, B. B[e] phenomenon extending to lower luminosities in the Magellanic Clouds. *Astron. Astrophys.* **1995**, *302*, 409.
21. McGregor, P.J.; Hillier, D.J.; Hyland, A.R. CO Overtone Emission from Magellanic Cloud Supergiants. *Astrophys. J.* **1988**, *334*, 639. [CrossRef]
22. Zickgraf, F.-J.; Kovacs, J.; Wolf, B.; Stahl, O.; Kaufer, A.; Appenzeller, I. R4 in the Small Magellanic Cloud: A spectroscopic binary with a B[e]/LBV-type component. *Astron. Astrophys.* **1996**, *309*, 505–514.
23. Zickgraf, F.-J.; Wolf, B.; Stahl, O.; Humphreys, R.M. S 18: A new B(e) supergiant in the Small Magellanic Cloud with evidence for an excretion disk. *Astron. Astrophys.* **1989**, *220*, 206–214.
24. Zickgraf, F.-J.; Stahl, O.; Wolf, B. IR survey of OB emission-line stars in the SMC: Detection of a new B E supergiant, AV 172. *Astron. Astrophys.* **1992**, *260*, 205–212.
25. Lamers, H.J.G.L.M.; Zickgraf, F.-J.; de Winter, D.; Houziaux, L.; Zorec, J. An improved classification of B[e]-type stars. *Astron. Astrophys.* **1998**, *340*, 117–128.
26. Appenzeller, I.; Klare, G.; Stahl, O.; Wolf, B.; Zickgraf, F.-J. CASPEC/Cassegrain Echelle Spectrograph/and IUE—A Perfect Match. *Messenger* **1984**, *38*, 28.
27. Clark, J.S.; Bartlett, E.S.; Coe, M.J.; Dorda, R.; Haberl, F.; Lamb, J.B.; Negueruela, I.; Udalski, A. The supergiant B[e] star LHA 115-S 18—Binary and/or luminous blue variable? *Astron. Astrophys.* **2013**, *560*, A10. [CrossRef]
28. Massey, P.; Neugent, K.F.; Morrell, N.; Hillier, D.J. A Modern Search for Wolf-Rayet Stars in the Magellanic Clouds: First Results. *Astrophys. J.* **2014**, *788*, 83. [CrossRef]
29. Bartlett, E.S.; Clark, J.S. Observational Constraints on the Nature of the X-ray bright supergiant B[e] stars LHA 115-S18 & LHA 120- S134. In Proceedings of the SALT Science Conference, Stellenbosch, South Afrcia, 1–5 June 2015.
30. Magalhaes, A.M. Polarization and the Envelopes of B[e] Supergiants in the Magellanic Clouds. *Astrophys. J.* **1992**, *398*, 286. [CrossRef]
31. Melgarejo, R.; Magalhães, A.M.; Carciofi, A.C.; Rodrigues, C.V. S 111 and the polarization of the B[e] supergiants in the Magellanic Clouds. *Astron. Astrophys.* **2001**, *377*, 581–588. [CrossRef]
32. Magalhães, A.M.; Melgarejo, R.; Pereyra, A.; Carciofi, A.C. Polarimetry and the Envelopes of Magellanic B[e] Supergiants. In *Stars with the B[e] Phenomenon*; ASP Conference Series; Astronomical Society of the Pacific: San Francisco, CA, USA, 2006; Volume 355, p. 147.
33. Seriacopi, D.B.; Carciofi, A.C.; Magalhães, A.M. The Envelopes of B[e] Supergiants in the Magellanic Clouds as Seen by Polarimetry. In *The B[e] Phenomenon: Forty Years of Studies*; Miroshnichenko, A., Zharikov, S., Korčáková, D., Wolf, M., Eds.; ASP Conference Series; Astronomical Society of the Pacific: San Francisco, CA, USA, 2017; Volume 508, p. 109.
34. Zickgraf, F.-J.; Schulte-Ladbeck, R.E. Polarization characteristics of galactic Be stars. *Astron. Astrophys.* **1989**, *214*, 274–284.
35. McGregor, P.J.; Hyland, A.R.; Hillier, D.J. Atomic and Molecular Line Emission from Early-Type High-Luminosity Stars. *Astrophys. J.* **1988**, *324*, 1071. [CrossRef]
36. McGregor, P.J.; Hyland, A.R.; McGinn, M.T. Emission-line stars in the Magellanic Clouds: Infrared spectroscopy of Be and Ofpe/WN9 stars. *Astron. Astrophys.* **1989**, *223*, 237–240.
37. Morris, P.W.; Eenens, P.R.J.; Hanson, M.M.; Conti, P.S.; Blum, R.D. Infrared Spectra of Massive Stars in Transition: WNL, Of, Of/WN, Be, B[e], and Luminous Blue Variable Stars. *Astrophys. J.* **1996**, *470*, 597. [CrossRef]
38. Wheelwright, H.E.; de Wit, W.J.; Weigelt, G.; Oudmaijer, R.D.; Ilee, J.D. AMBER and CRIRES observations of the binary sgB[e] star HD 327083: Evidence of a gaseous disc traced by CO bandhead emission. *Astron. Astrophys.* **2012**, *543*, A77. [CrossRef]

39. Muratore, M.F.; Kraus, M.; de Wit, W.J. Near-infrared spectroscopic survey of galactic B[e] stars. *Boletín Asoc. Argent. Astron.* **2012**, *55*, 123–127.
40. Oksala, M.E.; Kraus, M.; Arias, M.L.; Borges Fernandes, M.; Cidale, L.; Muratore, M.F.; Curé, M. The sudden appearance of CO emission in LHA 115-S 65. *Mon. Not. R. Astron. Soc.* **2012**, *426*, L56–L60. [CrossRef]
41. Oksala, M.E.; Kraus, M.; Cidale, L.S.; Muratore, M.F.; Borges Fernandes, M. Probing the ejecta of evolved massive stars in transition. A VLT/SINFONI K-band survey. *Astron. Astrophys.* **2013**, *558*, A17. [CrossRef]
42. Kraus, M.; Oksala, M.E.; Nickeler, D.H.; Muratore, M.F.; Borges Fernandes, M.; Aret, A.; Cidale, L.S.; de Wit, W.J. Molecular emission from GG Carinae's circumbinary disk. *Astron. Astrophys.* **2013**, *549*, A28. [CrossRef]
43. Kraus, M.; Cidale, L.S.; Arias, M.L.; Maravelias, G.; Nickeler, D.H.; Torres, A.F.; Borges Fernandes, M.; Aret, A.; Curé, M.; Vallverdú, R.; et al. Inhomogeneous molecular ring around the B[e] supergiant LHA 120-S 73. *Astron. Astrophys.* **2016**, *593*, A112. [CrossRef]
44. Torres, A.F.; Cidale, L.S.; Kraus, M.; Arias, M.L.; Barbá, R.H.; Maravelias, G.; Borges Fernandes, M. Resolving the clumpy circumstellar environment of the B[e] supergiant LHA 120-S 35. *Astron. Astrophys.* **2018**, *612*, A113. [CrossRef]
45. Kraus, M.; Oksala, M.E.; Cidale, L.S.; Arias, M.L.; Torres, A.F.; Borges Fernandes, M. Discovery of SiO Band Emission from Galactic B[e] Supergiants. *Astrophys. J. Lett.* **2015**, *800*, L20. [CrossRef]
46. Torres, A.F.; Kraus, M.; Cidale, L.S.; Barba, R.; Borges Fernandes, M.; Brandi, E. Discovery of Raman-scattered lines in the massive luminous emission-line star LHA 115-S 18. *Mon. Not. R. Astron. Soc.* **2012**, *427*, L80–L84. [CrossRef]
47. Domiciano de Souza, A.; Driebe, T.; Chesneau, O.; Hofmann, K.-H.; Kraus, S.; Miroshnichenko, A.S.; Ohnaka, K.; Petrov, R.G.; Preisbisch, T.; Stee, P.; et al. AMBER/VLTI and MIDI/VLTI spectro-interferometric observations of the B[e] supergiant CPD-57^{deg}2874. Size and geometry of the circumstellar envelope in the near- and mid-IR. *Astron. Astrophys.* **2007**, *464*, 81–86. [CrossRef]
48. Domiciano de Souza, A.; Bendjoya, P.; Niccolini, G.; Chesneau, O.; Borges Fernandes, M.; Carciofi, A.C.; Spang, A.; Stee, P.; Driebe, T. Fast ray-tracing algorithm for circumstellar structures (FRACS). II. Disc parameters of the B[e] supergiant CPD-57^{deg}2874 from VLTI/MIDI data. *Astron. Astrophys.* **2011**, *525*, A22. [CrossRef]
49. Millour, F.; Meilland, A.; Chesneau, O.; Stee, P.; Kanaan, S.; Petrov, R.; Mourard, D.; Kraus, S. Imaging the spinning gas and dust in the disc around the supergiant A[e] star HD 62623. *Astron. Astrophys.* **2011**, *526*, A107. [CrossRef]
50. Cidale, L.S.; Borges Fernandes, M.; Andruchow, I.; Arias, M.L.; Kraus, M.; Chesneau, O.; Kanaan, S.; Curé, M.; de Wit, W.J.; Muratore, M.F. Observational constraints for the circumstellar disk of the B[e] star CPD-52 9243. *Astron. Astrophys.* **2012**, *548*, A72. [CrossRef]
51. Wang, Y.; Weigelt, G.; Kreplin, A.; Hofmann, K.-H.; Kraus, S.; Miroshnichenko, A.S.; Schertl, D.; Chelli, A.; Domiciano de Souza, A.; Massi, F.; et al. AMBER/VLTI observations of the B[e] star MWC 300. *Astron. Astrophys.* **2012**, *545*, L10. [CrossRef]
52. Wheelwright, H.E.; de Wit, W.J.; Oudmaijer, R.D.; Vink, J.S. VLTI/AMBER observations of the binary B[e] supergiant HD 327083. *Astron. Astrophys.* **2012**, *538*, A6. [CrossRef]
53. Aret, A.; Kraus, M.; Muratore, M.F.; Borges Fernandes, M. A new observational tracer for high-density disc-like structures around B[e] supergiants. *Mon. Not. R. Astron. Soc.* **2012**, *423*, 284–293. [CrossRef]
54. Condori, C.A.H.; Borges Fernandes, M.; Kraus, M.; Panoglou, D.; Guerrero, C.A. The study of unclassified B[e] stars and candidates in the Galaxy and Magellanic Clouds. *Mon. Not. R. Astron. Soc.* **2019**, *488*, 1090–1110. [CrossRef]
55. Kraus, M.; Liimets, T.; Cappa, C.E.; Cidale, L.S.; Nickeler, D.H.; Duronea, N.U.; Arias, M.L.; Gunawan, D.S.; Oksala, M.E.; Borges Fernandes, M.; et al. Resolving the Circumstellar Environment of the Galactic B[e] Supergiant Star MWC 137 from Large to Small Scales. *Astron. J.* **2017**, *154*, 186. [CrossRef]
56. Kraus, M.; Borges Fernandes, M.; de Araújo, F.X. On the hydrogen neutral outflowing disks of B[e] supergiants. *Astron. Astrophys.* **2007**, *463*, 627–634. [CrossRef]
57. Kraus, M.; Borges Fernandes, M.; de Araújo, F.X. Neutral material around the B[e] supergiant star LHA 115-S 65. An outflowing disk or a detached Keplerian rotating disk? *Astron. Astrophys.* **2010**, *517*, A30. [CrossRef]

58. Maravelias, G.; Kraus, M.; Cidale, L.S.; Borges Fernandes, M.; Arias, M.L.; Curé, M.; Vasilopoulos, G. Resolving the kinematics of the discs around Galactic B[e] supergiants. *Mon. Not. R. Astron. Soc.* **2018**, *480*, 320–344. [CrossRef]
59. Muratore, M.F.; Kraus, M.; Oksala, M.E.; Arias, M.L.; Cidale, L.; Borges Fernandes, M.; Liermann, A. Evidence of the Evolved Nature of the B[e] Star MWC 137. *Astron. J.* **2015**, *149*, 13. [CrossRef]
60. Aret, A.; Kraus, M.; Šlechta, M. Spectroscopic survey of emission-line stars—I. B[e] stars. *Mon. Not. R. Astron. Soc.* **2016**, *456*, 1424–1437. [CrossRef]
61. Kraus, M.; Krügel, E.; Thum, C.; Geballe, T.R. CO band emission from MWC 349. I. First overtone bands from a disk or from a wind? *Astron. Astrophys.* **2000**, *362*, 158–168.
62. Kraus, M.; Arias, M.L.; Cidale, L.S.; Torres, A.F. Evidence of an evolved nature of MWC 349A. **2019**, in preparation.
63. Miroshnichenko, A.S.; Bjorkman, K.S.; Grosso, M.; Hinkle, K.; Levato, H.; Marang, F. Properties of galactic B[e] supergiants. IV. Hen 3-298 and Hen 3-303. *Astron. Astrophys.* **2005**, *436*, 653–659. [CrossRef]
64. Liermann, A.; Schnurr, O.; Kraus, M.; Kreplin, A.; Arias, M.L.; Cidale, L.S. A K-band spectral mini-survey of Galactic B[e] stars. *Mon. Not. R. Astron. Soc.* **2014**, *443*, 947–956. [CrossRef]
65. Miroshnichenko, A.S.; Bjorkman, K.S.; Chentsov, E.L.; Klochkova, V.G.; Ezhkova, O.V.; Gray, R.O.; García-Lario, P.; Perea Calderón, J.V.; Rudy, R.J.; Lynch, D.K.; et al. The luminous B[e] binary AS 381. *Astron. Astrophys.* **2002**, *383*, 171–181. [CrossRef]
66. Andrillat, Y.; Jaschek, C. B[e] stars. VIII. MWC 342. *Astron. Astrophys. Suppl.* **1999**, *136*, 59–63. [CrossRef]
67. Borges Fernandes, M.; Kraus, M.; Lorenz Martins, S.; de Araújo, F.X. On the evolutionary stage of the unclassified B[e] star CD-42°11721. *Mon. Not. R. Astron. Soc.* **2007**, *377*, 1343–1362. [CrossRef]
68. Kraus, M.; Borges Fernandes, M.; Kubát, J.; de Araújo, F.X. From B[e] to A[e]. On the peculiar variations of the SMC supergiant LHA 115-S 23 (AzV 172). *Astron. Astrophys.* **2008**, *487*, 697–707. [CrossRef]
69. Chentsov, E.L.; Klochkova, V.G.; Miroshnichenko, A.S. Spectral variability of the peculiar A-type supergiant 3Pup. *Astrophys. Bull.* **2010**, *65*, 150–163. [CrossRef]
70. Kraus, M.; Lamers, H.J.G.L.M. Ionization structure in the winds of B[e] supergiants. I. Ionization equilibrium calculations in a H plus He wind. *Astron. Astrophys.* **2003**, *405*, 165–174. [CrossRef]
71. Kraus, M. Ionization structure in the winds of B[e] supergiants. II. Influence of rotation on the formation of equatorial hydrogen neutral zones. *Astron. Astrophys.* **2006**, *456*, 151–159. [CrossRef]
72. Zsargó, J.; Hillier, D.J.; Georgiev, L.N. Axi-symmetric models of B[e] supergiants. I. The effective temperature and mass-loss dependence of the hydrogen and helium ionization structure. *Astron. Astrophys.* **2008**, *478*, 543–551. [CrossRef]
73. Glatzel, W.; Kiriakidis, M. Stability of Massive Stars and the Humphreys-Davidson Limit. *Mon. Not. R. Astron. Soc.* **1993**, *263*, 375–384. [CrossRef]
74. Kiriakidis, M.; Fricke, K.J.; Glatzel, W. The stability of massive stars and its dependence on metallicity and opacity. *Mon. Not. R. Astron. Soc.* **1993**, *264*, 50–62. [CrossRef]
75. Glatzel, W.; Mehren, S. Non-radial pulsations and stability of massive stars. *Mon. Not. R. Astron. Soc.* **1996**, *282*, 1470–1482. [CrossRef]
76. Kurfürst, P.; Feldmeier, A.; Krtička, J. Time-dependent modeling of extended thin decretion disks of critically rotating stars. *Astron. Astrophys.* **2014**, *569*, A23. [CrossRef]
77. Kurfürst, P.; Feldmeier, A.; Krtička, J. Two-dimensional modeling of density and thermal structure of dense circumstellar outflowing disks. *Astron. Astrophys.* **2018**, *613*, A75. [CrossRef]
78. Pelupessy, I.; Lamers, H.J.G.L.M.; Vink, J.S. The radiation driven winds of rotating B[e] supergiants. *Astron. Astrophys.* **2000**, *359*, 695–706.
79. Curé, M. The Influence of Rotation in Radiation-driven Wind from Hot Stars: New Solutions and Disk Formation in Be Stars. *Astrophys. J.* **2004**, *614*, 929–941. [CrossRef]
80. Curé, M.; Rial, D.F.; Cidale, L. Outflowing disk formation in B[e] supergiants due to rotation and bi-stability in radiation driven winds. *Astron. Astrophys.* **2005**, *437*, 929–933. [CrossRef]
81. Kraus, M. Spectroscopic Diagnostics for Circumstellar Disks of B[e] Supergiants. In *The B[e] Phenomenon: Forty Years of Studies*; Miroshnichenko, A., Zharikov, S., Korčáková, D., Wolf, M., Eds.; ASP Conference Series; Astronomical Society of the Pacific: San Francisco, CA, USA, 2017; Volume 508, pp. 219–228.
82. Kastner, J.H.; Buchanan, C.; Sahai, R.; Forrest, W.J.; Sargent, B.A. The Dusty Circumstellar Disks of B[e] Supergiants in the Magellanic Clouds. *Astron. J.* **2010**, *139*, 1993–2002. [CrossRef]

83. Mehner, A.; de Wit, W.J.; Groh, J.H.; Oudmaijer, R.D.; Baade, D.; Rivinius, T.; Selman, F.; Boffin, H.M.J.; Martayan, C. VLT/MUSE discovers a jet from the evolved B[e] star MWC 137. *Astron. Astrophys.* **2016**, *585*, A81. [CrossRef]
84. Clark, J.S.; Negueruela, I.; Crowther, P.A.; Goodwin, S.P. On the massive stellar population of the super star cluster Westerlund 1. *Astron. Astrophys.* **2005**, *434*, 949–969. [CrossRef]
85. Clark, J.S.; Ritchie, B.W.; Negueruela, I. The circumstellar environment and evolutionary state of the supergiant B[e] star Wd1-9. *Astron. Astrophys.* **2013**, *560*, A11. [CrossRef]
86. Liermann, A.; Kraus, M.; Schnurr, O.; Borges Fernandes, M. The ^{13}Carbon footprint of B[e] supergiants. *Mon. Not. R. Astron. Soc.* **2010**, *408*, L6–L10. [CrossRef]
87. Kraus, M. The pre- versus post-main sequence evolutionary phase of B[e] stars. Constraints from 13CO band emission. *Astron. Astrophys.* **2009**, *494*, 253–262. [CrossRef]
88. Ekström, S.; Georgy, C.; Eggenberger, P.; Meynet, G.; Mowlavi, N.; Wyttenbach, A.; Granada, A.; Decressin, T.; Hirschi, R.; Frischknecht, U.; et al. Grids of stellar models with rotation. I. Models from 0.8 to 120 $M_\odot$ at solar metallicity (Z = 0.014). *Astron. Astrophys.* **2012**, *537*, A146. [CrossRef]
89. Berkhuijsen, E.M.; Humphreys, R.M.; Ghigo, F.D.; Zumach, W. A catalogue of the brightest stars in the field of M 31. *Astron. Astrophys. Suppl.* **1988**, *76*, 65–99.
90. Massey, P.; Armandroff, T.E.; Conti, P.S. Massive stars in M31. *Astron. J.* **1986**, *92*, 1303–1333. [CrossRef]
91. Freedman, W.L. The Young Stellar Content of Nearby Resolved Galaxies. Ph.D. Thesis, University of Toronto, Toronto, ON, Canada, 1984.
92. Humphreys, R.M.; Massey, P.; Freedman, W.L. Spectroscopy of Luminous Blue Stars in M31 and M33. *Astron. J.* **1990**, *99*, 84. [CrossRef]
93. Massey, P.; Olsen, K.A.G.; Hodge, P.W.; Strong, S.B.; Jacoby, G.H.; Schlingman, W.; Smith, R.C. A Survey of Local Group Galaxies Currently Forming Stars. I. UBVRI Photometry of Stars in M31 and M33. *Astron. J.* **2006**, *131*, 2478–2496. [CrossRef]
94. Massey, P.; Olsen, K.A.G.; Hodge, P.W.; Jacoby, G.H.; McNeill, R.T.; Smith, R.C.; Strong, S.B. A Survey of Local Group Galaxies Currently Forming Stars. II. UBVRI Photometry of Stars in Seven Dwarfs and a Comparison of the Entire Sample. *Astron. J.* **2007**, *133*, 2393–2417. [CrossRef]
95. Massey, P.; McNeill, R.T.; Olsen, K.A.G.; Hodge, P.W.; Blaha, C.; Jacoby, G.H.; Smith, R.C.; Strong, S.B. A Survey of Local Group Galaxies Currently Forming Stars. III. A Search for Luminous Blue Variables and Other Hα Emission-Line Stars. *Astron. J.* **2007**, *134*, 2474–2503. [CrossRef]
96. Massey, P.; Waterhouse, E.; DeGioia-Eastwood, K. The Progenitor Masses of Wolf-Rayet Stars and Luminous Blue Variables Determined from Cluster Turnoffs. I. Results from 19 OB Associations in the Magellanic Clouds. *Astron. J.* **2000**, *119*, 2214–2241. [CrossRef]
97. Clark, J.S.; Castro, N.; Garcia, M.; Herrero, A.; Najarro, F.; Negueruela, I.; Ritchie, B.W.; Smith, K.T. On the nature of candidate luminous blue variables in M 33. *Astron. Astrophys.* **2012**, *541*, A146. [CrossRef]
98. Humphreys, R.M.; Weis, K.; Davidson, K.; Bomans, D.J.; Burggraf, B. Luminous and Variable Stars in M31 and M33. II. Luminous Blue Variables, Candidate LBVs, Fe II Emission Line Stars, and Other Supergiants. *Astrophys. J.* **2014**, *790*, 48. [CrossRef]
99. Humphreys, R.M.; Gordon, M.S.; Martin, J.C.; Weis, K.; Hahn, D. Luminous and Variable Stars in M31 and M33. IV. Luminous Blue Variables, Candidate LBVs, B[e] Supergiants, and the Warm Hypergiants: How to Tell Them Apart. *Astrophys. J.* **2017**, *836*, 64. [CrossRef]
100. Gvaramadze, V.V.; Kniazev, A.Y.; Fabrika, S. Revealing evolved massive stars with Spitzer. *Mon. Not. R. Astron. Soc.* **2010**, *405*, 1047–1060. [CrossRef]
101. Wachter, S.; Mauerhan, J.C.; Van Dyk, S.D.; Hoard, D.W.; Kafka, S.; Morris, P.W. A Hidden Population of Massive Stars with Circumstellar Shells Discovered with the Spitzer Space Telescope. *Astron. J.* **2010**, *139*, 2330–2346. [CrossRef]
102. Aadland, E.; Massey, P.; Neugent, K.F.; Drout, M.R. Shedding Light on the Isolation of Luminous Blue Variables. *Astron. J.* **2018**, *156*, 294. [CrossRef]
103. Campagnolo, J.C.N.; Borges Fernandes, M.; Drake, N.A.; Kraus, M.; Guerrero, C.A.; Pereira, C.B. Detection of new eruptions in the Magellanic Clouds luminous blue variables R 40 and R 110. *Astron. Astrophys.* **2018**, *613*, A33. [CrossRef]

104. Cutri, R.M.; Skrutskie, M.F.; van Dyk, S.; Beichman, C.A.; Carpenter, J.M.; Chester, T.; Cambresy, L.; Evans, T.; Fowler, J.; Gizis, J.; et al. VizieR Online Data Catalog: 2MASS All-Sky Catalog of Point Sources. *VizieR On-line Data Catalog* **2003**, II/246.
105. Cutri, R.M. VizieR Online Data Catalog: WISE All-Sky Data Release. *VizieR On-line Data Catalog* **2012**, II/311.
106. Worthey, G.; Lee, H. An Empirical UBV RI JHK Color-Temperature Calibration for Stars. *Astrophys. J. Suppl.* **2011**, *193*, 1. [CrossRef]
107. Morris, P.W.; Voors, R.H.M.; Lamers, H.J.G.L.M.; Eenens, P.R.J. Near-Infrared Spectra of LBV; Be and B[e] Stars: Does Axisymmetry Provide a Morphological Link? In *Luminous Blue Variables: Massive Stars in Transition*; Nota, A., Lamers, H., Eds.; ASP Conference Series; Astronomical Society of the Pacific: San Francisco, CA, USA, 1997; Volume 120, p. 20.
108. Bonanos, A.Z.; Massa, D.; Sewilo, M.; Lennon, D.J.; Panagia, N.; Smith, L.J.; Meixner, M.; Babler, B.L.; Bracker, S.; Meade, M.R.; et al. Spitzer SAGE Infrared Photometry of Massive Stars in the Large Magellanic Cloud. *Astron. J.* **2009**, *138*, 1003–1021. [CrossRef]
109. Stahl, O.; Smolinski, J.; Wolf, B.; Zickgraf, F.-J. High-Dispersion Spectroscopy of the B[e] Supergiant S 111. In *Physics of Luminous Blue Variables*; Davidson, K., Moffat, A.F.J., Lamers, H.J.G.L.M., Eds.; Astrophysics and Space Science Library: Dordrecht, The Netherlands; Kluwer Academic Publishers: Boston, MA, USA, 1989; Volume 157, p. 295.
110. Miroshnichenko, A.S.; Manset, N.; Polcaro, F.; Rossi, C.; Zharikov, S. The B[e] phenomenon in the Milky Way and Magellanic Clouds. *Proc. Int. Astron. Union* **2011**, *6*, 260–264. [CrossRef]
111. Evans, C.J.; Taylor, W.D.; Hénault-Brunet, V.; Sana, H.; de Koter, A.; Simón-Díaz, S.; Carraro, G.; Bagnoli, T.; Bastian, N.; Bestenlehner, J.M.; et al. The VLT-FLAMES Tarantula Survey. I. Introduction and observational overview. *Astron. Astrophys.* **2011**, *530*, A108. [CrossRef]
112. Evans, C.J.; Kennedy, M.B.; Dufton, P.L.; Howarth, I.D.; Walborn, N.R.; Markova, N.; Clark, J.S.; de Mink, S.E.; de Koter, A.; Dunstall, P.R.; et al. The VLT-FLAMES Tarantula Survey. XVIII. Classifications and radial velocities of the B-type stars. *Astron. Astrophys.* **2015**, *574*, A13. [CrossRef]
113. Kalari, V.M.; Vink, J.S.; Dufton, P.L.; Evans, C.J.; Dunstall, P.R.; Sana, H.; Clark, J.S.; Ellerbroek, L.; de Koter, A.; Lennon, D.J.; et al. The VLT-FLAMES Tarantula Survey. XV. VFTS 822: A candidate Herbig B[e] star at low metallicity. *Astron. Astrophys.* **2014**, *564*, L7. [CrossRef]
114. Kamath, D.; Wood, P.R.; Van Winckel, H. Optically visible post-AGB stars, post-RGB stars and young stellar objects in the Large Magellanic Cloud. *Mon. Not. R. Astron. Soc.* **2015**, *454*, 1468–1502. [CrossRef]
115. González-Fernández, C.; Dorda, R.; Negueruela, I.; Marco, A. A new survey of cool supergiants in the Magellanic Clouds. *Astron. Astrophys.* **2015**, *578*, A3. [CrossRef]
116. Hernández, J.; Calvet, N.; Briceño, C.; Hartmann, L.; Berlind, P. Spectral Analysis and Classification of Herbig Ae/Be Stars. *Astron. J.* **2004**, *127*, 1682–1701. [CrossRef]
117. Cardelli, J.A.; Clayton, G.C.; Mathis, J.S. The Relationship between Infrared, Optical, and Ultraviolet Extinction. *Astrophys. J.* **1989**, *345*, 245. [CrossRef]
118. Levato, H.; Miroshnichenko, A.S.; Saffe, C. New objects with the B[e] phenomenon in the Large Magellanic Cloud. *Astron. Astrophys.* **2014**, *568*, A28. [CrossRef]
119. Dunstall, P.R.; Fraser, M.; Clark, J.S.; Crowther, P.A.; Dufton, P.L.; Evans, C.J.; Lennon, D.J.; Soszyński, I.; Taylor, W.D.; Vink, J.S. The VLT-FLAMES Tarantula Survey. V. The peculiar B[e]-like supergiant, VFTS698, in 30 Doradus. *Astron. Astrophys.* **2012**, *542*, A50. [CrossRef]
120. Whitney, B.A.; Sewilo, M.; Indebetouw, R.; Robitaille, T.P.; Meixner, M.; Gordon, K.; Meade, M.R.; Babler, B.L.; Harris, J.; Hora, J.L.; et al. Spitzer Sage Survey of the Large Magellanic Cloud. III. Star Formation and 1000 New Candidate Young Stellar Objects. *Astron. J.* **2008**, *136*, 18–43. [CrossRef]
121. Campbell, M.A.; Evans, C.J.; Mackey, A.D.; Gieles, M.; Alves, J.; Ascenso, J.; Bastian, N.; Longmore, A.J. VLT-MAD observations of the core of 30 Doradus. *Mon. Not. R. Astron. Soc.* **2010**, *405*, 421–435. [CrossRef]
122. Evans, C.J.; Howarth, I.D.; Irwin, M.J.; Burnley, A.W.; Harries, T.J. A 2dF survey of the Small Magellanic Cloud. *Mon. Not. R. Astron. Soc.* **2004**, *353*, 601–623. [CrossRef]
123. Graus, A.S.; Lamb, J.B.; Oey, M.S. Discovery of New, Dust-poor B[e] Supergiants in the Small Magellanic Cloud. *Astrophys. J.* **2012**, *759*, 10. [CrossRef]
124. Bonanos, A.Z.; Lennon, D.J.; Köhlinger, F.; van Loon, J.T.; Massa, D.L.; Sewilo, M.; Evans, C.J.; Panagia, N.; Babler, B.L.; Block, M.; et al. Spitzer SAGE-SMC Infrared Photometry of Massive Stars in the Small Magellanic Cloud. *Astron. J.* **2010**, *140*, 416–429. [CrossRef]

125. Wisniewski, J.P.; Bjorkman, K.S.; Bjorkman, J.E.; Clampin, M. Discovery of a New Dusty B[e] Star in the Small Magellanic Cloud. *Astrophys. J.* **2007**, *670*, 1331–1336. [CrossRef]
126. Meyssonnier, N.; Azzopardi, M. A new catalogue of H-alpha emission-line stars and small nebulae in the Small Magellanic Cloud. *Astron. Astrophys. Suppl.* **1993**, *102*, 451–593.
127. Keller, S.C.; Wood, P.R.; Bessell, M.S. Be stars in and around young clusters in the Magellanic Clouds. *Astron. Astrophys. Suppl.* **1999**, *134*, 489–503. [CrossRef]
128. Keller, L.D.; Sloan, G.C.; Forrest, W.J.; Ayala, S.; D'Alessio, P.; Shah, S.; Calvet, N.; Najita, J.; Li, A.; Hartmann, L.; et al. PAH Emission from Herbig Ae/Be Stars. *Astrophys. J.* **2008**, *684*, 411–429. [CrossRef]
129. Peeters, E.; Hony, S.; Van Kerckhoven, C.; Tielens, A.G.G.M.; Allamandola, L.J.; Hudgins, D.M.; Bauschlicher, C.W. The rich 6 to 9 µm spectrum of interstellar PAHs. *Astron. Astrophys.* **2002**, *390*, 1089–1113. [CrossRef]
130. Ruffle, P.M.E.; Kemper, F.; Jones, O.C.; Sloan, G.C.; Kraemer, K.E.; Woods, P.M.; Boyer, M.L.; Srinivasan, S.; Antoniou, V.; Lagadec, E.; et al. Spitzer infrared spectrograph point source classification in the Small Magellanic Cloud. *Mon. Not. R. Astron. Soc.* **2015**, *451*, 3504–3536. [CrossRef]
131. Whelan, D.G.; Lebouteiller, V.; Galliano, F.; Peeters, E.; Bernard-Salas, J.; Johnson, K.E.; Indebetouw, R.; Brandl, B.R. An In-depth View of the Mid-infrared Properties of Point Sources and the Diffuse ISM in the SMC Giant H II Region, N66. *Astrophys. J.* **2013**, *771*, 16. [CrossRef]
132. Kamath, D.; Wood, P.R.; Van Winckel, H. Optically visible post-AGB/RGB stars and young stellar objects in the Small Magellanic Cloud: Candidate selection, spectral energy distributions and spectroscopic examination. *Mon. Not. R. Astron. Soc.* **2014**, *439*, 2211–2270. [CrossRef]
133. Heydari-Malayeri, M. Discovery of a low mass B[e] supergiant in the Small magellanic Cloud. *Astron. Astrophys.* **1990**, *234*, 233.
134. Corral, L.J. LBV-Type Stars in M33. *Astron. J.* **1996**, *112*, 1450. [CrossRef]
135. Fabrika, S.; Sholukhova, O. A survey of blue—Hα objects in the galaxy M 33. *Astron. Astrophys. Suppl.* **1999**, *140*, 309–326. [CrossRef]
136. Kraus, M.; Cidale, L.S.; Arias, M.L.; Oksala, M.E.; Borges Fernandes, M. Discovery of the First B[e] Supergiants in M 31. *Astrophys. J. Lett.* **2014**, *780*, L10. [CrossRef]
137. Sholukhova, O.; Bizyaev, D.; Fabrika, S.; Sarkisyan, A.; Malanushenko, V.; Valeev, A. New luminous blue variables in the Andromeda galaxy. *Mon. Not. R. Astron. Soc.* **2015**, *447*, 2459–2467. [CrossRef]
138. Kourniotis, M.; Kraus, M.; Arias, M.L.; Cidale, L.; Torres, A.F. On the evolutionary state of massive stars in transition phases in M33. *Mon. Not. R. Astron. Soc.* **2018**, *480*, 3706–3717. [CrossRef]
139. Sholukhova, O.; Fabrika, S.; Valeev, A. Classification of LBV-Star Candidates in the Galaxy M31. In *Stars: From Collapse to Collapse*; Balega, Y.Y., Kudryavtsev, D.O., Romanyuk, I.I., Yakunin, I.A., Eds.; ASP Conference Series; Astronomical Society of the Pacific: San Francisco, CA, USA, 2017; Volume 510, p. 468.
140. Humphreys, R.M.; Davidson, K.; Grammer, S.; Kneeland, N.; Martin, J.C.; Weis, K.; Burggraf, B. Luminous and Variable Stars in M31 and M33. I. The Warm Hypergiants and Post-red Supergiant Evolution. *Astrophys. J.* **2013**, *773*, 46. [CrossRef]
141. Fabrika, S.; Sholukhova, O.; Becker, T.; Afanasiev, V.; Roth, M.; Sanchez, S.F. Crowded field 3D spectroscopy of LBV candidates in M 33. *Astron. Astrophys.* **2005**, *437*, 217–226. [CrossRef]
142. Humphreys, R.M.; Davidson, K.; Hahn, D.; Martin, J.C.; Weis, K. Luminous and Variable Stars in M31 and M33. V. The Upper HR Diagram. *Astrophys. J.* **2017**, *844*, 40. [CrossRef]
143. Gordon, M.S.; Humphreys, R.M.; Jones, T.J. Luminous and Variable Stars in M31 and M33. III. The Yellow and Red Supergiants and Post-red Supergiant Evolution. *Astrophys. J.* **2016**, *825*, 50. [CrossRef]
144. Stahl, O.; Wolf, B.; de Groot, M.; Leitherer, C. Atlas of hig-dispersion spectra of peculiar emission-line stars in the Magellanic Clouds. *Astron. Astrophys. Suppl.* **1985**, *61*, 237–258.
145. Sholukhova, O.; Fabrika, S.; Roth, M.; Becker, T. B 416—A B[e]-SUPERGIANT in Interacting Binary? *Balt. Astron.* **2004**, *13*, 156–158.
146. Sandage, A. The brightest stars in nearby galaxies. II. The color-magnitude diagram for the brightest red and blue stars in M101. *Astron. J.* **1983**, *88*, 1569–1578. [CrossRef]
147. Sandage, A. The brightest stars in nearby galaxies. III. The color-magnitude diagram for the brightest red and blue stars in M 81 and Holmberg IX. *Astron. J.* **1984**, *89*, 621–629. [CrossRef]
148. Sandage, A. The brightest stars in nearby galaxies. IV. The color-magnitude diagram for the brightest red and blue stars in NGC 2403. *Astron. J.* **1984**, *89*, 630–635. [CrossRef]

149. Zickgraf, F.-J.; Humphreys, R.M. A Stellar Content Survey of NGC 2403 and M81. *Astron. J.* **1991**, *102*, 113. [CrossRef]
150. Grammer, S.; Humphreys, R.M. The Massive Star Population in M101. I. The Identification and Spatial Distribution of the Visually Luminous Stars. *Astron. J.* **2013**, *146*, 114. [CrossRef]
151. Humphreys, R.M. Studies of luminous stars in nearby galaxies. VII—The brightest blue stars in the spiral galaxies M101 and NGC 2403. *Astrophys. J.* **1980**, *241*, 598–601. [CrossRef]
152. Humphreys, R.M.; Aaronson, M. The Visually Brightest Early-Type Supergiants in the Spiral Galaxies NGC 2403, M81, and M101. *Astron. J.* **1987**, *94*, 1156, [CrossRef]
153. Sholukhova, O.N.; Fabrika, S.N.; Vlasyuk, V.V.; Dodonov, S.N. Spectroscopy of stars in the galaxy M 81. *Astron. Lett.* **1998**, *24*, 507–515.
154. Sholukhova, O.N.; Fabrika, S.N.; Vlasyuk, V.V. Spectroscopy of stars in the Galaxy NGC 2403. *Astron. Lett.* **1998**, *24*, 603–610.
155. Grammer, S.H.; Humphreys, R.M.; Gerke, J. The Massive Star Population in M101. III. Spectra and Photometry of the Luminous and Variable Stars. *Astron. J.* **2015**, *149*, 152. [CrossRef]
156. Humphreys, R.M.; Stangl, S.; Gordon, M.S.; Davidson, K.; Grammer, S.H. Luminous and Variable Stars in NGC 2403 and M81. *Astron. J.* **2019**, *157*, 22. [CrossRef]
157. Pustilnik, S.A.; Makarova, L.N.; Perepelitsyna, Y.A.; Moiseev, A.V.; Makarov, D.I. The extremely metal-poor galaxy DDO 68: The luminous blue variable, Hα shells and the most luminous stars. *Mon. Not. R. Astron. Soc.* **2017**, *465*, 4985–5002. [CrossRef]
158. Spetsieri, Z.T.; Bonanos, A.Z.; Kourniotis, M.; Yang, M.; Lianou, S.; Bellas-Velidis, I.; Gavras, P.; Hatzidimitriou, D.; Kopsacheili, M.; Moretti, M.I.; et al. Massive variable candidates with the Hubble Space Telescope. *Astron. Astrophys.* **2018**, *618*, A185. [CrossRef]
159. Spetsieri, Z.T.; Bonanos, A.Z.; Yang, M.; Kourniotis, M.; Hatzidimitriou, D. The HST Key Project galaxies NGC 1326A, NGC 1425 and NGC 4548: New variable stars and massive star population. *Astron. Astrophys.* **2019**, *629*, A3. [CrossRef]
160. Ababakr, K.M.; Oudmaijer, R.D.; Vink, J.S. A statistical spectropolarimetric study of Herbig Ae/Be stars. *Mon. Not. R. Astron. Soc.* **2017**, *472*, 854–868. [CrossRef]
161. Bik, A.; Thi, W.F. Evidence for an inner molecular disk around massive Young Stellar Objects. *Astron. Astrophys.* **2004**, *427*, L13–L16. [CrossRef]
162. Blum, R.D.; Barbosa, C.L.; Damineli, A.; Conti, P.S.; Ridgway, S. Accretion Signatures from Massive Young Stellar Objects. *Astrophys. J.* **2004**, *617*, 1167–1176. [CrossRef]
163. Ilee, J.D.; Wheelwright, H.E.; Oudmaijer, R.D.; de Wit, W.J.; Maud, L.T.; Hoare, M.G.; Lumsden, S.L.; Moore, T.J.T.; Urquhart, J.S.; Mottram, J.C. CO bandhead emission of massive young stellar objects: Determining disc properties. *Mon. Not. R. Astron. Soc.* **2013**, *429*, 2960–2973. [CrossRef]
164. Ilee, J.D.; Fairlamb, J.; Oudmaijer, R.D.; Mendigutía, I.; van den Ancker, M.E.; Kraus, S.; Wheelwright, H.E. Investigating the inner discs of Herbig Ae/Be stars with CO bandhead and Brγ emission. *Mon. Not. R. Astron. Soc.* **2014**, *445*, 3723–3736. [CrossRef]
165. Ilee, J.D.; Oudmaijer, R.D.; Wheelwright, H.E.; Pomohaci, R. Blinded by the light: On the relationship between CO first overtone emission and mass accretion rate in massive young stellar objects. *Mon. Not. R. Astron. Soc.* **2018**, *477*, 3360–3368. [CrossRef]
166. Esteban, C.; Fernandez, M. S266: A ring nebula around a Galactic B[e] supergiant? *Mon. Not. R. Astron. Soc.* **1998**, *298*, 185–192. [CrossRef]
167. Gvaramadze, V.V.; Menten, K.M. Discovery of a parsec-scale bipolar nebula around MWC 349A. *Astron. Astrophys.* **2012**, *541*, A7. [CrossRef]
168. Cohen, M.; Bieging, J.H.; Dreher, J.W.; Welch, W.J. The binary system MWC 349. *Astrophys. J.* **1985**, *292*, 249–256. [CrossRef]
169. Marchiano, P.; Brandi, E.; Muratore, M.F.; Quiroga, C.; Ferrer, O.E.; García, L.G. The spectroscopic orbits and physical parameters of GG Carinae. *Astron. Astrophys.* **2012**, *540*, A91. [CrossRef]
170. Miroshnichenko, A.S.; Levato, H.; Bjorkman, K.S.; Grosso, M. Properties of galactic B[e] supergiants II. HDE 327083. *Astron. Astrophys.* **2003**, *406*, 673–683. [CrossRef]
171. Wolf, B.; Stahl, O. The absorption spectrum of the Be star MWC 300. *Astron. Astrophys.* **1985**, *148*, 412–416.

172. Miroshnichenko, A.S.; Levato, H.; Bjorkman, K.S.; Grosso, M.; Manset, N.; Men'shchikov, A.B.; Rudy, R.J.; Lynch, D.K.; Mazuk, S.; Venturini, C.C.; et al. Properties of galactic B[e] supergiants. III. MWC 300. *Astron. Astrophys.* **2004**, *417*, 731–743. [CrossRef]
173. Miroshnichenko, A.; Corporon, P. Revealing the nature of the B[e] star MWC 342. *Astron. Astrophys.* **1999**, *349*, 126–134.
174. Millour, F.; Chesneau, O.; Borges Fernandes, M.; Meilland, A.; Mars, G.; Benoist, C.; Thiébaut, E.; Stee, P.; Hofmann, K.-H.; Baron, F.; et al. A binary engine fuelling HD 87643's complex circumstellar environment. Determined using AMBER/VLTI imaging. *Astron. Astrophys.* **2009**, *507*, 317–326. [CrossRef]
175. Plets, H.; Waelkens, C.; Trams, N.R. The peculiar binary supergiant 3 Puppis. *Astron. Astrophys.* **1995**, *293*, 363–370.
176. Meilland, A.; Kanaan, S.; Borges Fernandes, M.; Chesneau, O.; Millour, F.; Stee, P.; Lopez, B. Resolving the dusty circumstellar environment of the A[e] supergiant HD 62623 with the VLTI/MIDI. *Astron. Astrophys.* **2010**, *512*, A73. [CrossRef]
177. de Wit, W.J.; Oudmaijer, R.D.; Vink, J.S. Dusty Blue Supergiants: News from High-Angular Resolution Observations. *Adv. Astron.* **2014**, *2014*, 270848. [CrossRef]
178. Hendry, M.A.; Smartt, S.J.; Skillman, E.D.; Evans, C.J.; Trundle, C.; Lennon, D.J.; Crowther, P.A.; Hunter, I. The blue supergiant Sher 25 and its intriguing hourglass nebula. *Mon. Not. R. Astron. Soc.* **2008**, *388*, 1127–1142. [CrossRef]
179. Smith, N.; Arnett, W.D.; Bally, J.; Ginsburg, A.; Filippenko, A.V. The ring nebula around the blue supergiant SBW1: Pre-explosion snapshot of an SN 1987A twin. *Mon. Not. R. Astron. Soc.* **2013**, *429*, 1324–1341. [CrossRef]
180. Smith, N.; Tombleson, R. Luminous blue variables are antisocial: Their isolation implies that they are kicked mass gainers in binary evolution. *Mon. Not. R. Astron. Soc.* **2015**, *447*, 598–617. [CrossRef]
181. Richardson, N.D.; Mehner, A. The 2018 Census of Luminous Blue Variables in the Local Group. *Res. Notes Am. Astron. Soc.* **2018**, *2*, 121. [CrossRef]
182. Langer, N.; Heger, A. B[e] Supergiants: What is Their Evolutionary Status? In *B[e] Stars*; Hubert, A.M., Jaschek, C., Eds.; Astrophysics and Space Science Library: Dordrecht, The Netherlands; Kluwer Academic Publishers: Boston, MA, USA, 1998; Volume 233, p. 235.
183. Pasquali, A.; Nota, A.; Langer, N.; Schulte-Ladbeck, R.E.; Clampin, M. R4 and Its Circumstellar Nebula: Evidence for a Binary Merger? *Astron. J.* **2000**, *119*, 1352–1358. [CrossRef]
184. Martin, J.C.; Humphreys, R.M. Multi-epoch BVRI Photometry of Luminous Stars in M31 and M33. *Astron. J.* **2017**, *154*, 81. [CrossRef]
185. Stanek, K.Z.; Kaluzny, J.; Krockenberger, M.; Sasselov, D.D.; Tonry, J.L.; Mateo, M. Detached Eclipsing Binaries and Cepheids. III. Variables in the Field M31C. *Astron. J.* **1999**, *117*, 2810–2830. [CrossRef]
186. Vilardell, F.; Ribas, I.; Jordi, C. Eclipsing binaries suitable for distance determination in the Andromeda galaxy. *Astron. Astrophys.* **2006**, *459*, 321–331. [CrossRef]
187. Bartlett, E.S.; Clark, J.S.; Negueruela, I. CI Camelopardalis: The first sgB[e]-high mass X-ray binary twenty years on: A supernova imposter in our own Galaxy? *Astron. Astrophys.* **2019**, *622*, A93. [CrossRef]
188. Filliatre, P.; Chaty, S. The Optical/Near-Infrared Counterpart of the INTEGRAL Obscured Source IGR J16318-4848: An sgB[e] in a High-Mass X-Ray Binary? *Astrophys. J.* **2004**, *616*, 469–484. [CrossRef]
189. Lau, R.M.; Kasliwal, M.M.; Bond, H.E.; Smith, N.; Fox, O.D.; Carlon, R.; Cody, A.M.; Contreras, C.; Dykhoff, D.; Gehrz, R.; et al. Rising from the Ashes: Mid-infrared Re-brightening of the Impostor SN 2010da in NGC 300. *Astrophys. J.* **2016**, *830*, 142. [CrossRef]
190. Lau, R.M.; Heida, M.; Kasliwal, M.M.; Walton, D.J. First Detection of Mid-infrared Variability from an Ultraluminous X-Ray Source Holmberg II X-1. *Astrophys. J. Lett.* **2017**, *838*, L17. [CrossRef]
191. Villar, V.A.; Berger, E.; Chornock, R.; Margutti, R.; Laskar, T.; Brown, P.J.; Blanchard, P.K.; Czekala, I.; Lunnan, R.; Reynolds, M.T. The Intermediate Luminosity Optical Transient SN 2010da: The Progenitor, Eruption, and Aftermath of a Peculiar Supergiant High-mass X-ray Binary. *Astrophys. J.* **2016**, *830*, 11. [CrossRef]

© 2019 by the author. Licensee MDPI, Basel, Switzerland. This article is an open access article distributed under the terms and conditions of the Creative Commons Attribution (CC BY) license (http://creativecommons.org/licenses/by/4.0/).

Review

Luminous Blue Variables

Kerstin Weis [1,*] and Dominik J. Bomans [1,2,3]

1. Astronomical Institute, Faculty for Physics and Astronomy, Ruhr University Bochum, 44801 Bochum, Germany
2. Research Department Plasmas with Complex Interactions, Ruhr University Bochum, 44801 Bochum, Germany
3. Ruhr Astroparticle and Plasma Physics (RAPP) Center, 44801 Bochum, Germany
* Correspondence: kweis@astro.rub.de

Received: 29 October 2019; Accepted: 18 February 2020; Published: 29 February 2020

Abstract: Luminous Blue Variables are massive evolved stars, here we introduce this outstanding class of objects. Described are the specific characteristics, the evolutionary state and what they are connected to other phases and types of massive stars. Our current knowledge of LBVs is limited by the fact that in comparison to other stellar classes and phases only a few "true" LBVs are known. This results from the lack of a unique, fast and always reliable identification scheme for LBVs. It literally takes time to get a true classification of a LBV. In addition the short duration of the LBV phase makes it even harder to catch and identify a star as LBV. We summarize here what is known so far, give an overview of the LBV population and the list of LBV host galaxies. LBV are clearly an important and still not fully understood phase in the live of (very) massive stars, especially due to the large and time variable mass loss during the LBV phase. We like to emphasize again the problem how to clearly identify LBV and that there are more than just one type of LBVs: The giant eruption LBVs or η Car analogs and the S Dor cycle LBVs.

Keywords: Luminous Blue Variables; giant eruption; massive stars; stellar population; Wolf-Rayet stars; Eddington limit; mass loss rate; nebulae of Luminous Blue Variable; Supernova impostors; bistability limit

---

1. Historic Background and Naming

Studying the brightest stars in M 31 and M 33 Hubble and Sandage [1] found irregular variable stars that defined a new object class: Var 19 in M 31 and Var 2, Var A, Var B and Vary C in M 33. The variability of Var 2 has been recognized already 1922 by Duncan [2] and 1923 by Wolf [3]. All irregular variable stars Hubble and Sandage found showed the three common characteristics: high luminosity, blue color indice and at the date of observation an intermediate F-type spectrum. Objects of this class became known as *Hubble-Sandage Variables*. In 1974 Sandage and Tammann [4] observed bright stars in NGC 2366, NGC 4236, IC 2574, Ho I, Ho II, and NGC 2403 originally to further constrain the Hubble constant using Cepheids. In some of these galaxies however they identified stars they designated as *Irregular Luminous Blue Variables*. At the same time Humphreys [5] published additional spectral analysis on the M 31 and M 33 Variables and put them into context to the η Carina-like objects. Few years later Humphreys and Davidson [6] studied our galaxy and the LMC and identified the most luminous and massive stars. In that work it became more and more obvious that a certain region in the HRD is not populated: very luminous cool stars seems to not exist or more likely stay for only for a very short time in this region. The boundary to that area was defined by the authors and has been referred to as the Humphreys-Davidson limit. Shortly after in his publication entitled "The stability limit of hypergiant photospheres" de Jager [7] was the first to addressed the presence of such a limit and related possible instabilities from a more theoretical perspective. His argumentation was based on turbulent pressure initiating an instability. Lamers & Fitzpatrick [8] however showed in a 1986 publication that —as still accepted now—radiation and not turbulent pressure is the driver. This linked Humphreys and Davidson observations to the fact that stars will become unstable in this cool and luminous state.

The variability of S Doradus in the LMC was first notices by Pickering in 1897 [9], he also found the star to be bright in H_β, H_γ and H_δ. Later further studies of its variability [10] showed that S Dor characteristics are very similar to those of the Hubble-Sandage Variables. In our own galaxy, a class known as the P Cygni type stars also showed the same behavior. Note in that context that not all stars that show P Cygni line profiles were automatically members of this historically defined class. Humphreys noted already in her 1975 paper [5] that: "The spectral and photometric properties of these extragalactic variables suggest that they may all be related to stars like η Car in our Galaxy and S Dor in the Large Magellanic Cloud.". This hints to the fact that all are only samples of one larger class of variable stars.

In 1984 Peter Conti [11] used the term Luminous Blue Variable during a talk at the IAU Symposium 105 on Observational Tests of the Stellar Evolution Theory. Herewith he finally united—as Humphreys already suggested in her 1975 paper—the earlier defined stellar subgroups of Hubble-Sandage Variables, S Dor Variables, P Cygni and η Car type stars, and explicitly excluded Wolf-Rayet stars and normal blue supergiants from LBVs.

2. Characteristic of Luminous Blue Variables

The name already suggests that features that LBVs seem to have in common are being blue and luminous stars that are variable. This however is a rather weak constraint and not even true for a LBV all the time.

It is not simple to disentangle a LBV from a blue O B supergiant and even cooler supergiant of spectral type A of F. A significant number of LBVs have at least temporarily an Of/WN type spectrum [12,13], indicating the presence of emission line and in particular a larger amount of nitrogen in their photosphere. Others were detected with a Be or B[e] spectrum.

It is not possible to identify and classify an LBV by its spectrum or analog its color. It is the a specific variability or an eruption that distinguishes LBVs from "normal stars". The variability of LBVs is a combination of a photometric brightness and color change, caused and accompanied by changes in the stellar spectrum. During such a S Dor variability or S Dor cycle which lasts years or decades [14,15] the star varies from a optically fainter to a brighter star and back. This variability is therefore caused by the star changing from an early (hot) to a late (cool) spectral type, it implies also that not only brightens up but also goes from a blue to a redder color. Historically the brightening of a LBV in the bright (cool) phase during an cycle has also been called an eruption (or S Dor eruption). As we will see later this term is confusing.

With S Doradus in the Large Magellanic Cloud as the first to show this and therefore the prototype, this alternation from hot to cool and back was accordingly named a S Dor cycle and is observed in LBVs only!. The S Dor variability is the one and only clear distinction of LBVs from other massive evolved stars. An example of a long term lightcurve is given for the LBV Var B in M 33 in Figure 1, the analog version for Var C was published by Burggraf [16]. Also plotted here are the changes of the spectral type for the star, that mark an S Dor cycle.

Bernhard Wolf [17] noticed that the change of the spectrum (or equivalent T_{eff}) within an S Dor cycle from a hot to a cool type is larger for more luminous LBVs. This became known as the *amplitude-luminosity-relation*. His plot as well as a new version we made to visualize this relation is given in Figure 2. Instead of a classical HRD we plotted the change of Temperature (ΔT_{eff}) versus the Luminosity L by using the LBVs given in the HRD in Figure 3. The new plot visualizes nicely how tight this relation really is.

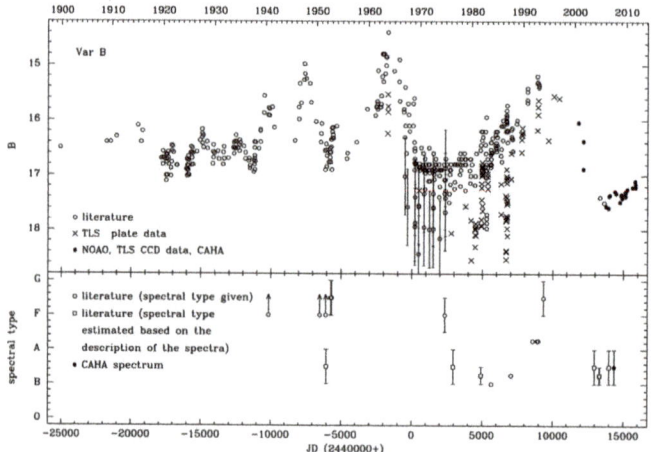

Figure 1. This figure taken from Burggraf (2015) [18] shows a lightcurve spanning more than 100 years of the LBV and original Hubble Sandage Variable Var B in M 33. In addition to the B magnitudes upper section the spectral type if know for the same date is plotted in the lower section. Note the for S Dor cycles typical changes in the spectral type.

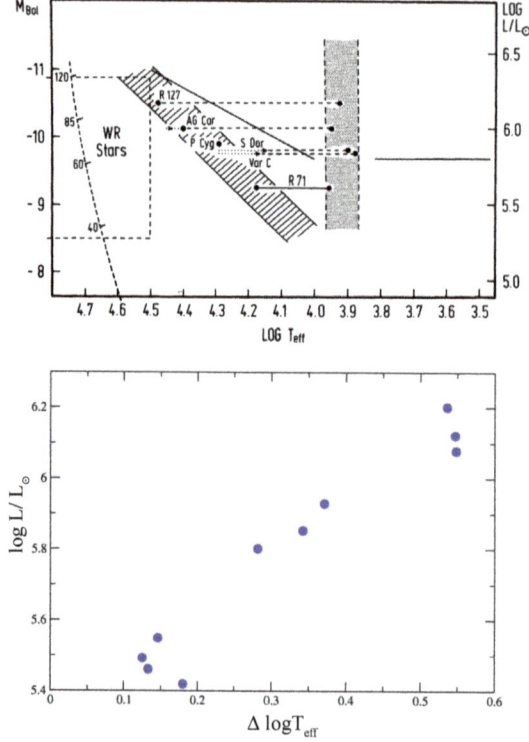

Figure 2. This figure shows the classical plot by Wolf [17] (**top**) and in a new version (**bottom**) we plotted the luminosity L and change in Temperature ΔT_{eff} for a new way to visualize the amplitude-luminosity-relation.

A more elaborate photometric classification, based on the duration of the S Dor cycle was made by van Genderen in 2001 [19]. He subdivided the phase and thereby objects into long S Dor (L-SD), here the cycle lasts ≤ 20 years and the short S Dor (S-SD) with the cycle being less than 10 yrs. Beside that he added a group he designated as ex-/dormant for those that currently (within the last 100 years) showed only a weak or no activity at all. Note that these variation are much larger as the microvariability which is common for supergiants in general [20].

In contrast to the more ordered S Dor variability (or eruption) LBVs can undergo more energetic events. In spontaneous *giant eruptions* the visible brightness increases spontaneously by several magnitudes [21]. The best known and well documented event is the giant eruption of the LBV η Carinae around 1843. During the eruption the star (or rather outburst) was with -1^m the second brightest star in the sky, surpassed only by Sirius with -1.46^m [21,22]. Other known and documented historic and present giant eruptions of LBVs are those of P Cygni around ~1600 [23], SN1954J (=V12) in NGC 2403 ([21,24], and SN1961V in NGC 1058 ([25,26]). It is really important to distinguish between the "S Dor eruption" and a giant eruption. The latter being much more energetic and have changes of $\Delta \sim$ 5mag. With that different strength of the "eruptions" both are most likely caused by very different physical mechanism.

LBVs that showed a giant eruption are referred to as *giant eruption LBVs* or η *Car Variables*, to distinct them from LBVs that show only S Dor variations. Or more precisely for which we at least do not know if they have had a giant eruption, since we are limited to historic records of the last centuries, several giant eruption could have passed unnoticed. See the contribution by Kris Davidson in this volume for more details on giant eruption LBVs and there important distinction from LBVs with S Dor variability only. Concerning these two very different variabilities it has so far not been observed and therefore is not clear if the S Dor variability and the giant eruptions occur separately or a LBV can show both variations.

Beside their variability LBVs stand out by having a rather high mass loss rate. In 1997 Leitherer [27] gave a first list for the mass loss rates of LBVs. They range from $7\ 10^{-7}$ to $6.6\ 10^{-4}$ with a typical values around 10^{-5} $M_\odot$/yr^{-1}. Stahl et al. [28] used the H_α line to determined the mass loss rate the during one complete S Dor cycle of AG Car. They find that the derive mass-loss rates in the visual maximum is about a factor five higher as in the visual minimum. More recent studies of the same object by Groh et al. [29] support and extend this study. The authors associate the changes with the bistability limit. Lamers et al. [30] first discussed that while evolving from hot to cool temperatures stars will pass the bistability limit at roughly 21000 K. At this limit a change in the stellar wind occurs. On the hot side the wind velocities are higher and the mass loss rates lower (see also [31]). The cool side of the bistability limit matches in the HRD to the region of LBVs in their cool state and causes a high mass loss in that phase. The closeness to the Eddington limit [32,33] of LBV in their cool phase also favors a high mass loss. This is even more so if the stars rotate fast and the modified Eddington limit the $\Omega\Gamma$ limit applies [33] lowering the gravitational force even further. And indeed AG Car [34] and HR Car [35] are fast rotating LBVs.

3. The Evolutionary Status of LBVs

LBVs are massive evolved stars. The LBV phase is in comparison to other phases massive stars will pass with roughly 25,000 years rather short [36]. Originally, in the classical Conti scenario [37,38], only stars above roughly 50 $M_\odot$ were thought to turn into LBVs. Observations however identified LBVs that have a significantly lower mass. The position of LBVs in the HRD, see Figure 3 is associated with bright and generally blue stars (like AG Car, R 127, S 61, P Cyg, WRA 751), but an additional area is populated with LBVs that are fainter and somewhat cooler (HR Car, R 71, HD 160529). These maybe indeed hint for two subclasses the first group being massive LBVs and the latter less massive LBVs. Figure 3 shows the position of galactic and LMC LBVs and LBV candidates in the HRD. If known the position is given for both the cool (open circles) and the hot phase (filled circles).

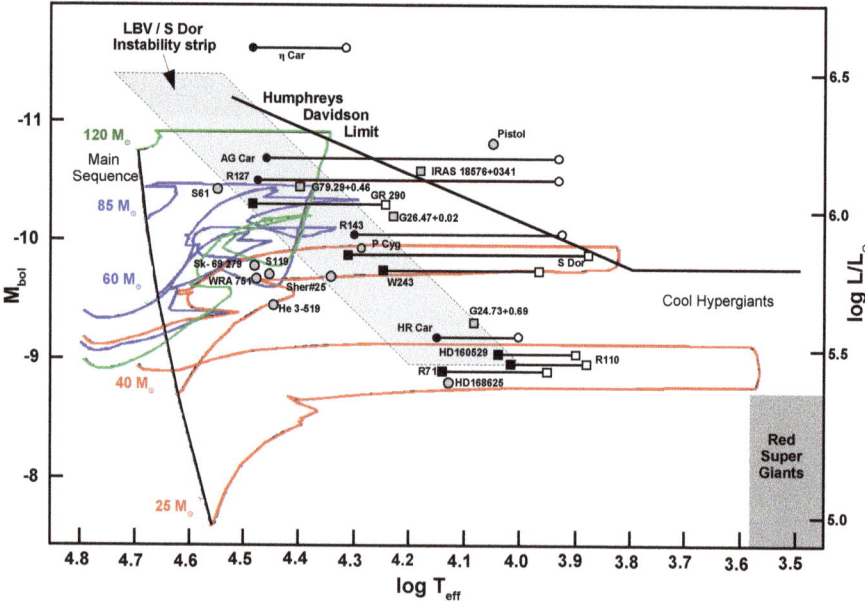

Figure 3. HRD with Galactic and LMC LBVs and LBV candidates. Circles are used for LBVs with an emission line (optical/NIR) nebulae, squares for all others. If an S Dor cycle has been observed both the cool (open symbol) and the hot phase (filled symbol) are marked. Otherwise an open grayish symbol is used. In color evolutionary tracks for different masses are added. The tracks are based on the data from the Geneva code for Z = 0.02 and v_{rot} = 300 km/s, colors code the generally three different evolutionary scenarios, see text for details.

Stellar evolution models by the Geneva group [39] that include rotation also shows the position of stars with lower mass matching both location of LBVs in the HRD Figure 3. Also plotted in this figure are tracks of the Geneva group, the Humphreys Davidson limit as well as the LBV/S Dor instability strip, the area of LBV in the hot phase.

The Geneva models [39] yield the following evolutionary scenarios:
least massive stars (red color code in Figure 3):

$M < M_{WR}$: O – BSG/RSG

intermediate massive stars (blue color code in Figure 3):

$M_{WR} < M < M_{OWR}$: O – LBV

or alternatively: O – RSG – eWNL – eWNE – WC/WO

most massive stars (green color code in Figure 3):

$M > M_{OWR}$: O – eWNL – eWNE – WC/WO

The authors define that mass limits as follows: "M_{OWR} is the minimum initial mass of a single star entering the WR phase during the MS phase...M_{WR} is the minimum initial mass of a single star entering the WR phase at any point in the course of its lifetime." Both limits M_{WR} and M_{OWR} depend on the rotation rate and metallicity. For a rotation rate of 300 km/s and solar metallicity M_{WR} = 22 $M_\odot$ and M_{OWR} = 45 $M_\odot$. Both values are higher for lower metallicity and lower for higher metallicity. This leads to a mass as low as 21 $M_\odot$ for LBVs at Z = 0.04. Depending on the mass and mass loss LBVs either evolve into Wolf-Rayet or directly turn supernovae. Figure 3 with the tracks and LBV positions also yield a clue to Wolfs amplitude-luminosity-relation: In their evolution the point in temperature (open circle) the stars start to turned back around towards hotter temperatures is relatively independent of the stars mass. The more massive, luminous stars start with a hotter temperature, so for them the crossing in the HRD (or change in T_{eff}) to the turning point is larger. This cool limit is caused by the stars forming an "extended envelope" or pseudo-photosphere in an opaque stellar wind [40,41]. An analysis of this concept using NLTE expanding atmosphere models showed that the formation of a pseudo-photosphere due to strongly increased mass-loss alone does not explain large brightness excursions [42,43]. A later discussion in context of the bi-stability jump implied that the formation of a pseudo-photosphere might work for rotating, relatively low mass LBVs [44,45]. Still, the idea of pseudo-photospheres may explain the power-law shape of the variability spectrum at higher frequencies [46]. An promising alternative idea to explain S Dor variability is envelope inflation [47] potentially induced by changes of the stars rotation. A similar idea based on an instability induced by the lowering of the effective stellar mass by rotation was also suggested [35,48].

4. Nebulae around LBVs

4.1. Emission Line Nebulae

One consequence of the high mass loss rate that LBVs posses and if present giant eruption is the formation of circumstellar nebulae. Many, however apparently not all LBVs are surrounded by a small nebula. Nebulae form by wind wind interaction of faster and slower winds during a S Dor cycle, while giant eruption LBVs nebula are the result of mass ejection in the eruption.

LBV nebulae predominantly contain stellar material, noticeable by the presence of stronger [N II] emission lines as a result of CNO processed material that was mixed up into the wind and/or ejecta of the star. During one of the first conferences devoted to LBVs in 1988, Stahl [49] reviewed on what was known about the nebulae around LBVs. Our current knowledge of LBV nebula is however still restricted mainly to nebula in our own galaxy and the Magellanic Clouds, only these nebulae are are spatially resolved and can be studied in detail.

A more recent study by Weis [50] show that the morphologies of the nebula are manifold. A signification fraction (on average 60%, 75% for galactic LBVs) show bipolarity. This bipolarity is either strong with a hourglass shape (i.e., η Car, HR Car, AG Car) or more weak in bipolar attachments, like Caps as seen in (i.e., WRA 751, R 127). Figure 4 shows one example of all so far known types of morphologies of either a galactic or LMC nebula. The true bipolar nature of the nebulae around AG Car has been identified by Weis [51]. Its hourglass structure is seen pol on and appears more spherical or rather boxy. Only by using high resolution Echelle spectra the kinematics revealed the true bipolar nature. Only one, the nebula around the LMC LBV R 143 is really irregular [52], this however is not surprising given the stars is situated in the middle of the 30 Doradus HII region. Spherical are S 61 and S 119 the latter showing signs of an outflow [53].

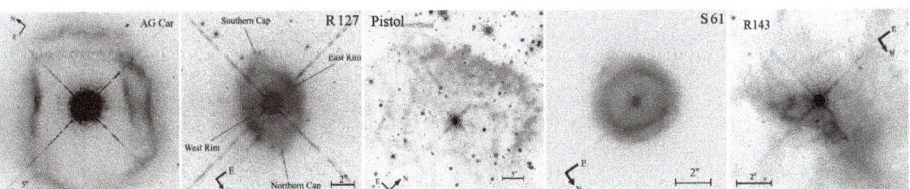

Figure 4. HST images of LBV nebulae sorted by morphology: hourglass AG Car [54], R 127 with bipolar attachments, weakly bipolar He 3-519 [55], spherical S 61 [52] and last in row irregular R 143 [52].

The list with parameters in Table 1 reveals that LBV nebulae are with only a few parsec rather small. The largest is with a diameter of about 4.5 pc the nebula around Sk-69° 279 in the LMC, the nebula shows an 1.7 pc extension in one direction, enlarging the nebula size to a dimension of 4.5 × 6.2 pc [56].

The the smallest (detected so far) are the Homunculus around η Carinae (see Section 4.4), the inner nebula around P Cygni (see section below) and the nebula around HD 168625 ([57], Weis et al.

in prep) all with sizes of roughly 0.2 parsec. Note in that context that Weis [52] found for S Dor(LBV) nebula emission in the spectrum but it's physically to small to be spatially resolved (so far). The same is true for GR 290 in M 33 (see Maryeva this volume) and the galactic LBV W243 in Westerlund 1. They are therefore excluded from Table 1 and not marked bold for LBVs with nebulae in Table 2. The expansion velocities of LBV nebulae are a few km/s to 100 km/s [50,58,59] They are higher for η Car see section below.

Table 1. Parameters of Galactic and LMC LBVs and LBV candidates with an line emission (optical/NIR) nebulae. LBVs with dust nebulae only have been excluded here. In case the nebula has several spatially distinct parts (inner and outer regions) a slash is used for separation between them.

LBV	Host Galaxy	Maximum Size [pc]	v_{exp} [km/s]	Morphology	Reference
η Carinae	Milky Way	0.2/0.67	300/10 – 3200	bipolar	[58,60]
AG Carinae	Milky Way	1.4 × 2	25/43	bipolar	[61]
HD 168625	Milky Way	0.13 × 0.17	40	bipolar	[57]
He 3-519	Milky Way	2.1	61	spherical	[55]
HR Carinae	Milky Way	1.3 × 0.65	75	bipolar	[62]
P Cygni	Milky Way	0.2/0.8	110 – 150/185	spherical	see text
WRA 751	Milky Way	0.5	26	bipolar	[63]
Pistol star	Milky Way	0.8 × 1.2	60	spherical	[64]
Sher 25	Milky Way	0.4×1	20/83	bipolar	[65]
R 127	LMC	1.3	32	bipolar	[52]
R 143	LMC	1.2	24 (line split)	irregular	[52]
S 61	LMC	0.82	27	spherical	[52]
S 119	LMC	1.8	26	spherical plus outflow	[53]
Sk-69° 279	LMC	4.5×6.2	14	spherical plus outflow	[56]

4.2. Dust Nebulae

With the SPITZER MIPSGal survey more than 400 small ($\sim 1'$) single bubbles were detected in 24 μm emission [66,67]. An extended sample was even derived using citizen-science and machine-learning methods [68]. For most of these bubbles no optical counterpart is known, making them heavy obscured gas and/or pure dust bubbles. Some of the bubbles contain central, NIR bright stars (some even faintly visible in the optical), while others do not show central sources at all, not even in SPITZER IRAC images 3.6 & 4.5 μm from the GLIMPSE surveys. The nature of these small bubbles were an enigma, until first classification spectra of some the bright central sources were taken [69,70]. Several of those turned out to be massive evolved stars, like blue supergiants, LBV candidates and Wolf-Rayet stars. Others were red supergiants and AGB stars. The nature and origin of the emission of the bubbles however remained uncertain. It could be hot dust, or MIR lines of ionized gas, or both. Taking SPITZER IRS spectra of several of the bubble revealed that all cases exist [71,72]: bubbles for which no central stars are detected seems to be dominated by line emission (mostly the high ionization [OIV] λ25.9 μm line), and are therefore most likely planetary nebulae. Bubbles with NIR visible central stars that show dust dominated IRS spectra are even less frequent. Stellar NIR spectroscopic classification again prove that the central stars are dominated by evolved massive stars [73] of various types, with several Wolf-Rayet stars, e.g., [74,75], and a number of LBV candidates or related stars [76–78]. Two of these candidates can now be seen as established LBVs (see Table 2, WS1 [79] and MN48 [80]). Also, several previously known LBVs show MIR nebulae, e.g., MWC 930 [81] or HR Car [82].

Still there are some problems with interpreting the small MIR bubbles and especially the LBV candidate interpretation. Most of these bubbles are round/spherical, e.g., [69], and bipolar structures are rare among the 24 μm bubbles. While this is consistent with the morphology of circumstellar gas nebulae of Wolf-Rayet stars, it seems to contradict the results found for LBV nebulae [52] which have as reported above a preference for bipolar morphologies. From the 24 µm images alone it is not clear, whether the nebulae are (a) dust only (b) partly dust, partly gas, or (c) dominated by ionized gas. The distribution and kinematics of the dust and gas are often different for circumstellar nebulae of massive stars, as e.g., shown for the nebula of the classical LBV AG Carinae [61]. Gail et al. [83] were among the first to investiage the problem of dust formation in an CNO precessed material like LBV envelops. Later hydrodynamical simulations of gas and dust nebulae, e.g., [84], show the small dust grains follow the gas quite well, but the larger grains show their own unique distribution. The morphologies problem may therefore be dominantly a wavelength bias. Last but not least one can speculate about the bubbles being signposts of a more general, previously overlooked short high mass-loss phase in the evolution of many massive stars. We are currently running a program with the LBT infrared spectrograph LUCI to classify more of the central stars using HK band spectroscopy.

4.3. P Cygni

As was mentioned above P Cygni is one of the classical giant eruption LBVs, with an eruption observed in 1600. In the Van Genderen classification it is currently a weakly active LBV. The nebula around P Cygni has at least, two distinct parts. One larger structure with a diameter of 0.8 to 0.9 pc named the outer shell or OS which is rather spherical and a much smaller and clumpy structure the inner shell or IS which is less than 0.2 pc across [85]. Beside the IS and OS Meaburn et al. [86] reported 1999 a giant lobe to associated with the stars. Its has a PA = 50° to the other nebulae and stretches to an extend of 7 arcminutes or up 3.6 pc. He finds expansion velocities around 110 to 140 km/s (depending on which line he uses) associate with the inner shell and structures as high 185 km/s in the outer shell.

With a diameter of only 0.2 pc the IS is in most images barely resolved. A new LBT/LUCI AO image (Figure 5) we made recently shows the large amount of fine structure and details of inner nebula for the very first time. With a resolution down to 85 AU size structures it is an improvement from the previously published LBT image by Arcidiacono et al. [87].

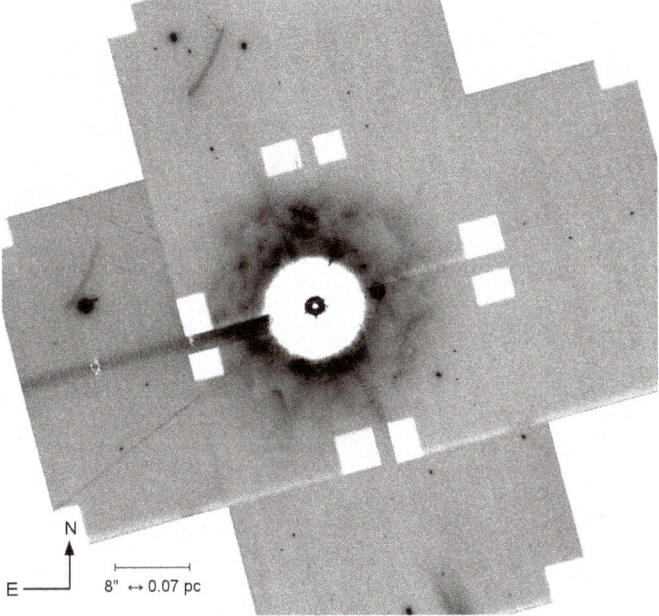

Figure 5. An LBT LUCI AO [FeII] image of the inner nebula (or inner shell) of P Cygni (Weis et al., in prep). The images has pixel scale of 0.015"/pixel and resolve scales down to 85 AU.

Mapping the nebulae with KPNO high resolution longslit Echelle Spectra we measured the expansion velocity of the inner nebula is 100–150 km/s, this is well in agreement to Meaburns values. This would assuming no larger acceleration or deceleration match to the inner shell having been ejected during the 1600 giant eruption. With the spectra we can also associate velocities to distinct clumps that appear in the spectra and can be identified on the image (Weis et al. in prep.).

4.4. η Carinae — the Most Peculiar LBV?

η Carinae used to be the most classical giant eruption LBV or η Carinae variable. With the discovery of many unique and unusual characteristics η Carinae or the η Carinae system is not the LBV par excellence anymore. A book devoted to η Carina and the Supernova impostors [88] can be consulted for all details on this object. Here a short summary of the characteristics:

- A giant eruption that took place in 1843
- A binary system with two massive components one with 60 $M_\odot$ the second with 30 $M_\odot$ [89]
- A nebula that has at least three section: *The little Homunculus, The Homunculus, The outer ejecta*.

The Homunculus was identified first and photographed in 1950 by Gaviola [90], the name of the nebula was motivated by the first images showing a man like morphology. The little Homunculus resides within the Homunculus and was revealed only using HST STIS long slit spectroscopy by Ishibashi et al. [91]. The outer ejecta as the name implies surrounds the Homunculus. It consists of a countless number of clumps and filaments. A first report and catalog with designation of several part of the outer ejecta was made 1976 by Walborn [92]. A summary of more recent optical, x-ray and kinematic studies of the outer ejecta is given by Weis [59]. Today we know that all three sections of the nebula are of bipolar morphology. The expansion velocities are with up to 3000 km/s faster than in any other LBV nebula. Shocks of these extremely fast structures in the outer ejecta create X-ray emission [93,94]. This emission is shown in the right section of Figure 6, here a a CHANDRA image is color coded and in indicates in red soft Xray emission of the outer ejecta is, in blue the more central emission results mark shocks of the central stellar system not the Homunculus nebula!

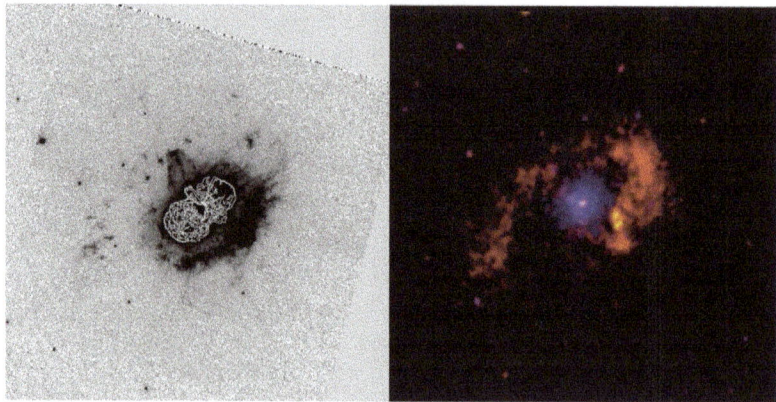

Figure 6. The nebula around η Carinae in the optical and X-ray. **Left**: An optical F658N HST image in greyscale, the Homunculus nebula additionally marked in contour to distinguish it from the outer ejecta, shown only in grey scale [58]. **Right**: A CHANDRA Xray image with color coded energy regimes, green:0.2-0.6 keV, red: 0.6–1.2 keV and blue 1.2–12 keV color version of Figure 1 in [95].

5. Instabilities and the Origin of Variability

What are possible origins of the LBV variability. First we have to differentiate between the S Dor variability and giant eruptions. The latter are in need of much larger energy being released. Already in their 1994 paper Humphreys and Davidson [41] discussed what could cause the variabilities and whether one or more mechanism are at work. They argue and cite several works showing that a classical κ-mechanism seems not to work. A more likely cause also discussed in that paper is the proximity of LBV to the Eddington, or in case of rotation $\Omega\Gamma$ limit. This limit indeed lies in the HRD in the same region as the Humphreys Davison limit, which also resembles the cool position of an LBV in the S Dor cycle. Clearly the properties of the stellar winds and their dependence of metallicity have to be a major contributor to the mechanisms creating the variability. For various more detailed theoretical works the reader is referred a review by Glatzel [96] a newer overviews by Vink [97] and Owocki [98]. Alternative models like non-radial gravity mode oscillations have been proposed by Guzik [99]. The potential importance of pulsations for the driving of the S Dor mechanism was discussed in [100]. The analysis of long, well sampled lightcurves may provide more information about the properties of the S Dor process [20,101]. Still, the link of low amplitude variability patterns with LBV nature is far from clear. Kalari et al. [101] showed that SMC blue supergiant AzV 261 exhibits variability patterns consistent with other LBVs in a 3 year time span and nearly nightly photometric observations, but their detailed analysis of high dispersion spectra showed no temperature changes typical for an S Dor cycle over a decade, precluding a classification as LBV.

Recently, new 3d radiation hydrodynamic simulations of 80 and 35 $M_\odot$ performed and the results point at variations of the He opacity as a possible cause of the S Dor variability and link the shorter time scale irregular oscillations to convection [102]. Other ideas for the origin of the S Dor variability were already discussed in Section 3.

6. The LBV Wolf-Rayet Star Connection

In the last years it has been found that the masses of Wolf-Rayet stars in that state (not their initial mass) is much lower as can be explained by the stellar winds only. Furthermore the empirical mass loss rates for hot, massive stars have also been seriously questioned, mainly because of the effects of wind clumping [103,104]. Wind clumping will reduce the mass loss rate and leave us with even higher mass in evolved stars. Stellar evolution models use theoretical mass loss rates, generally lower that the empirical ones even without clumping. A phase of enhanced mass loss with a different mechanism may be needed [105].

The LBVs phase would just fit. It is passed right before the WR phase and is known for high mass loss as well as the formation of massive nebulae. A LBV phase therefore might be mandatory to explain at least some WR classes and the lower WR star masses. One might even speculate that WR nebula are only by fast WR winds blown up, enlarged former LBV nebulae. Indications for such a hypothesis are the somewhat larger size of WR nebulae in combination with a N enhancement. The latter being a well known attribute of LBV nebula. Most WR nebula are not found around WO or WC but WN type stars, the natural and direct predecessor of LBVs.

First hints for such a scenario have been shown for the Wolf-Rayet stars WR 124 with its nebula M1-67 and the LBV He 3-519 [55]. The M1-67 WR nebula is one of few if not the only one that has a bipolar morphology and a size of only 2pc. As described above these are very typical values for LBV nebula. One might picture WR 124 as an old LBV that has just left the LBV and entered the WR phase, matching well to its current WN 8 spectrum. The scenario for He 3-519 might be just reversed, the stars is an LBV that is turning into an WR right now. This would also explain why no S Dor variability is seen for that star. Its current spectra type is already that of a WN 11. The nebula is only weakly bipolar and with 2 × 2.5 pc rather large for an galactic LBV see HST image in Figure 4. It looks more like an old LBV nebula that by inflation via the strong WR stars wind has already increased its size. Doing so also caused bipolarity to fade o[55]. For a more general review about WR stars see the contribution by Kathryn Neugent and Philip Massey in this volume.

7. Links of SN Impostors and LBVs

In recent years several projects and monitoring surveys that search for supernovae found what has become known as *SN impostors*. These transients show spectra similar to core-collapse SN, especially of the type SNIIn, but are generally significant fainter than core-collapse SN. SN impostors show lightcurves quite different from all core collapse SN, sometimes even showing strong fluctuations on short timescales some time after the initial eruption, see e.g., [106]. It is interesting to note, that the brightest impostors events even overlap in energy with the faint SN IIP, e.g., [107,108]. A very tempting and likely explanation is to identify at least a subset of these SN impostors with giant eruptions or even S Dor variabilities of LBVs [109] in distant galaxies. Note in this context that while a LBV giant eruption will look like a SN impostor, not all SN impostors might indeed be LBV giant eruptions!

With the current list of about 40 SN impostors [110], light curves and spectra during the eruption, and especially the pre- and post-eruption behavior imply at least two different object classes are summarized in the name Impostor: the transients with strong narrow emission lines and erratic lightcurves with secondary, smaller outbursts following the first eruption, and the transients, which are followed with less than a decade by a true supernova explosion, e.g., [111]. This diversity of the lightcurves and spectra of the transients denoted as SN impostor was also noted by Smith [112]. It is potentially important, that the rise of the eruptions can be very steep [106,110,113], putting interesting limits on the kinetic energy and size scales evolved.

SN impostors will be discussed in detail in this volume by Kris Davidson.

8. Multiplicity of LBVs

In recent years it became clear, that a significant part of massive stars are born in double (or multiple) systems. A detailed analysis of the results from the FLAMES-Tarantula survey lead Sana et al. [114] to the following percentages for massive stars: effective single (real single stars or wide binaries without significant interaction ~29%, stellar merger ~24%, accretion and spin up or common envelope evolution ~14%, and envelope stripping ~33% [114]. Therefore, about ~71% are affected by binary interaction. Alternatively, if one sees the result of mergers as apparent single stars for most of their lifetime, then only ~47% of the massive stars should show a companion. The same data also imply that the numbers of equal mass binaries are lower than unequal mass pairs. The ratio goes up to ~50% at $M_2/M_1 = 0.3$, the lowest mass ratio probed by their data [114].

This immediately implies that a sizable number of LBVs should have binary companions. The idea, that binary star evolution is linked to the LBV phenomenon is quite old, see e.g., Gallagher (1989) [115]. More recently, the idea of mergers triggering giant eruptions (or being one path to SN impostors) gained some interest, e.g., [116]. Still, the observation of binary companions of LBVs are difficult, due to the large luminosity of the primary, and its strong stellar wind, which both limits spectroscopic searches. Direct imaging searches only cover relatively large separations, and only few LBVs are analyzed with stellar interferometers, yet. A search for X-ray only covers situations in which colliding winds can occur, and may be in part contaminated by the X-ray emission of circumstellar nebula.

The current state on observed stellar companions to LBVs is the following: As shown in Section 4.4 η Carinae show strong signs of being a binary star with a massive, hot companion stars. HD 5980 was first reported as an exlipsing LBV Wolf-Rayet binary system that showed an LBV like eruption [117]. Koenigsberger et al. [118] report new analysis which is consistent with the system being more complex and multiple: a double binary scenario and manifests a quadruple system. The LBV candidate [KMN95] Star A (= 1806-20) showes double He lines [119] but single emission lines, implying a dense stellar wind for the primary, similar to the case of η Carinae. MCW 314 shows clear indications in its lightcurve and its radial velocity curve for having a lower luminosity supergiant companion [120]. If the wide companion candidate [121] is truly bound, than MWC 314 would be a hierarchical triple star. The LBV HR Car was observed with stellar interferometry and strong indications of a companion was found [122]. The companion star appears to be relatively low mass (below ~$15 M_\odot$).

A search for wide companions based on natural seeing, AO assisted imaging, and archival HST imaging of 7 galactic LBVs, LBV candidates, and some related objects yielded one star with potential companion (MWC 314) and no apparent bound companions for the 5 other LBVs and LBV candidates (the Pistol star, HD 168625, HD 168607, MWC 930, and [KMN95] Star A (= 1806-20)) [121]. The PSF subtracted HST images used in the study of LBV nebulae in the LMC by Weis [52] also showed no apparent companion stars, but only relatively large projected orbital distances could be probed (>0.1 pc).

A X-ray archival survey (using XMM-Newton and CHANDRA X-ray satellites) of 31 LBVs, LBV candidates, and related objects was performed by Naze et al. [123]. X-ray emission may indicated colliding winds in a binary, but (softer) X-ray could also be created in a circumstellar nebula, see e.g., Weis [94] for the case of η Carinae. The survey of Naze et al. yielded 4 detection (η Carinae, W243 (= Westerlund 1 #243), MSX6C G026.4700+00.0207 (= GAL 026.47+00.02), and Schulte #12 (= Cyg OB2 #12). Two more are labeled doubtful candidates (GCIRS 34W, and GCIRS 33SE) by the authors. This result also implies a long list of 25 non-detections, which includes confirmed LBVs like P Cygni, the Pistol star, and FMM 362. While acknowledging their rather heterogenous data base, the authors suggest that their detection rate is consistent with a binary fraction between 26% and 69%, roughly consistent with that of other classes of hot, massive stars.

Given the very different methods used, and the therefore very different orbital radii and mass (and luminosity) ratios probed up to now, it is hard to derive a reliable result on the binary fraction for LBVs as a class. An additional problem are the very different LBV input lists used in the different searches. There are clearly several good cases for binary companions of LBV stars. Still, we regard the actual binary fraction of LBVs as currently very uncertain, but most likely around ~20% for the confirmed LBVs. This would be somewhat lower than the binary fraction for other classes of massive stars like O supergiants or Wolf-Rayet stars. If this estimate of the binary fraction is correct, it may hold important clues for the evolutionary pathways leading to LBVs.

9. LBV and Their Neighborhood

Smith & Tombleson[124] analyzed the location of LBVs in comparison to their surrounding and concluded that LBVs in MW and LMC are isolated, and not spatially associated with young O-type stars. This would imply a complete change of the standard view of the evolution of LBVs, clearly a far reaching claim, which needed further investigation. Humphreys et al. [125] analyzed the location of a sample of LBVs in M 31, M 33, and the LMC in comparison too other massive main sequence and supergiant stars. With this large and more coherently selected sample,Humphreys et al. [125] concluded that LBVs are associated with supergiant stars and are neither isolated or preferentially run-away stars. Separating the more massive classical and the less luminous LBVs, the classical LBVs have a distribution similar to the late O-type stars, while the less luminous LBVs have a distribution like the red supergiants. Smith [126] questioned the results of this analysis and reiterated the results of his analysis. Davidson et al. [127] shortly after showed that the statistical analysis methods use in [126] are flawed. Independently, Aadland et al. [128] performed a very similar analysis and came to similar conclusions as Humphreys et al. [125], that the stellar environment of LBVs is the same as for supergiants. It is still be worth noting, that the Aadland et al. sample is not a clean LBV sample, but contains many B[e] supergiants. Note that this point was also pointed out by Kraus in her review paper on B[e] in this volume. In a recent paper Smith [129] gravitated to the interpretation by Humphreys et al. of LBV locations within (or near) their birth association. Just lately with an analysis of GAIA data [130], strong evidence was presented, that OB stars form not preferentially in bound clusters, but in a continuous distribution of gas densities, at many locations of the birth cloud. This view is also supported by recent simulations which also favor a hierarchical formation model for the formation of OB stars as a result of the fractal structure of the birth clouds, contrary to a monolithic collapse. In this picture many different stellar neighborhoods of massive stars would be natural, also consistent with our results.

10. The Population of LBVs

As mention above the first reports on Var 2 in M 33 was already in the 1920ties marks the first identification of an LBV-at that time without the knowledge that it is and what LBV are.

Since the Studies by Hubble & Sandage 1953 [1] and Sandage & Tamman 1974[4] we know that both M 31 and M 33, as well as NGC 2403 host several LBV and LBV candidates. S Dor added the LMC to the LBV host galaxies. The LMC has a remarkable population of LBVs [6,131]. Bernhard Wolf and his group in Heidelberg studied various LBVs and LBV candidates in several galaxies and with this first larger sample was able to identify the above mentioned amplitude luminosity relation. Beside that they found several LMC LBV candidates and confirmed many LBVs by observing their S Dor Cycles like R 127 [132], R 110 [133]. They also noticed an inverse P Cygni profile in the spectrum of S Dor [134]. and added HD 160529 to the galactic LBVs. Last but not least the group also identified with R 40 the very first LBV in the SMC [135,136]. Other LBV host galaxies now known are locally IC 10 and further out are the M 1 group members M 1, NGC 2366. LBV and LBV candidates are reported also in M 101, NGC 300, NGC 247, NGC 6822, NGC 4414, and IC 1613, just to name the most important galaxies.

In Table 2 a list of known LBV and LBV candidates is given. True LBVs are those stars where the membership is clear since a complete S Dor cycle has been observed, this is not the case for the LBV candidates. For LBVs that had a giant eruption, those are classified separately and named giant eruption LBVs (or η Car Variables) to distinct them from LBVs with S Dor variability only.

Several more stars have for the one or other reason be classified as LBVs by one or more authors, but show no clear hints like S Dor cycle or giant eruption. For the Milky Way the objects HD 80077 and Schulte 12 the new GAIA parallax moves both to a closer distance and to a lower luminosity. Still, the GAIA parallaxes are at this time (GAIA DR2) prone to several systematics [137,138]. Therefore the distance of at least the Schulte 12 is still not settled yet [139]. According to Humphreys et al. [125] in the LMC R 66, R 74, R 123 are B[e], R 149 is an Of star, HD 269604 an A supergiant and HD 34664 as well as HD 38489 are B[e]sg. Neither are R 81, R 84, R 99, R 126 LBVs. The SMC object R 50 is a B[e]sg while R 4 is a spectroscopic binary system with one B[e]sg. Finally the activity HD 5980 is more like a giant eruption but this most likely due to a binary interaction, see chapter on Multiplicity of LBVs below. Therefore its seen as a giant eruption LBV candidate with the above caveat. Just recently Humphreys [140] report that the following objects are not LBVs: I 8 in M 81 is an F supergiant, furthermore V 52 in NGC 2403 and I 3 in M 81 are foreground objects and not even part of those galaxies! HD 168625, He 3-519, Pistol star, and Sher 25 are LBV candidates due to the fact that they posses a circumstellar nebula. They however might indeed be LBVs, as members of what van Genderen classified as a group ex/dormant LBVs, just currently not showing any variability. Besides the variability searches, there is also a consistent search of luminous emission line stars using two or more broad band colors, Hα as detection and [OIII] as veto filter, done as part of the NOAO Local Group Survey [141,142] and independently by our group [143]. Both searches covered M 31, M 33, NGC 6822, IC 10, Wolf-Lundmark-Melotte, Sextans A, and Sextans B and finding very few candidates in the dwarf galaxies.

We searched also NGC 3109, a low metallicity galaxy forming a subgroup with Sextans A and Sextans B at the fringes of the Local Group. We found one candidate [144], similar to the low candidate numbers for the low metallicity dwarfs in the Local Group. An earlier attempt with the same idea to detect very luminous stars, which are strong Hα line emitters (either from strong mass loss, or from a circumstellar nebula), which are faint or absent in [OIII] (no stellar emission line, and faint for circumstellar nebulae of CNO processed material) was done by the Heidelberg group [145] for M 33, M 81, NGC 2403, and M 101, but was not published. We used e.g., these data to complement our list of good candidates for spectroscopy in M 33 [146,147]. It is interesting to note here, that coordinated searches for variable stars (in particular not only analyzing the Cepheids) is done only for small number of massive local galaxies since the photographic plate area. A new effort is ongoing with the LBT and yielded already interesting results [140]. Our group is currently working on a search for LBV and related objects in several nearby galaxies.

Table 2. LBVs and LBV candidates in alphabetic order. Giant eruption LBVs are italic. Objects marked bold have (optical) emission LBV nebula. Except for the Milky Way and LMC which have a to large number of objects, references are given.

Galaxy	LBVs	LBV Candidates	References
Milky Way	**AG Car**, *η Car*, FMM 362, [GKF2010] MN44, [GKM2012] WS1, HD 168607, HD 160529, HD 193237, **HR Car**, **LBV G0.120-0.048**, MWC 930, *P Cygni*, V481 Sct, W243, **WRA 751**	BD+143887, BD-13 5061, B[B61] 2, G025.520+0.216, G79.29+0.46, GCIRS 16C, GCIRS 16NE, GCIRS 16NW, GCIRS 33SE, GCIRS 16SW, [GKF2010] MN58, [GKF2010] MN61, [GKF2010] MN76, [GKF2010] MN 80, [GKF2010] MN83, [GKF2010] MN96, [GKF2010] MN112, [GKF2010] WS2, **HD 168625**, HD 316285, HD 326823, **He 3-519**, IRAS16278-4808, IRAS19040+0817, J17082913-3925076, [KMN95] Star A, MSX6C G026.4700+00.0207, **Pistol star**, **Sher 25**, WR 102ka, WRAY 16-137, WRAY 16-232	
LMC	HD 269216, R 71, R 85, **R 127**, **R 143**, R 110, S Dor,	HDE 269582, **S 61**, **S 119**, **Sk-69° 279**	
SMC	R 40	(R 4), (R 50)	[148]
M 31	AE And, AF And, LAMOST J0037+4016, UCAC4 660-00311, Var A-1, Var 15	J003910.85+403622.4, J00441132+4132568, M 31-004425.18, M31-004051.59	[147,149] [150]
M 33	Var B, Var C, Var 2, Var 83,	GR 290, [HS80] B48, [HS80] B416, [HS80] B517, J013228.99+302819.3, J013235.21+303017.4, J013317.01 + 305329.87, J013317.22+303201.6, J013334.11+304744.6, J013337.31336+303328.8, J013351.46+304057.0, J013354.85+303222.8, J01342475+3033061, J01342718+3045599, J013432.76+304717.2, J013459.36+304201.0, J01350971+3041565, M33C-5916, M33C-10788, M33C-15235, M33C-16364, M33C-21386, UIT 008	[147] [18]
NGC 2403	*SN 1954J=V12*, SN 2002kg=V37	V 22, V 35, V 38	[24] [140,151]
NGC 1058	*SN 1961V*		[25,26]
NGC 2366	NGC 2363 V1		[152]
M 101		J140220.98+542004.38, V 1, V 2, V 4, V 9, V 10	[153,154]
M 81		I1, I2	[140,155]
IC 10		unnamed	[156]
NGC 300		B 16	[157]
NGC 6822		unnamed	[158]
NGC 4414		unnamed	[159]
IC 1613		V 39, V1835, V2384, V3072, V3120, V0416, V0530,	[158] [160,161]
UGC 5340		unnamed	[162–164]
NGC 3109		unnamed	[144]

Another different approach was used by Khan et al. [165,166] to search for analogs of η Carinae. They applied SED fitting to HST and Spitzer IRAC data of resolved stars in nearby massive galaxies, and found 5 promising candidates (one each in M 51, M 101, NGC 6946, and two in M 83). Again, these are at best only candidate LBV, mainly due to the missing variability information.

11. LBVs in Low Metallicity Systems

The situation is even worse for LBVs in lower mass galaxies. Detections are rare as metal-poor also implies low mass and even in actively starforming dwarf galaxies the numbers of massive stars are more limited as in large, massive spirals. The SMC, for example, on has only one confirmed LBV, see Table 2. An interesting LBV candidate is V 39 in the low metallicity Local Group dwarf galaxy IC 1613. Detailed analysis of its spectrum shows some patterns similar to other LBVs, but is also consistent with that of a sgB[e] star [161].

Besides the aforementioned Local Group galaxies, there are only chance detections up to now, including the exceptional case of NGC 2366 V1. NGC 2366 V1 [152,167,168] is located in a dwarf galaxy with a metallicity below 1/10 solar. Its "outburst", with a change of only ~3 mag [167] was probably not really a giant eruption). Neither did it follow the classical S Dor pattern since it turned bluer (not redder) with increasing brightness.

The transient in UGC 5340 (DDO 68) was again a chance detection [162]. The brightening is 1 mag, and here again a blueing during the bright state is visible [163,164]. The galaxy is a morphologically peculiar, low mass system, which has with ~1/30 solar (log (O/H) = 7.12) one of the lowest gas-phase metalicities in the nearby universe (distance 12.6 Mpc [169]. The transient in PHL 293B (= SDSS J223036.79-000636.9) [170] is difficult to study mainly due to its distance of ~25 Mpc. The host galaxy is a dwarf galaxy and more metal-poor than the SMC (log O/H = 7.72). The transient discovery spectrum shows clear P Cygni profiles, but no details on the temporal variability were known, only 2 spectra (one without and one with P Cygni profiles). An additional spectrum brought the time baseline to 8 years and proved temporal variations of the broad stellar lines [171]. While being an interesting object, which may acquire LBV candidate status with a longer term photometric and spectroscopic monitoring, but the currently limited data makes the label LBV for this object a bit premature. As similar problem is the transient in the galaxy SDSS J094332.35+332657.6 [172], an apparent stellar transient in an very low mass and extremely low metallicity (log (O/H) = 7.03) galaxy at a distance of ~ 8 Mpc. Only a very limited historical record is available, and therefore the LBV nature of the transient is quite unclear. It may be interesting to note here that the LBV GR 290 (= Romano's star) in M 33 also shows spectra variability, but not consistent with an S Dor pattern, see Maryeva et al., this volume. The star is located in the outer regions of the disk of M 33 (r= 4.3 kpc from the center of M 33. The observed metallicity gradient [173] therefore implies a low metallicity of log (O/H) = 8.2 (roughly between LMC and SMC [174]) for the star. Note in that context that metallicity gradients are a common feature in spiral galaxies, e.g., [175,176], so large spiral galaxies do not have one fixed metallicity.

Another intriguing object was detected by as a point source with high velocity dispersion in Hα Fabry-Perot observations of the local (D~2.6 Mpc), low metallicity dwarf galaxy UGC 8508 [177]. An intermediate dispersion spectrum of the source shows a bright Hα line with broad wings, a relatively strong Fe II λ4924 line, but also a strong He II λ4686 line. The classification of the authors as a massive star with strong mass loss is convincing, but if it is indeed a good LBV candidate is more uncertain, given the high temperature (and/or hard radiation field) implied by the presence of the strong, narrow He II line.

We detected another unusual point source [178] in NGC 1705, a starburst dwarf galaxy at D~ 5 Mpc with a metallicity similar to the LMC. The spectrum shows several very strong (and split) forbidden emission lines, all showing an expansion velocity of 50 km s^{-1}, and an underlying spectrum of the source is that of an A supergiant. Again this is a massive star with an expanding circumstellar bubble, but its exact nature is not determined yet.

The starter for the question, how many galaxies do we know in the local universe (e.g., the Local Volume = D < 11 Mpc) based on the classic compilation of [179]. There is an obvious distance limit when using photometry from the ground, especially historic photographic plate material for long term light curves to identify LBV candidates. This limit is depending on seeing, size of the telescope used, and the detector. The limits for photographic plate work is about 7 Mpc (the distance of M 101) [24], and is for most telescopes more like ~4 Mpc (the M 81 and IC 342 groups in the north, and Sculptor and Fornax groups in the south). Obviously with CCDs and good seeing this can be extended (and/or the quality of the photometry improved), but access of older CCD data is tricky, if the observatory does not run a well maintained archive. Clearly, HST and in the near future EUCLID and JWST, can go much farther out, but it gets hard beyond 20 Mpc (especially due to the crowding of stars).

Low metallicity LBVs are especially interesting, since the metallicity can influence opacity in the interior of the stars and in the wind. Metallicity also affects the path of the evolutionary tracks (at which mass stars still go RSG, return to the blue, or go through a LBV phase with S Dor-like variability and instability that caused these, etc...) Furthermore rotation rate, binary fraction, and potentially IMF as well as magnetic fields are important.

Several of this markers of LBV candidates are directly, or indirectly influenced by metallicity. Mass loss e.g., [180–182], emission lines of heavy elements (e.g., photospheric or wind emission lines of FeII, FeIII, [FeII], then HeI, and [NII] in a circumstellar nebula) [146,147], and variability due to the metallicity dependence of the instabilities involved (see above). It could be that at low enough metallicity, massive stars behave differently, e.g., not showing an typical S Dor variability pattern anymore. The cases of V1 in NGC 2366 [167] and the transient in UGC 5340 (DDO 68) [163,164] hints towards and seem to support such a scenario. No coordinated search for luminous variable sources in a sample of low metallicity dwarf galaxies outside the Local Group was done yet. A pilot search on a few selected very low metallicity galaxies was reported by [183] using HST archival data. While there are several interesting candidates of luminous stars with signs of variability and in some cases Hα emission, the data yield not enough proofs to claim LBV candidates. Figure 7 demonstrates one of the problem, very low metallicity galaxies are rare and spatial resolution poses severe problems for ground based studies beyond ~5 Mpc, requiring HST time. This aspect may improve with the upcoming EUCLID mission and more in the future by WFIRST, and is alleviated somewhat by the improving image quality of the large survey instruments, link e.g., SUBARU SuprimeCAM, DECam, and hopefully LSST. Another problem is the metallicity-luminosity relation, which implies that low metallicity is in the local universe the exclusive regime of dwarf galaxies. Therefore, even in a burst of starformation the absolute number of massive stars produced is, during a short time frame only, still comparable to the production rate of a massive spiral galaxy. With the current data situation it is to early to speculated on trends of LBV numbers and LBV nature at low metalicities, but as noted above, it is intriguing to see so many LBVs and LBV candidates in the LMC. With at the same time nearly non in the SMC.

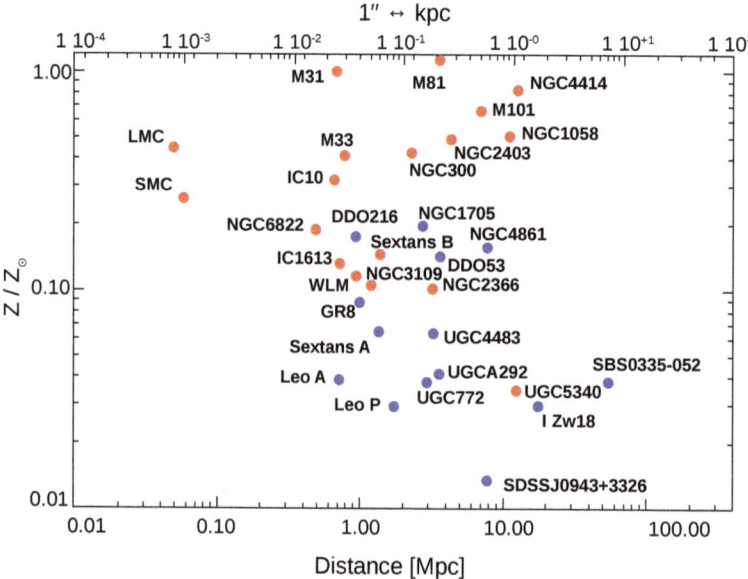

Figure 7. Plot of the gas-phase metallicity of nearby galaxies versus their distance and spatial resolution. Only a selection of the galaxies in the Local volume are plotted, but the sample is complete for the significantly starforming galaxies in the Local Group. Metalicities of the inner disk are chosen for the spiral galaxies with metallicity gradients, the metalicities of stars in the outer disk of these galaxies can be a a few faction of tens solar lower. Galaxies with LBVs and/or LBV candidates are plotted as red dots, the other galaxies are plotted as blue dots. Plot was adapted and updated from [163].

12. Summary and Conclusions

The Luminous Blue Variable phase is a short phase in the life of massive stars. It may be passed by stars with an initial mass as low as 21 $M_\odot$. LBVs have a specific variability the S Dor variable, can undergo giant eruption and have very high mass loss rate. The one and only way to pinpoint and truly classify LBVs is by the variability and/or giant eruption This asks for the detection of at least one S Dor cycle the star passes or to catch it in a giant eruption. These variabilities also subdivide the LBV class in classical (S Dor variable) LBVs and giant eruption LBVs. With the variability as the only clear classification method many LBVs in a quiescence state might be overlooked and not be identified as such. It is therefore not trivial to describe the LBV population in a galaxy. In that connection not knowing the true amount of LBVs and non-LBVs makes it hard to give an estimate for the real duration of the LBV phase. This again is directly linked to uncertainties of the total mass loss rate of massive stars. Even small changes of the phase length are linked to large changes in the mass total loss of the stars, given LBVs have very high mass loss rates. Last but not least that implies that the final mass of stars that pass a LBV phase could be much lower as thought so far. In that case this would even effect amounts and ratios of different SN types.

The path is therefore clear, to better characterize the LBV population and the underlying physics more long-term variability studies of nearby galaxies are needed. Spanning the parameter space especially towards lower metalicities will potentially clarify the importance of opacity effects and rotation for the S Dor variability. Also analyses of the long-term variability of massive stars in all the most metal-rich spiral galaxies in the Local Universe are not really done yet. First attempts are already ongoing, partly using data from well maintained archives, and the time-domain section of future large survey projects like LSST will be a major step forward. This will also be true for a better understanding of eruption LBVs. Another promising avenue will be the "archaeology" of the mass-loss of LBVs and related stars using their circumstellar nebulae. In this way, information on energy, mass, and chemical composition of earlier mass-loss of the stars can be investigated, again providing clues about the underlying mechanism of instability and the evolutionary state of the stars. With the rise of integral field spectrographs, even with AO support (e.g., MUSE at the ESO/VLT), such analyses should be possible in all Local Group galaxies and the nearest galaxy groups. First such analyses are already appearing for galaxies in the Sculptor group: NGC 300 [184] and NGC 7793 [185]. An unfortunate weakness in the currently available instrumentation are high-dispersion spectrographs fed by long-slits and IFUs, an important capability for kinematics/energetics of nebulae, which is becoming rare [186] at the intermediate and large telescopes. High-multiplex spectroscopic survey instruments at large telescopes, like e.g., Hectospec at the MMT, and soon MOONS and 4MOST at ESO telescopes, as well as WEAVE at the WHT, can be very useful tools to set LBVs in context to their massive star environment, as they are capable of providing good quality spectra for many photometrically selected LBV candidates (as well as other supergiants). This still requires that starforming, nearby galaxies will be targeted in the upcoming large surveys at these facilities.

Taking this all together, one can be optimistic, that in the coming years many more good quality observational data will be available to improve our understanding of the LBV phenomenon and its importance for the evolution of massive stars.

Author Contributions: K.W. planned the review structure. Both authors contributed then roughly equally to writing this review, with slightly different relative amounts depending on the topics covered in each chapter. All authors have read and agreed to the published version of the manuscript.

Funding: This work was supported through the Astronomical Institute of the Ruhr University Bochum, the Ruhr University Research Department Plasmas with Complex Interactions, and DFG Research Unit FOR 1254.

Acknowledgments: The authors thank Roberta Humphreys for many discussions, and her many comments and suggestions for this text; Kris Davidson for several helpful comments; Jochen Heidt and Alexander Becker for their contributions to the LBT AO observations of P Cygni; and our students for their contributions to many aspects of LBV related research here at Bochum during the last years. Thanks also got to two anonymous referees whose suggestions improved this paper.

Conflicts of Interest: The authors declare no conflict of interest.

Abbreviations

The following abbreviations are used in this manuscript:

BSG	Blue Supergiant
ESO	European Southern Observatory
HRD	Hertzsprung-Russell Diagram
HST	Hubble Space Telescope
JWST	James Webb Space Telescope
LBV	Luminous Blue Variable
LMC	Large Magellanic Cloud
LSST	Large Synoptic Survey Telescope
MMT	(converted) Multi Mirror Telescope
NOAO	National Optical Astronomy Observatory
RSG	Red Supergiant
SMC	Small Magellanic Cloud
SN	supernova
WFIRST	Wide Field Infrared Survey Telescope
WHT	William Herschel Telescope
WR	Wolf-Rayet star

References

1. Hubble, E.; Sandage, A. The Brightest Variable Stars in Extragalactic Nebulae. I. M31 and M33. *Astrophys. J.* **1953**, *118*, 353. [CrossRef]
2. Duncan, J.C. Three Variable Stars and Suspected Nova in the Spiral Nebula M 33 Trianguli. *Publ. Astron. Soc. Pac.* **1922**, *34*, 290. [CrossRef]
3. Wolf, M. Zwei neue Veränderliche. *Astronomische Nachrichten* **1923**, *217*, 475. [CrossRef]
4. Sandage, A.; Tammann, G.A. Steps toward the Hubble constant. IV. Distances to 39 galaxies in the general field leading to a calibration of the galaxy luminosity classes and a first hint of the value of H_0. *Astrophys. J.* **1974**, *194*, 559–568. [CrossRef]
5. Humphreys, R.M. The spectra of AE Andromedae and the Hubble-Sandage variables in M31 and M33. *Astrophys. J.* **1975**, *200*, 426–429. [CrossRef]
6. Humphreys, R.M.; Davidson, K. Studies of luminous stars in nearby galaxies. III. Comments on the evolution of the most massive stars in the Milky Way and the Large Magellanic Cloud. *Astrophys. J.* **1979**, *232*, 409–420. [CrossRef]
7. De Jager, C. The stability limit of hypergiant photospheres. *Astron. Astrophys.* **1984**, *138*, 246–252.
8. Lamers, H.J.G.L.M.; Fitzpatrick, E.L. The Relationship between the Eddington Limit, the Observed Upper Luminosity Limit for Massive Stars, and the Luminous Blue Variables. *Astrophys. J.* **1988**, *324*, 279. [CrossRef]
9. Pickering, E.C. Large Magellanic Clouds. *Popular Astronomy* **1897**, *5*, 411–413. [CrossRef]
10. Feast, M.W.; Thackeray, A.D.; Wesselink, A.J. The brightest stars in the Magellanic Clouds. *Mon. Not. R. Astron. Soc.* **1960**, *121*, 337. [CrossRef]
11. Conti, P.S. Basic Observational Constraints on the Evolution of Massive Stars. In *Observational Tests of the Stellar Evolution Theory*; Maeder, A., Renzini, A., Eds.; Cambridge University Press: Cambridge, UK, 1984; Volume 105, p. 233.
12. Bohannan, B.; Walborn, N.R. The Ofpe/WN9 Class in the Large Magellanic Cloud. *Publ. Astron. Soc. Pac.* **1989**, *101*, 520. [CrossRef]
13. Walborn, N.R.; Fitzpatrick, E.L. The OB Zoo: A Digital Atlas of Peculiar Spectra. *Publ. Astron. Soc. Pac.* **2000**, *112*, 50–64. [CrossRef]
14. Van Genderen, A.M.; de Groot, M.; Sterken, C. New perceptions on the S Doradus phenomenon and the micro variations of five Luminous Blue Variables (LBVs). *Astron. Astrophys. Suppl.* **1997**, *124*, 517–531. [CrossRef]
15. Van Genderen, A.M.; Sterken, C.; de Groot, M. New discoveries on the S DOR phenomenon based on an investigation of the photometric history of the variables AG Car, S Dor and η Car. *Astron. Astrophys.* **1997**, *318*, 81–98.
16. Burggraf, B.; Weis, K.; Bomans, D.J.; Henze, M.; Meusinger, H.; Sholukhova, O.; Zharova, A.; Pellerin, A.; Becker, A. Var C: Long-term photometric and spectral variability of a luminous blue variable in M 33. *Astron. Astrophys.* **2015**, *581*, A12.
17. Wolf, B. Empirical amplitude-luminosity relation of S Doradus variables and extragalactic distances. *Astron. Astrophys.* **1989**, *217*, 87–91.
18. Burggraf, B. LBVs in M33: Variability and Evolutionary State. Ph.D. thesis, Ruhr University Bochum, Bochum, Germany, 2015.
19. Van Genderen, A.M. S Doradus variables in the Galaxy and the Magellanic Clouds. *Astron. Astrophys.* **2001**, *366*, 508–531. [CrossRef]
20. Dorn-Wallenstein, T.Z.; Levesque, E.M.; Davenport, J.R.A. Short-term Variability of Evolved Massive Stars with TESS. *Astrophys. J.* **2019**, *878*, 155.
21. Humphreys, R.M.; Davidson, K.; Smith, N. η Carinae's Second Eruption and the Light Curves of the η Carinae Variables. *Publ. Astron. Soc. Pac.* **1999**, *111*, 1124–1131.[CrossRef]
22. Davidson, K.; Humphreys, R.M. Eta Carinae and Its Environment. *Annu. Rev. Astron. Astrophys.* **1997**, *35*, 1–32. [CrossRef]
23. De Groot, M. The most luminous stars in the universe. *Irish Astron. J.* **1988**, *18*, 163–170.
24. Tammann, G.A.; Sandage, A. The Stellar Content and Distance of the Galaxy NGC 2403 IN the M81 Group. *Astrophys. J.* **1968**, *151*, 825. [CrossRef]
25. Zwicky, F. NGC 1058 and its Supernova 1961. *Astrophys. J.* **1964**, *139*, 514. [CrossRef]

26. Goodrich, R.W.; Stringfellow, G.S.; Penrod, G.D.; Filippenko, A.V. SN 1961V: an Extragalactic Eta Carinae Analog? *Astrophys. J.* **1989**, *342*, 908. [CrossRef]
27. Leitherer, C. Mass Loss from LBVs: Observational Constraints. In *Luminous Blue Variables: Massive Stars in Transition*; Nota, A.; Lamers, H., Eds.; Astronomical Society of the Pacific: San Francisco, CA, USA, 1997; Volume 120, p. 58.
28. Stahl, O.; Jankovics, I.; Kovács, J.; Wolf, B.; Schmutz, W.; Kaufer, A.; Rivinius, T.; Szeifert, T. Long-term spectroscopic monitoring of the Luminous Blue Variable AG Carinae. *Astron. Astrophys.* **2001**, *375*, 54–69. [CrossRef]
29. Groh, J.H.; Damineli, A.; Hillier, D.J. LBVs and the nature of the S Dor cycles: the case of AG Carinae. In *Revista Mexicana de Astronomía y Astrofísica Conference Series*, Instituto de Astronomía: Ciudad de México, México, 2008; Volume 33, pp. 132–134.
30. Lamers, H.J.G.L.M.; Snow, T.P.; Lindholm, D.M. Terminal Velocities and the Bistability of Stellar Winds. *Astrophys. J.* **1995**, *455*, 269. [CrossRef]
31. Vink, J.S.; de Koter, A.; Lamers, H.J.G.L.M. On the nature of the bi-stability jump in the winds of early-type supergiants. *Astron. Astrophys.* **1999**, *350*, 181–196.
32. Eddington, A.S. On the radiative equilibrium of the stars. *Mon. Not. R. Astron. Soc.* **1916**, *77*, 16–35. [CrossRef]
33. Maeder, A.; Meynet, G. Stellar evolution with rotation. VI. The Eddington and Omega -limits, the rotational mass loss for OB and LBV stars. *Astron. Astrophys.* **2000**, *361*, 159–166.
34. Groh, J.H.; Hillier, D.J.; Damineli, A. AG Carinae: A Luminous Blue Variable with a High Rotational Velocity. *Astrophys. J. Lett.* **2006**, *638*, L33–L36. [CrossRef]
35. Groh, J.H.; Damineli, A.; Hillier, D.J.; Barbá, R.; Fernández-Lajús, E.; Gamen, R.C.; Moisés, A.P.; Solivella, G.; Teodoro, M. Bona Fide, Strong-Variable Galactic Luminous Blue Variable Stars are Fast Rotators: Detection of a High Rotational Velocity in HR Carinae. *Astrophys. J. Lett.* **2009**, *705*, L25–L30.
36. Humphreys, R.M. The Wolf-Rayet Connection - Luminous Blue Variables and Evolved Supergiants (review). In *Wolf-Rayet Stars and Interrelations with Other Massive Stars in Galaxies*; van der Hucht, K.A., Hidayat, B., Eds.; Cambridge University Press: Cambridge, UK, 1991; Volume 143, p. 485.
37. Conti, P.S. On the relationship between Of and WR stars. *Memoires of the Societe Royale des Sciences de Liege* **1975**, *9*, 193–212.
38. Maeder, A.; Conti, P.S. Massive Star Populations in Nearby Galaxies. *Annu. Rev. Astron. Astrophys.* **1994**, *32*, 227–275. [CrossRef]
39. Meynet, G.; Maeder, A. Stellar evolution with rotation. XI. Wolf-Rayet star populations at different metallicities. *Astron. Astrophys.* **2005**, *429*, 581–598. [CrossRef]
40. Davidson, K. The Relation between Apparent Temperature and Mass-Loss Rate in Hypergiant Eruptions. *Astrophys. J.* **1987**, *317*, 760. [CrossRef]
41. Humphreys, R.M.; Davidson, K. The Luminous Blue Variables: Astrophysical Geysers. *Publ. Astron. Soc. Pac.* **1994**, *106*, 1025. [CrossRef]
42. Leitherer, C.; Schmutz, W.; Abbott, D.C.; Hamann, W.R.; Wessolowski, U. Atmospheric Models for Luminous Blue Variables. *Astrophys. J.* **1989**, *346*, 919. [CrossRef]
43. De Koter, A.; Lamers, H.J.G.L.M.; Schmutz, W. Variability of luminous blue variables. II. Parameter study of the typical LBV variations. *Astron. Astrophys.* **1996**, *306*, 501.
44. Smith, N.; Vink, J.S.; de Koter, A. The Missing Luminous Blue Variables and the Bistability Jump. *Astrophys. J.* **2004**, *615*, 475–484. [CrossRef]
45. Vink, J.S., Mass-Loss Rates of Very Massive Stars. In *Very Massive Stars in the Local Universe*; Vink, J.S., Ed.; Astrophysics and Space Science Library, Springer: Basel, Switzerland, 2015; Volume 412, p. 77. [CrossRef]
46. Abolmasov, P. Stochastic variability of luminous blue variables. *New A* **2011**, *16*, 421–429.
47. Gräfener, G.; Owocki, S.P.; Vink, J.S. Stellar envelope inflation near the Eddington limit. Implications for the radii of Wolf-Rayet stars and luminous blue variables. *Astron. Astrophys.* **2012**, *538*, A40.
48. Groh, J.H.; Hillier, D.J.; Damineli, A. On the Nature of the Prototype Luminous Blue Variable AG Carinae. II. Witnessing a Massive Star Evolving Close to the Eddington and Bistability Limits. *Astrophys. J.* **2011**, *736*, 46.
49. Stahl, O. Circumstellar ejecta around LBV's. In *IAU Colloq. 113: Physics of Luminous Blue Variables*; Davidson, K., Moffat, A.F.J., Lamers, H.J.G.L.M., Eds.; Springer: Boston, MA, USA, 1989; Volume 157, pp. 149–155.

50. Weis, K. Nebulae around Luminous Blue Variables—large bipolar variety. In *Active OB Stars: Structure, Evolution, Mass Loss, and Critical Limits*; Neiner, C., Wade, G., Meynet, G., Peters, G., Eds.; Cambridge University Press: Cambridge, UK, 2011; Volume 272, pp. 372–377. [CrossRef]
51. Weis, K. Gone with the wind: Nebulae around LBVs. *Bull. Soc. R. Sci. Liege* **2011**, *80*, 440–444.
52. Weis, K. On the structure and kinematics of nebulae around LBVs and LBV candidates in the LMC. *Astron. Astrophys.* **2003**, *408*, 205–229. [CrossRef]
53. Weis, K.; Duschl, W.J.; Bomans, D.J. An outflow from the nebula around the LBV candidate S 119. *Astron. Astrophys.* **2003**, *398*, 1041–1048. [CrossRef]
54. Weis, K. 28 years of Luminous Blue Variables. In *370 Years of Astronomy in Utrecht*; Astronomical Society of the Pacific Conference Seriesl; Pugliese, G., de Koter, A., Wijburg, M., Eds.; Astronomical Society of the Pacific: San Francisco, CA, USA, 2013; Volume 470, p. 129.
55. Weis, K. Family ties of WR to LBV nebulae yielding clues for stellar evolution. In Proceedings of the Wolf-Rayet Stars: Proceedings of an International Workshop, Potsdam, Germany, 1–5 June 2015; pp. 167–170.
56. Weis, K.; Duschl, W.J. Outflow from and asymmetries in the nebula around the LBV candidate Sk-$69°279$. *Astron. Astrophys.* **2002**, *393*, 503–510. [CrossRef]
57. Nota, A.; Pasquali, A.; Clampin, M.; Pollacco, D.; Scuderi, S.; Livio, M. The Nebula around HD 168625: Morphology, Dynamics, and Physical Properties. *Astrophys. J.* **1996**, *473*, 946. [CrossRef]
58. Weis, K. The outer ejecta of η Carinae. In *The Fate of the Most Massive Stars*; Astronomical Society of the Pacific Conference Series; Humphreys, R., Stanek, K., Eds.; Astronomical Society of the Pacific: San Francisco, CA, USA, 2005; Volume 332, p. 275.
59. Weis, K. The Outer Ejecta. In *Eta Carinae and the Supernova Impostors*; Astrophysics and Space Science Library; Davidson, K., Humphreys, R.M., Eds.; Springer: Basel, Switzerland, 2012; Volume 384, p. 171. [CrossRef]
60. Weis, K. The outer ejecta of Eta Carinae. In *Eta Carinae and Other Mysterious Stars: The Hidden Opportunities of Emission Spectroscopy*; Astronomical Society of the Pacific Conference Series; Gull, T.R., Johannson, S., Davidson, K., Eds.; Astronomical Society of the Pacific: San Francisco, CA, USA, 2001; Volume 242, p. 129.
61. Weis, K. The AG Carinae Nebula—Bigger than ever? In *Mass Loss from Stars and the Evolution of Stellar Clusters*; Astronomical Society of the Pacific Conference Series; de Koter, A., Smith, L.J., Waters, L.B.F.M., Eds.; Astronomical Society of the Pacific: San Francisco, CA, USA, 2008; Volume 388, p. 231.
62. Weis, K.; Duschl, W.J.; Bomans, D.J.; Chu, Y.H.; Joner, M.D. The bipolar structure of the LBV nebula around HR Carinae. *Astron. Astrophys.* **1997**, *320*, 568–574.
63. Weis, K. A kinematic and morphological investigation of the asymmetric nebula around the LBV candidate WRA 751. *Astron. Astrophys.* **2000**, *357*, 938–944.
64. Figer, D.F.; Morris, M.; Geballe, T.R.; Rich, R.M.; Serabyn, E.; McLean, I.S.; Puetter, R.C.; Yahil, A. High-Resolution Infrared Imaging and Spectroscopy of the Pistol Nebula: Evidence for Ejection. *Astrophys. J.* **1999**, *525*, 759–771. [CrossRef]
65. Brandner, W.; Grebel, E.K.; Chu, Y.H.; Weis, K. Ring Nebula and Bipolar Outflows Associated with the B1.5 Supergiant Sher 25 in NGC 3603. *Astrophys. J. Lett.* **1997**, *475*, L45–L48. [CrossRef]
66. Mizuno, D.R.; Kraemer, K.E.; Flagey, N.; Billot, N.; Shenoy, S.; Paladini, R.; Ryan, E.; Noriega-Crespo, A.; Carey, S.J. A Catalog of MIPSGAL Disk and Ring Sources. *Astron. J.* **2010**, *139*, 1542–1552.
67. Simpson, R.J.; Povich, M.S.; Kendrew, S.; Lintott, C.J.; Bressert, E.; Arvidsson, K.; Cyganowski, C.; Maddison, S.; Schawinski, K.; Sherman, R.; et al. The Milky Way Project First Data Release: a bubblier Galactic disc. *Mon. Not. R. Astron. Soc.* **2012**, *424*, 2442–2460.
68. Beaumont, C.N.; Goodman, A.A.; Kendrew, S.; Williams, J.P.; Simpson, R. The Milky Way Project: Leveraging Citizen Science and Machine Learning to Detect Interstellar Bubbles. *Astrophys. J. Suppl.* **2014**, *214*, 3.
69. Gvaramadze, V.V.; Kniazev, A.Y.; Fabrika, S.; Sholukhova, O.; Berdnikov, L.N.; Cherepashchuk, A.M.; Zharova, A.V. MN112: a new Galactic candidate luminous blue variable. *Mon. Not. R. Astron. Soc.* **2010**, *405*, 520–524.
70. Wachter, S.; Mauerhan, J.C.; Van Dyk, S.D.; Hoard, D.W.; Kafka, S.; Morris, P.W. A Hidden Population of Massive Stars with Circumstellar Shells Discovered with the Spitzer Space Telescope. *Astron. J.* **2010**, *139*, 2330–2346.
71. Flagey, N.; Noriega-Crespo, A.; Billot, N.; Carey, S.J. Spitzer/InfraRed Spectrograph Investigation of MIPSGAL 24 μm Compact Bubbles. *Astrophys. J.* **2011**, *741*, 4.

72. Nowak, M.; Flagey, N.; Noriega-Crespo, A.; Billot, N.; Carey, S.J.; Paladini, R.; Van Dyk, S.D. Spitzer/Infrared Spectrograph Investigation of MIPSGAL 24 µm Compact Bubbles: Low-resolution Observations. *Astrophys. J.* **2014**, *796*, 116.
73. Flagey, N.; Noriega-Crespo, A.; Petric, A.; Geballe, T.R. Palomar/TripleSpec Observations of Spitzer/MIPSGAL 24 µm Circumstellar Shells: Unveiling the Natures of Their Central Sources. *Astron. J.* **2014**, *148*, 34.
74. Gvaramadze, V.V.; Kniazev, A.Y.; Fabrika, S. Revealing evolved massive stars with Spitzer. *Mon. Not. R. Astron. Soc.* **2010**, *405*, 1047–1060.
75. Mauerhan, J.C.; Wachter, S.; Morris, P.W.; Van Dyk, S.D.; Hoard, D.W. Discovery of Twin Wolf-Rayet Stars Powering Double Ring Nebulae. *Astrophys. J. Lett.* **2010**, *724*, L78–L83.
76. Gvaramadze, V.V.; Kniazev, A.Y.; Miroshnichenko, A.S.; Berdnikov, L.N.; Langer, N.; Stringfellow, G.S.; Todt, H.; Hamann, W.R.; Grebel, E.K.; Buckley, D.; et al. Discovery of two new Galactic candidate luminous blue variables with Wide-field Infrared Survey Explorer. *Mon. Not. R. Astron. Soc.* **2012**, *421*, 3325–3337.
77. Gvaramadze, V.V.; Kniazev, A.Y.; Berdnikov, L.N.; Langer, N.; Grebel, E.K.; Bestenlehner, J.M. Discovery of a new Galactic bona fide luminous blue variable with Spitzer. *Mon. Not. R. Astron. Soc.* **2014**, *445*, L84–L88.
78. Gvaramadze, V.V.; Kniazev, A.Y.; Berdnikov, L.N. Discovery of a new bona fide luminous blue variable in Norma. *Mon. Not. R. Astron. Soc.* **2015**, *454*, 3710–3721.
79. Kniazev, A.Y.; Gvaramadze, V.V.; Berdnikov, L.N. WS1: one more new Galactic bona fide luminous blue variable. *Mon. Not. R. Astron. Soc.* **2015**, *449*, L60–L64.
80. Kniazev, A.Y.; Gvaramadze, V.V.; Berdnikov, L.N. MN48: a new Galactic bona fide luminous blue variable revealed by Spitzer and SALT. *Mon. Not. R. Astron. Soc.* **2016**, *459*, 3068–3077.
81. Cerrigone, L.; Umana, G.; Buemi, C.S.; Hora, J.L.; Trigilio, C.; Leto, P.; Hart, A. Spitzer observations of a circumstellar nebula around the candidate luminous blue variable MWC 930. *Astron. Astrophys.* **2014**, *562*, A93.
82. Umana, G.; Buemi, C.S.; Trigilio, C.; Hora, J.L.; Fazio, G.G.; Leto, P. The Dusty Nebula Surrounding HR Car: A Spitzer View. *Astrophys. J.* **2009**, *694*, 697–703.
83. Gail, H.P.; Duschl, W.J.; Ferrarotti, A.S.; Weis, K. Dust formation in LBV envelopes. In *The Fate of the Most Massive Stars*; Humphreys, R.; Stanek, K., Eds.; Astronomical Society of the Pacific: San Francisco, CA, USA, 2005; Volume 332, p. 323.
84. van Marle, A.J.; Meliani, Z.; Keppens, R.; Decin, L. Computing the Dust Distribution in the Bow Shock of a Fast-moving, Evolved Star. *Astrophys. J. Lett.* **2011**, *734*, L26.
85. Barlow, M.J.; Drew, J.E.; Meaburn, J.; Massey, R.M. The Shock-Excited P-Cygni Nebula. *Mon. Not. R. Astron. Soc.* **1994**, *268*, L29. [CrossRef]
86. Meaburn, J.; López, J.A.; O'Connor, J. The Kinematical Association of a Giant Lobe with the Luminous Blue Variable Star P Cygni. *Astrophys. J. Lett.* **1999**, *516*, L29–L32. [CrossRef]
87. Arcidiacono, C.; Ragazzoni, R.; Morossi, C.; Franchini, M.; di Marcantonio, P.; Kulesa, C.; McCarthy, D.; Briguglio, R.; Xompero, M.; Busoni, L.; et al. A high-resolution image of the inner shell of the P Cygni nebula in the infrared [Fe II] line. *Mon. Not. R. Astron. Soc.* **2014**, *443*, 1142–1150.
88. Davidson, K.; Humphreys, R.M. *Eta Carinae and the Supernova Impostors*; Astrophysics and Space Science Library; Springer: Basel, Switzerland, 2012; Volume 384. [CrossRef]
89. Madura, T.I.; Gull, T.R.; Owocki, S.P.; Groh, J.H.; Okazaki, A.T.; Russell, C.M.P. Constraining the absolute orientation of η Carinae's binary orbit: a 3D dynamical model for the broad [Fe III] emission. *Mon. Not. R. Astron. Soc.* **2012**, *420*, 2064–2086.
90. Gaviola, E. Eta Carinae. I. The Nebulosity. *Astrophys. J.* **1950**, *111*, 408. [CrossRef]
91. Ishibashi, K.; Gull, T.R.; Davidson, K.; Smith, N.; Lanz, T.; Lindler, D.; Feggans, K.; Verner, E.; Woodgate, B.E.; Kimble, R.A.; et al. Discovery of a Little Homunculus within the Homunculus Nebula of η Carinae. *Astron. J.* **2003**, *125*, 3222–3236. [CrossRef]
92. Walborn, N.R. The complex outer shell of Eta Carinae. *Astrophys. J. Lett.* **1976**, *204*, L17–L19. [CrossRef]
93. Weis, K.; Duschl, W.J.; Bomans, D.J. High velocity structures in, and the X-ray emission from the LBV nebula around η Carinae. *Astron. Astrophys.* **2001**, *367*, 566–576. [CrossRef]
94. Weis, K.; Corcoran, M.F.; Bomans, D.J.; Davidson, K. A spectral and spatial analysis of η Carinae's diffuse X-ray emission using CHANDRA. *Astron. Astrophys.* **2004**, *415*, 595–607. [CrossRef]

95. Weis, K.; Corcoran, M.F.; Davidson, K., On the X-ray Emission of η Carinae's Outer Ejecta. In *The High Energy Universe at Sharp Focus: Chandra Science*; Astronomical Society of the Pacific Conference Series; Schlegel, E.M., Vrtilek, S.D., Eds.; Astronomical Society of the Pacific: San Francisco, CA, USA, 2002; Volume 262, p. 275.
96. Glatzel, W. Instabilities in the Most Massive Evolved Stars. In *The Fate of the Most Massive Stars*; Astronomical Society of the Pacific Conference Series; Humphreys, R., Stanek, K., Eds.; Astronomical Society of the Pacific: San Francisco, CA, USA, 2005; Volume 332, p. 22.
97. Vink, J.S. Eta Carinae and the Luminous Blue Variables. In *Eta Carinae and the Supernova Impostors; Astrophysics and Space Science Series Library*; Davidson, K., Humphreys, R.M., Eds.; Springer: Boston, MA, USA, 2012; Volume 384, p. 221. [CrossRef]
98. Owocki, S.P. Instabilities in the Envelopes and Winds of Very Massive Stars. In *Very Massive Stars in the Local Universe*; Vink, J.S., Ed.; Springer: Boston, MA, USA, 2015; Volume 412, p. 113. [CrossRef]
99. Guzik, J.A. Instability Considerations for Massive Star Eruptions. In *The Fate of the Most Massive Stars*; Humphreys, R., Stanek, K., Eds.; Springer: Basel, Switzerland 2005; Volume 332, p. 208.
100. Lovekin, C.C.; Guzik, J.A. Pulsations as a driver for LBV variability. *Mon. Not. R. Astron. Soc.* **2014**, *445*, 1766–1773.
101. Kalari, V.M.; Vink, J.S.; Dufton, P.L.; Fraser, M. How common is LBV S Doradus variability at low metallicity? *Astron. Astrophys.* **2018**, *618*, A17.
102. Jiang, Y.F.; Cantiello, M.; Bildsten, L.; Quataert, E.; Blaes, O.; Stone, J. Outbursts of luminous blue variable stars from variations in the helium opacity. *Nature* **2018**, *561*, 498–501. [CrossRef] [PubMed]
103. Hamann, W.R.; Feldmeier, A.; Oskinova, L.M. Clumping in hot-star winds. In Proceedings of the an International Workshop, Potsdam, Germany, 18–22 June 2007.
104. Sundqvist, J.O.; Puls, J.; Feldmeier, A.; Owocki, S.P. Mass loss from inhomogeneous hot star winds. II. Constraints from a combined optical/UV study. *Astron. Astrophys.* **2011**, *528*, A64.
105. Puls, J.; Vink, J.S.; Najarro, F. Mass loss from hot massive stars. *Astron. Astrophys. Rev.* **2008**, *16*, 209–325.
106. Pastorello, A.; Botticella, M.T.; Trundle, C.; Taubenberger, S.; Mattila, S.; Kankare, E.; Elias-Rosa, N.; Benetti, S.; Duszanowicz, G.; Hermansson, L.; et al. Multiple major outbursts from a restless luminous blue variable in NGC 3432. *Mon. Not. R. Astron. Soc.* **2010**, *408*, 181–198.
107. Arnett, W.D.; Bahcall, J.N.; Kirshner, R.r.P.; Woosley, S.E. Supernova 1987A. *Annu. Rev. Astron. Astrophys.* **1989**, *27*, 629–700. [CrossRef]
108. Pastorello, A.; Della Valle, M.; Smartt, S.J.; Zampieri, L.; Benetti, S.; Cappellaro, E.; Mazzali, P.A.; Patat, F.; Spiro, S.; Turatto, M.; et al. A very faint core-collapse supernova in M85. *Nature* **2007**, *449*, 1–2.
109. Weis, K.; Bomans, D.J. SN 2002kg - the brightening of LBV V37 in NGC 2403. *Astron. Astrophys.* **2005**, *429*, L13–L16. [CrossRef]
110. Bomans, D.J.; Weis, K. SN Impostors, a complex mix-bag of transients. *Astron. Astrophys.* **2019**, in prep.
111. Pastorello, A.; Fraser, M. Supernova impostors and other gap transients. *Nature Astronomy* **2019**, *3*, 676–679.
112. Smith, N.; Li, W.; Silverman, J.M.; Ganeshalingam, M.; Filippenko, A.V. Luminous blue variable eruptions and related transients: diver sity of progenitors and outburst properties. *Mon. Not. R. Astron. Soc.* **2011**, *415*, 773–810.
113. Bomans, D.J.; Mueller, A.; Becker, A.; Weis, K.; Granzer, T. Gaia16ada: the most recent outburst of the supernova impostor in NGC 4559. *The Astronomer's Telegram* **2016**, *8755*, 1.
114. Sana, H.; de Mink, S.E.; de Koter, A.; Langer, N.; Evans, C.J.; Gieles, M.; Gosset, E.; Izzard, R.G.; Le Bouquin, J.B.; Schneider, F.R.N. Binary Interaction Dominates the Evolution of Massive Stars. *Science* **2012**, *337*, 444.
115. Gallagher, J.S. Close Binary Models for Luminous Blue Variables Stars. In *IAU Colloq. 113: Physics of Luminous Blue Variables*; Davidson, K., Moffat, A.F.J., Lamers, H.J.G.L.M., Eds.; Cambridge University Press: Cambridge, UK, 1989; Volume 157, p. 185. [CrossRef]
116. Justham, S.; Podsiadlowski, P.; Vink, J.S. Luminous Blue Variables and Superluminous Supernovae from Binary Mergers. *Astrophys. J.* **2014**, *796*, 121.
117. Barba, R.; Niemela, V. HD 5980. IAU Circular: Cambridge, MA, USA, 1994.
118. Koenigsberger, G.; Morrell, N.; Hillier, D.J.; Gamen, R.; Schneider, F.R.N.; González-Jiménez, N.; Langer, N.; Barbá, R. The HD 5980 Multiple System: Masses and Evolutionary Status. *Astron. J.* **2014**, *148*, 62.
119. Figer, D.F.; Najarro, F.; Kudritzki, R.P. The Double-lined Spectrum of LBV 1806-20. *Astrophys. J. Lett.* **2004**, *610*, L109–L112. [CrossRef]

120. Lobel, A.; Groh, J.H.; Martayan, C.; Frémat, Y.; Torres Dozinel, K.; Raskin, G.; Van Winckel, H.; Prins, S.; Pessemier, W.; Waelkens, C.; et al. Modelling the asymmetric wind of the luminous blue variable binary MWC 314. *Astron. Astrophys.* **2013**, *559*, A16.
121. Martayan, C.; Lobel, A.; Baade, D.; Mehner, A.; Rivinius, T.; Boffin, H.M.J.; Girard, J.; Mawet, D.; Montagnier, G.; Blomme, R.; et al. Luminous blue variables: An imaging perspective on their binarity and near environment. *Astron. Astrophys.* **2016**, *587*, A115.
122. Boffin, H.M.J.; Rivinius, T.; Mérand, A.; Mehner, A.; LeBouquin, J.B.; Pourbaix, D.; de Wit, W.J.; Martayan, C.; Guieu, S. The luminous blue variable HR Carinae has a partner. Discovery of a companion with the VLTI. *Astron. Astrophys.* **2016**, *593*, A90.
123. Nazé, Y.; Rauw, G.; Hutsemékers, D. The first X-ray survey of Galactic luminous blue variables. *Astron. Astrophys.* **2012**, *538*, A47.
124. Smith, N.; Tombleson, R. Luminous blue variables are antisocial: their isolation implies that they are kicked mass gainers in binary evolution. *Mon. Not. R. Astron. Soc.* **2015**, *447*, 598–617.
125. Humphreys, R.M.; Weis, K.; Davidson, K.; Gordon, M.S. On the Social Traits of Luminous Blue Variables. *Astrophys. J.* **2016**, *825*, 64.
126. Smith, N. The isolation of luminous blue variables: on subdividing the sample. *Mon. Not. R. Astron. Soc.* **2016**, *461*, 3353–3360.
127. Davidson, K.; Humphreys, R.M.; Weis, K. LBVs and Statistical Inference. *arXiv* **2016**, arXiv:1608.02007.
128. Aadland, E.; Massey, P.; Neugent, K.F.; Drout, M.R. Shedding Light on the Isolation of Luminous Blue Variables. *Astron. J.* **2018**, *156*, 294.
129. Smith, N. The isolation of luminous blue variables resembles aging B-type supergiants, not the most massive unevolved stars. *Mon. Not. R. Astron. Soc.* **2019**, *489*, 4378–4388.
130. Ward, J.L.; Kruijssen, J.M.D.; Rix, H.W. Not all stars form in clusters – Gaia-DR2 uncovers the origin of OB associations. *arXiv* **2019**, arXiv:1910.06974.
131. Lortet, M.C. A provisory catalogue of S-Dor candidate stars in the Magellanic Clouds. *Bull. Soc. R. Sci. Liege* **1988**, *35*, 145–154.
132. Stahl, O.; Wolf, B. The Spectral Evolution of the LMC S DOR Variable R127 during Outburst. In *The Impact of Very High S/N Spectroscopy on Stellar Physics*; Cayrel de Strobel, G., Spite, M., Eds.; Cambridge University Press: Cambridge, UK, 1988; Volume 132, p. 557.
133. Stahl, O.; Wolf, B.; Klare, G.; Jüttner, A.; Cassatella, A. Observation of the new luminous blue variable R 110 of the Large Magellanic cloud during an F star-phase. *Astron. Astrophys.* **1990**, *228*, 379–386.
134. Wolf, B.; Stahl, O. Inverse P Cygni-type profiles in the spectrum of the luminous blue variable S Doradus. *Astron. Astrophys.* **1990**, *235*, 340.
135. Stahl, O.; Wolf, B.; de Groot, M.; Leitherer, C. Atlas of hig-dispersion spectra of peculiar emission-line stars in the Magellanic Clouds. *Astron. Astrophys. Suppl.* **1985**, *61*, 237–258.
136. Szeifert, T.; Stahl, O.; Wolf, B.; Zickgraf, F.J.; Bouchet, P.; Klare, G. R 40 : the first luminous blue variable in the Small Magellanic Cloud. *Astron. Astrophys.* **1993**, *280*, 508–518.
137. Lindegren, L. The Tycho-Gaia Astrometric Solution. In *Astrometry and Astrophysics in the Gaia Sky*; Recio-Blanco, A.; de Laverny, P.; Brown, A.G.A.; Prusti, T., Eds.; Cambridge University Press: Cambridge, UK, 2018, Volume 330; pp. 41–48. [CrossRef]
138. Lindegren, L.; Hernández, J.; Bombrun, A.; Klioner, S.; Bastian, U.; Ramos-Lerate, M.; de Torres, A.; Steidelmüller, H.; Stephenson, C.; Hobbs, D.; et al. Gaia Data Release 2. The astrometric solution. *Astron. Astrophys.* **2018**, *616*, A2.
139. Berlanas, S.R.; Wright, N.J.; Herrero, A.; Drew, J.E.; Lennon, D.J. Disentangling the spatial substructure of Cygnus OB2 from Gaia DR2. *Mon. Not. R. Astron. Soc.* **2019**, *484*, 1838–1842.
140. Humphreys, R.M.; Stangl, S.; Gordon, M.S.; Davidson, K.; Grammer, S.H. Luminous and Variable Stars in NGC 2403 and M81. *Astron. J.* **2019**, *157*, 22.
141. Massey, P.; Olsen, K.A.G.; Hodge, P.W.; Jacoby, G.H.; McNeill, R.T.; Smith, R.C.; Strong, S.B. A Survey of Local Group Galaxies Currently Forming Stars. II. UBVRI Photometry of Stars in Seven Dwarfs and a Comparison of the Entire Sample. *Astron. J.* **2007**, *133*, 2393–2417. [CrossRef]
142. Massey, P.; McNeill, R.T.; Olsen, K.A.G.; Hodge, P.W.; Blaha, C.; Jacoby, G.H.; Smith, R.C.; Strong, S.B. A Survey of Local Group Galaxies Currently Forming Stars. III. A Search for Luminous Blue Variables and Other Hα Emission-Line Stars. *Astron. J.* **2007**, *134*, 2474–2503.

143. Burggraf, B.; Weis, K.; Bomans, D.J. LBVs in Local Group Galaxies. *Astron. Nachr.* **2007**, *328*, 716.
144. Bomans, D.J.; Weis, K.; Wittkowski, M. Massive Variable Stars at Low Metallicity: The Case of NGC 3109. In *Proceedings of a Scientific Meeting in Honor of Anthony F. J. Moffat*; Drissen, L., Robert, C., St-Louis, N., Moffat, A.F.J., Eds.; Astronomical Society of the Pacific: San Francisco, CA, USA, 2012; Volume 465, p. 508.
145. Spiller, F. Suche nach Leuchtkräftigen Blauen Veränderlichen und verwandten Objekten in M33, NGC 2403, M81 und M101. Ph.D. Thesis, University Heidelberg, Heidelberg, Germany, 1992.
146. Humphreys, R.M.; Weis, K.; Davidson, K.; Bomans, D.J.; Burggraf, B. Luminous and Variable Stars in M31 and M33. II. Luminous Blue Variables, Candidate LBVs, Fe II Emission Line Stars, and Other Supergiants. *Astrophys. J.* **2014**, *790*, 48.
147. Humphreys, R.M.; Gordon, M.S.; Martin, J.C.; Weis, K.; Hahn, D. Luminous and Variable Stars in M31 and M33. IV. Luminous Blue Variables, Candidate LBVs, B[e] Supergiants, and the Warm Hypergiants: How to Tell Them Apart. *Astrophys. J.* **2017**, *836*, 64.
148. Szeifert, T.; Stahl, O.; Wolf, B.; Zickgraf, F.J., R40: First Luminous Blue Variable in the Small Magellanic Cloud. In *New Aspects of Magellanic Cloud Research*; Baschek, B., Klare, G., Lequeux, J., Eds.; Springer: Berlin/Heidelberg, Germany, 1993; Volume 416, p. 280. [CrossRef]
149. Humphreys, R.M.; Martin, J.C.; Gordon, M.S. A New Luminous Blue Variable in M31. *Publ. Astron. Soc. Pac.* **2015**, *127*, 347.
150. Huang, Y.; Zhang, H.W.; Wang, C.; Chen, B.Q.; Zhang, Y.W.; Guo, J.C.; Yuan, H.B.; Xiang, M.S.; Tian, Z.J.; Li, G.X.; et al. A New Luminous Blue Variable in the Outskirts of the Andromeda Galaxy. *Astrophys. J. Lett.* **2019**, *884*, L7.
151. Humphreys, R.M.; Davidson, K.; Van Dyk, S.D.; Gordon, M.S. A Tale of Two Impostors: SN2002kg and SN1954J in NGC 2403. *Astrophys. J.* **2017**, *848*, 86.
152. Drissen, L.; Crowther, P.A.; Smith, L.J.; Robert, C.; Roy, J.R.; Hillier, D.J. Physical Parameters of Erupting Luminous Blue Variables: NGC 2363-V1 Caught in the Act. *Astrophys. J.* **2001**, *546*, 484–495. [CrossRef]
153. Sandage, A. The brightest stars in nearby galaxies. II. The color-magnitude diagram for the brightest red and blue stars in M101. *Astron. J.* **1983**, *88*, 1569–1578. [CrossRef]
154. Grammer, S.H.; Humphreys, R.M.; Gerke, J. The Massive Star Population in M101. III. Spectra and Photometry of the Luminous and Variable Stars. *Astron. J.* **2015**, *149*, 152.
155. Sandage, A. The brightest stars in nearby galaxies. III. The color-magnitude diagram for the brightest red and blue stars in M 81 and Holmberg IX. *Astron. J.* **1984**, *89*, 621–629. [CrossRef]
156. Corral, L.J.; Herrero, A. Candidate LBVs in Local Group galaxies. In *A Massive Star Odyssey: From Main Sequence to Supernova*; van der Hucht, K., Herrero, A., Esteban, C., Eds.; Springer: Berlin, Germany, 2003; Volume 212, p. 160.
157. Bresolin, F.; Kudritzki, R.P.; Najarro, F.; Gieren, W.; Pietrzyński, G. Discovery and Quantitative Spectral Analysis of an Ofpe/WN9 (WN11) Star in the Sculptor Spiral Galaxy NGC 300. *Astrophys. J. Lett.* **2002**, *577*, L107–L110. [CrossRef]
158. Humphreys, R.M. Studies of luminous stars in nearby galaxies. V. The local group irregulars NGC 6822 and IC 1613. *Astrophys. J.* **1980**, *238*, 65. [CrossRef]
159. Turner, A.; Ferrarese, L.; Saha, A.; Bresolin, F.; Kennicutt, Robert C., J.; Stetson, P.B.; Mould, J.R.; Freedman, W.L.; Gibson, B.K.; Graham, J.A.; et al. The Hubble Space Telescope Key Project on the Extragalactic Distance Scale. XI. The Cepheids in NGC 4414. *Astrophys. J.* **1998**, *505*, 207–229. [CrossRef]
160. Antonello, E.; Fugazza, D.; Mantegazza, L.; Bossi, M.; Covino, S. Variable stars in nearby galaxies. III. White light observations of Field B of IC 1613. *Astron. Astrophys.* **2000**, *363*, 29–40.
161. Herrero, A.; Garcia, M.; Uytterhoeven, K.; Najarro, F.; Lennon, D.J.; Vink, J.S.; Castro, N. The nature of V39: an LBV candidate or LBV impostor in the very low metallicity galaxy IC 1613? *Astron. Astrophys.* **2010**, *513*, A70.
162. Pustilnik, S.A.; Tepliakova, A.L.; Kniazev, A.Y.; Burenkov, A.N. Discovery of a massive variable star with Z = Z_{solar}/36 in the galaxy DDO 68. *Mon. Not. R. Astron. Soc.* **2008**, *388*, L24–L28.
163. Bomans, D.J.; Weis, K. The nature of the massive stellar transient in DDO 68. *Bull. Soc. R. Sci. Liege* **2011**, *80*, 341–345.
164. Pustilnik, S.A.; Makarova, L.N.; Perepelitsyna, Y.A.; Moiseev, A.V.; Makarov, D.I. The extremely metal-poor galaxy DDO 68: the luminous blue variable, Hα shells and the most luminous stars. *Mon. Not. R. Astron. Soc.* **2017**, *465*, 4985–5002.

165. Khan, R.; Adams, S.M.; Stanek, K.Z.; Kochanek, C.S.; Sonneborn, G. Discovery of Five Candidate Analogs for η Carinae in Nearby Galaxies. *Astrophys. J. Lett.* **2015**, *815*, L18.
166. Khan, R.; Kochanek, C.S.; Stanek, K.Z.; Gerke, J. Finding η Car Analogs in Nearby Galaxies Using Spitzer. II. Identification of An Emerging Class of Extragalactic Self-Obscured Stars. *Astrophys. J.* **2015**, *799*, 187.
167. Drissen, L.; Roy, J.R.; Robert, C. A New Luminous Blue Variable in the Giant Extragalactic H II Region NGC 2363. *Astron. J.* **1997**, *474*, L35–L38. [CrossRef]
168. Petit, V.; Drissen, L.; Crowther, P.A. Spectral Evolution of the Luminous Blue Variable NGC 2363-V1. I. Observations and Qualitative Analysis of the Ongoing Giant Eruption. *Astron. J.* **2006**, *132*, 1756–1762. [CrossRef]
169. Sacchi, E.; Annibali, F.; Cignoni, M.; Aloisi, A.; Sohn, T.; Tosi, M.; van der Marel, R.P.; Grocholski, A.J.; James, B. Stellar Populations and Star Formation History of the Metal-poor Dwarf Galaxy DDO 68. *Astrophys. J.* **2016**, *830*, 3.
170. Izotov, Y.I.; Thuan, T.X. Luminous Blue Variable Stars in the two Extremely Metal-Deficient Blue Compact Dwarf Galaxies DDO 68 and PHL 293B. *Astrophys. J.* **2009**, *690*, 1797–1806.
171. Izotov, Y.I.; Guseva, N.G.; Fricke, K.J.; Henkel, C. VLT/X-shooter observations of the low-metallicity blue compact dwarf galaxy PHL 293B including a luminous blue variable star. *Astron. Astrophys.* **2011**, *533*, A25.
172. Filho, M.E.; Sánchez Almeida, J. An unusual transient in the extremely metal-poor Galaxy SDSS J094332.35+332657.6 (Leoncino Dwarf). *Mon. Not. R. Astron. Soc.* **2018**, *478*, 2541–2556.
173. Rosolowsky, E.; Simon, J.D. The M33 Metallicity Project: Resolving the Abundance Gradient Discrepancies in M33. *Astrophys. J.* **2008**, *675*, 1213–1222.
174. Toribio San Cipriano, L.; Domínguez-Guzmán, G.; Esteban, C.; García-Rojas, J.; Mesa-Delgado, A.; Bresolin, F.; Rodríguez, M.; Simón-Díaz, S. Carbon and oxygen in H II regions of the Magellanic Clouds: abundance discrepancy and chemical evolution. *Mon. Not. R. Astron. Soc.* **2017**, *467*, 3759–3774.
175. Sánchez, S.F.; Rosales-Ortega, F.F.; Iglesias-Páramo, J.; Mollá, M.; Barrera-Ballesteros, J.; Marino, R.A.; Pérez, E.; Sánchez-Blazquez, P.; González Delgado, R.; Cid Fernandes, R.; et al. A characteristic oxygen abundance gradient in galaxy disks unveiled with CALIFA. *Astron. Astrophys.* **2014**, *563*, A49. [CrossRef]
176. Bresolin, F. Metallicity gradients in small and nearby spiral galaxies. *Mon. Not. R. Astron. Soc.* **2019**, *488*, 3826–3843.
177. Moiseev, A.V.; Lozinskaya, T.A. Ionized gas velocity dispersion in nearby dwarf galaxies: looking at supersonic turbulent motions. *Mon. Not. R. Astron. Soc.* **2012**, *423*, 1831–1844.
178. Kleemann, B. Integral Field Spectroscopy of Late-Type Galaxies: Kinematics and Ionization Structure of Starbursts. Ph.D. thesis, Ruhr University Bochum, Bochum, Germany, 2019.
179. Tully, R.B.; Rizzi, L.; Shaya, E.J.; Courtois, H.M.; Makarov, D.I.; Jacobs, B.A. The Extragalactic Distance Database. *Astron. J.* **2009**, *138*, 323–331. [CrossRef]
180. Tramper, F.; Sana, H.; de Koter, A.; Kaper, L. On the Mass-loss Rate of Massive Stars in the Low-metallicity Galaxies IC 1613, WLM, and NGC 3109. *Astrophys. J. Lett.* **2011**, *741*, L8.
181. Tramper, F.; Sana, H.; de Koter, A.; Kaper, L.; Ramírez-Agudelo, O.H. The properties of ten O-type stars in the low-metallicity galaxies IC 1613, WLM, and NGC 3109. *Astron. Astrophys.* **2014**, *572*, A36.
182. Garcia, M.; Evans, C.J.; Bestenlehner, J.M.; Bouret, J.C.; Castro, N.; Cerviño, M.; Fullerton, A.W.; Gieles, M.; Herrero, A.; de Koter, A.; et al. Massive stars in extremely metal-poor galaxies: A window into the past. *arXiv* **2019**, arXiv:1908.04687.
183. Bomans, D.J.; Weis, K. Massive variable stars at very low metallicity? In *Active OB Stars: Structure, Evolution, Mass Loss, and Critical Limits*; Neiner, C., Wade, G., Meynet, G., Peters, G., Eds.; IAU symposium; Cambridge University Press: Cambridge, UK, 2011; Volume 272, pp. 265–270. [CrossRef]
184. Roth, M.M.; Sandin, C.; Kamann, S.; Husser, T.O.; Weilbacher, P.M.; Monreal-Ibero, A.; Bacon, R.; den Brok, M.; Dreizler, S.; Kelz, A.; et al. MUSE crowded field 3D spectroscopy in NGC 300. I. First results from central fields. *Astron. Astrophys.* **2018**, *618*, A3.

185. Wofford, A.; Ramirez, V.; Lee, J.C.; Thilker, D.A.; Della Bruna, L.; Adamo, A.; Van Dyk, S.D.; Herrero, A.; Kim, H.; Aloisi, A.; et al. Candidate LBV stars in galaxy NGC 7793 found via HST photometry + MUSE spectroscopy. *arXiv* **2020**, arXiv:2001.10113.
186. Bomans, D.J.; Weis, K. Long-slit échelle spectroscopy of galactic outflows: The case of NGC 4449. *Astron. Nachr.* **2014**, *335*, 99. [CrossRef]

© 2020 by the authors. Licensee MDPI, Basel, Switzerland. This article is an open access article distributed under the terms and conditions of the Creative Commons Attribution (CC BY) license (http://creativecommons.org/licenses/by/4.0/).

Review

The History Goes On: Century Long Study of Romano's Star [†]

Olga Maryeva [1,2,*], Roberto F. Viotti [3], Gloria Koenigsberger [4], Massimo Calabresi [5], Corinne Rossi [6] and Roberto Gualandi [7]

1. Astronomical Institute of the Czech Academy of Sciences, Fričova 298, 25165 Ondřejov, Czech Republic
2. Sternberg Astronomical Institute, Lomonosov Moscow State University, Universitetsky pr. 13, 119234 Moscow, Russia
3. INAF-Istituto di Astrofisica e Planetologia Spaziali di Roma (IAPS-INAF), Via del Fosso del Cavaliere 100, 00133 Roma, Italy; roberto.viotti@iaps.inaf.it
4. Instituto de Ciencias Físicas, Universidad Nacional Autónoma de México, Ave. Universidad S/N, 62210 Cuernavaca, Mexico; gloria@astro.unam.mx
5. Associazione Romana Astrofili, Via Carlo Emanuele I, n°12A, 00185 Roma, Italy; m.calabresi@mclink.it
6. Physics Department, Università di Roma "La Sapienza", Piazza le Aldo Moro 5, 00185 Roma, Italy; corinne.rossi@uniroma1.it
7. INAF-Osservatorio Astronomico di Bologna, Via Ranzani 1, I-40127 Bologna, Italy; roberto.gualandi@inaf.it
* Correspondence: olga.maryeva@gmail.com
† Based on observations made with the Gran Telescopio Canarias (GTC), installed at the Spanish Observatorio del Roque de los Muchachos of the Instituto de Astrofisica de Canarias, in the island of La Palma, and with the Cassini 1.52-m telescope of the Bologna Observatory (Italy), as well as data retrieved from the public archive of the Special Astrophysical observatory of Russian Academy of Sciences (SAO RAS).

Received: 31 July 2019; Accepted: 13 September 2019; Published: 18 September 2019

Abstract: GR 290 (M 33 V0532 = Romano's Star) is a unique variable star in the M 33 galaxy, which simultaneously displays variability typical for luminous blue variable (LBV) stars and physical parameters typical for nitrogen-rich Wolf-Rayet (WR) stars (WN). As of now, GR 290 is the first object which is confidently classified as a post-LBV star. In this paper, we outline the main results achieved from extensive photometric and spectroscopic observations of the star: the structure and chemical composition of its wind and its evolution over time, the systematic increase of the bolometric luminosity during the light maxima, the circumstellar environment. These results show that the current state of Romano's Star constitutes a fundamental link in the evolutionary path of very massive stars.

Keywords: galaxies: individual (M 33); stars: individual (GR 290, M 33 V0532); stars: variables: S Doradus; stars: Wolf-Rayet; stars: evolution; stars: winds, outflows

1. Introduction

GR 290 (M 33 V0532 = Romano's Star)[1] is a variable star in M 33 galaxy discovered by Giuliano Romano [1] who originally constructed its light curve and classified it as a Hubble-Sandage variable based on its photometric properties. Later, in 1984, Peter Conti [2] introduced a new class of objects which assimilated Hubble-Sandage variables—luminous blue variables (LBV), and thus GR 290 became an LBV candidate [3,4]. This classification has later been supported by the spectroscopic [5] and photometric [6] studies, as well as by its large bolometric luminosity [7]. However, some arguments suggest that the objects is rather on a post-LBV stage already [8–10].

---

1. The object has coordinates $\alpha = 01:35:09.701$, $\delta = +30:41:57.17$ at J2000 epoch.

Romano's star displays both strong spectral and photometric variability, with several significant (about 1.5–2 mag) increases of brightness detected during its long monitoring (Polcaro et al. [10] and references therein). Such variability is typical for LBV stars, while in the Hertzsprung-Russell (H-R) GR 290 lies in Wolf-Rayet (WR) stars region, beyond LBV instability strip [10]. GR 290 is presently in a short, and thus very rare, transition phase between the LBV evolutionary phase and the nitrogen rich WR stellar class (WN). It is an extremely important target for studies of massive star evolution, especially the evolutionary link between LBVs, WR stars and supernovae (SNe).

In this paper we summarise the main results achieved in the study of Romano's star. We combine new studies of GR 290's vicinity (Section 2) with its updated century-long photometric light curve (Section 3). Then, based on spectral data, numerical simulations of its stellar atmosphere (Section 4) and the nebula surrounding it (Section 5), we discuss the current evolutionary stage of the star in Section 6.

2. Stellar Vicinity of Romano's Star

GR 290 is located in the outer spiral arm of the M 33 galaxy, and lies to the east of the OB 88 and OB 89 associations [11–13], located at 0.5 and 0.125 kpc projected distances[2], respectively. The most detailed information about photometry of stars in this area may be found in Massey et al. [15]. Figure 1 shows the identification chart of the object and its vicinity, with red symbols corresponding to the stars which were spectrally classified by Massey et al. [16]. Coordinates and spectral classes of the stars are listed in Table 1.

Massey and Johnson [17] found a couple of carbon-rich Wolf-Rayet (WC) stars in these associations, J013458.89+304129.0 (WC4) in OB 88 and J013505.37+304114.9 (WC4-5) in OB 89. Moreover, the OB 88 association contains the star J013500.30+304150.9 classified as an LBV candidate by Massey et al. [18], and later reclassified by Humphreys et al. [9] as a FeII emission-line star. The presence of evolved massive stars in the associations indicates that their age is close to that of GR 290 and that they might have a common origin. Therefore, it is quite reasonable to suppose that GR 290 might have been originally ejected from the OB 89 association. Then, assuming a median escape velocity for runaway stars of 40–200 km/s [19], this ejection would have to have occurred 3.0–0.6 Myr ago, which is consistent with the evolutionary age of GR 290 and with the age of the OB 89 association.

---

[2] The adopted distance to M 33 is 847 ± 61 kpc (distance module 24.64 ± 0.15) from Galleti et al. [14].

Table 1. Stars in the vicinity of GR 290 spectrally classified by Massey et al. [16] (and references therein). Names and coordinates are given according to Massey et al. [15]. The three stars also included in Table 2 are marked by boldface.

Name	Coordinates		Spectral Class	V [mag]	(B − V)
	α	δ			
J013449.49+304127.2	01:34:49.46	+30:41:27.1	YSG:	16.468	0.854
J013453.20+304242.8	01:34:53.17	+30:42:42.7	G/KV	17.031	0.906
J013453.97+304043.4	01:34:53.94	+30:40:43.3	RSG:	19.492	1.558
J013454.31+304109.8	01:34:54.28	+30:41:09.7	RSG	18.450	2.045
J013455.06+304114.4	01:34:55.03	+30:41:14.3	B0I+Neb	18.246	−0.103
J013457.20+304146.1	01:34:57.17	+30:41:46.0	B3I	18.872	−0.088
J013458.77+304151.7	01:34:58.74	+30:41:51.6	RSG	19.121	1.605
J013458.89+304129.0	01:34:58.86	+30:41:28.9	WC4	20.662	0.238
J013459.07+304154.9	01:34:59.04	+30:41:54.8	RSG	19.030	1.986
J013459.08+304142.8	01:34:59.05	+30:41:42.7	B2I:	19.306	−0.163
J013459.29+304128.0	01:34:59.26	+30:41:27.9	B0.5:I	18.822	−0.112
J013459.39+304201.2	01:34:59.36	+30:42:01.1	O8Iaf [a]	18.254	−0.142
J013459.81+304156.9	**01:34:59.78**	**+30:41:56.8**	**RSG:**	**19.156**	**1.531**
J013500.30+304150.9	01:35:00.27	+30:41:50.8	cLBV [b]	19.298	−0.073
J013500.32+304147.3	01:35:00.29	+30:41:47.2	B0-2I	20.995	−0.183
J013501.36+304149.6	01:35:01.33	+30:41:49.5	Late O/Early B	19.346	−0.279
J013501.71+304159.2	01:35:01.68	+30:41:59.1	B1.5Ia	18.076	−0.099
J013502.06+304034.2	01:35:02.03	+30:40:34.1	RSG	18.500	1.365
J013502.30+304153.7	01:35:02.27	+30:41:53.6	B0.5Ia	18.933	−0.099
J013505.37+304114.9	01:35:05.34	+30:41:14.8	WC4-5	19.061	−0.293
J013505.74+304101.9	**01:35:05.71**	**+30:41:01.8**	**O6III(f)+Neb**	**18.218**	**−0.207**
J013506.87+304149.8	01:35:06.84	+30:41:49.7	B0.5Ib	18.655	−0.181
J013507.43+304132.6	**01:35:07.40**	**+30:41:32.5**	**RSG**	**18.582**	**1.991**
J013507.53+304208.4	01:35:07.50	+30:42:08.3	RSG	19.961	1.739

[a] later classified as Of/late-WN by Humphreys et al. [20]; [b] later classified as a FeII emission-line star by Humphreys et al. [9].

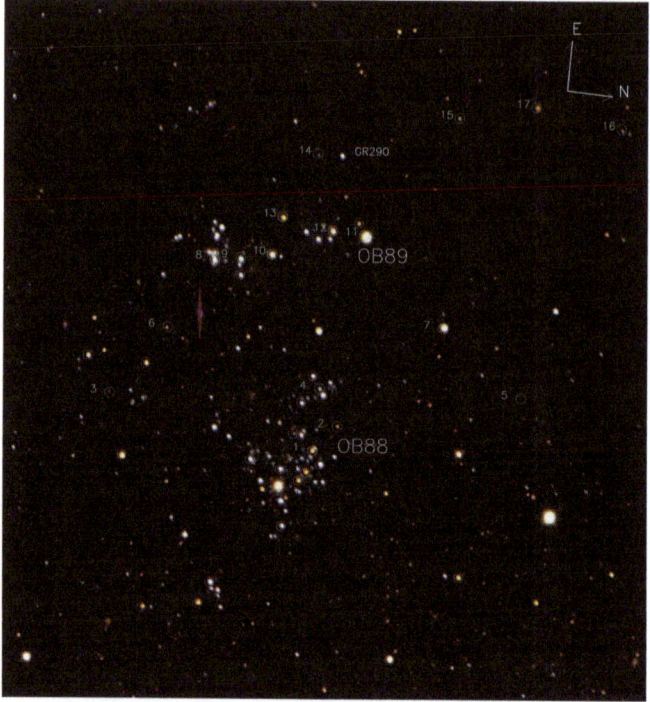

Figure 1. Identification chart of GR 290 vicinity and OB 88 and OB 89 associations. The colour picture is a combination of three direct images, with blue corresponding to B filter, green—to V and red—to R filter, all obtained with 2.5 m telescope of the Caucasian Mountain Observatory (CMO) of the Sternberg Astronomical Institute of Moscow State University. Green circles mark the stars studied in this work and red ones studied by Massey et al. [15,16]. Red squares are stars considered to be foreground objects by Massey et al. [15].

The field around GR 290 is not yet sufficiently explored as it consists mostly of faint ($V > 18$ mag) stars that require large telescopes for acquiring the spectra. Fortunately, some of the surrounding stars happened to lay on the slit during the long-slit observations of the object, thus that analysis of such data may provide additional information on the stellar contents and interstellar extinction in the vicinity of Romano's star. Therefore, we retrieved from General observational archive of Special Astrophysical observatory of Russian Academy of Sciences (SAO RAS)[3] all long-slit spectra of GR 290 obtained on Russian 6-m telescope with the Spectral Camera with Optical Reducer for Photometric and Interferometric Observations (SCORPIO) [21] during the years 2005–2016. We also utilised the spectra obtained with the OSIRIS spectrograph on the *Gran Telescopio Canarias (GTC)* and analysed by Maryeva et al. [22] and Maryeva et al. [23]. We reduced these spectra in a uniform way using the ScoRe package[4] initially created for the SCORPIO data reduction, and extracted the spectra of all stars crossed by the slit. To perform the spectral classification of these stars, we used an automatic code based on the χ^2 fitting with spectral standards from STELIB[5] (see Le Borgne et al. [24]) in the same way as used by Maryeva et al. [25]. The stars with spectra extracted and analysed in this way are

---

[3] General observational archive of Special Astrophysical observatory is available at https://www.sao.ru/oasis/cgi-bin/fetch?lang=en.
[4] ScoRE package available at http://www.sao.ru/hq/ssl/maryeva/score.htm.
[5] STELIB is availabte at http://webast.ast.obs-mip.fr/stelib.

marked with green circles in Figure 1, and their estimated spectral classes, measured positions and photometric magnitudes are listed in Table 2. The resulting spectra of the stars in flux units are shown in Figures A1–A5.

Table 2. The sample of stars in the field around GR 290 studied in this work. N corresponds to the labels in Figure 1. V and $(B-V)$ taken from Massey et al. [15].

N	Coordinates		Spectral Class	V [mag]	$(B-V)$	Instrument
	α	δ				
1 [a]	01:34:59.79	+30:41:56.9	RSG / M0-M1	19.156	1.531	OSIRIS
2	01:35:00.69	+30:42:07.5	RSG / K3-K4	20.507	1.613	OSIRIS
3	01:35:00.90	+30:40:18.2	RSG / K5-M0			SCORPIO
4	01:35:01.87	+30:41:57.3	B5-B7	19.980	−0.079	OSIRIS
5	01:35:02.37	+30:43:32.1	F:	21.773	0.059	OSIRIS
6	01:35:03.29	+30:40:42.5	RSG / K-M	20.339	1.088	SCORPIO
7	01:35:04.37	+30:42:53.1	G4-K1	16.459	0.733	SCORPIO
8 [b]	01:35:05.76	+30:41:02.2	star with em.lines			SCORPIO
9	01:35:05.87	+30:41:14.2	star with em.lines	18.522	−0.082	SCORPIO
10 [c]	01:35:06.14	+30:41:29.0	F4-F6 V	17.365	0.561	SCORPIO
11	01:35:07.09	+30:42:12.4	F4-F7 V	14.888	0.592	SCORPIO
12 [d]	01:35:07.15	+30:41:56.3	G8-K1	17.793	0.962	OSIRIS
13 [e]	01:35:07.40	+30:41:32.5	RSG / M0-M1	18.582	1.991	SCORPIO
14	01:35:09.63	+30:41:46.1	A9-F0	20.817	0.246	OSIRIS SCORPIO
15	01:35:11.40	+30:42:50.8	F4-G2 V	20.164	0.647	SCORPIO
16	01:35:11.66	+30:44:08.3	hot star	20.490	0.161	SCORPIO
17	01:35:12.05	+30:43:27.2	cool star	20.149	1.220	SCORPIO
	01:35:14.10	+30:44:23.3	hot star with abs.lines	17.654	−0.038	SCORPIO

[a,e] Stars classified as RSG by Massey et al. [16]; [b] star classified as O6III(f)+Neb by Massey et al. [16]; [c,d] stars classified as foreground objects by Massey et al. [15].

As we can see in Figure 1, our sample of stars partially intersect with the ones studied by Massey et al. [15,16]. We were able to refine the estimates of spectral classes for J013459.81+304156.9 and J013507.43+304132.6, classified earlier as just red supergiants (RSG) [16], as well as for J013506.17+304129.1 and J013507.18+304156.4 as foreground objects according to Massey et al. [15]. Our sample contains three more RSGs, which were not previously reported, and four hot stars, with only one (J013505.74+304101.9 with O6III(f)+Neb spectral class) known before [16]. Among three others, the spectrum of J013505.76+304102.21 displays the He I emission and strong nebular lines. The second one, J013514.1+304423.21, has a spectral slope corresponding to high temperature, and shows H and He absorption lines, while the last, J013501.87+304157.3, was preliminary classified as B5–B7 supergiant.

Knowing the spectral classes of these stars, and therefore their intrinsic colour indices, allows us to estimate the interstellar extinction around GR 290. Its value is comparable to the galactic foreground extinction value of $E_{(B-V)} = 0.052$ (according to the NED extinction calculator [26]). We did not register any star with higher reddening in the vicinity of GR 290.

3. Photometry

Photometric observations of GR 290 were initiated in the early 1960s by the Italian astronomer Giuliano Romano in the Asiago Observatory [1]. He obtained a light curve with the brightness of a star varying irregularly between $16^m\!.7$ and $18^m\!.1$, and classified it as a variable of the Hubble-Sandage type based on the shape of the light curve and GR 290's colour index.

Subsequent photometric investigations of GR 290 were undertaken by Kurtev et al. [6] and later by Zharova et al. [27]. The cumulative light curve derived in the latter work and covering half a century shows that GR 290 exhibits irregular light variations with different amplitudes and time scales [27]. The star shows large and intricate wave-like variations, with duration of the waves amounting to several years. In general, its variability is irregular, with the power spectrum fairly approximated by a red power-law spectrum [28] (i.e., the one dominated by a long timescale variations). Moreover, Kurtev et al. [6] discovered short-timescale variability with amplitude ∼0.^m5, which is also typical an LBV star.

Polcaro et al. [10] used various collections of photographic plates to further extend the historical light curve back to the beginning of the 20th century. The data between 1900 and 1950 suggest that no significant eruption took place during that half century. On the contrary, after 1960, two clear, long-term eruptions are evident (see Figure 2).

New photometric data, collected in Table 3 and shown as magenta dots in Figure 2, confirm the conclusion of Maryeva et al. [22] and Calabresi et al. [29] that the star has reached a long lasting visual minimum phase in 2013, and its brightness has been relatively stable since then.

Table 3. New photometric observations of GR 290 acquired by our group since Polcaro et al. [10].

Date	B		V		R		I		Obs [a]
31 July 2016	18.75	0.03	18.77	0.04	18.59	0.04	18.66	0.05	Loiano
4 August 2016	18.75	0.03	18.77	0.04	18.59	0.04	18.66	0.05	Loiano
29 October 2016			18.60	0.1					ARA
30 October 2016			18.80						Loiano
31 October 2016					18.68	0.15			ARA
28 December 2016			18.78	0.15	18.71	0.15			ARA
16 February 2017	18.70	0.04	18.68	0.04	18.59	0.05	18.73	0.06	Loiano
28 July 2017			18.80	0.04	18.62	0.05			Loiano
30 July 2017			18.87	0.12					ARA
17 December 2017			18.67	0.10					ARA
14 February 2018	18.69	0.05	18.77	0.06	18.64	0.08			RTT-150
19 August 2018			18.81	0.10					ARA
4 September 2018	18.53	0.04	18.67	0.04	18.54	0.04	19.48	0.04	CMO
11 September 2018			18.84	0.04	18.64	0.04			Loiano
13 September 2018	18.65	0.04	18.73	0.04	18.63	0.04	19.51	0.04	CMO
19 September 2018	18.74	0.05	18.83	0.04	18.70	0.04			Loiano
10 January 2019			18.84	0.06	18.74	0.05			Loiano

[a] observatories: Loiano: 1.52 m telescope at the Loiano station of the Bologna Astronomical; Observatory-INAF. ARA: 37 cm telescope of the Associazione Romana Astrofili at Frasso Sabino (Rieti); RTT-150: 1.5 m Russian–Turkish telescope. CMO: 2.5 m telescope of the Caucasian Mountain Observatory.

It is generally observed that, during the S Dor cycle, the colour of a typical LBV is bluer at the light minimum than close to the light maximum. In contrast, Polcaro et al. [10] demonstrated that $(B-V)$ colour of Romano's star is constant over time, within the error bars. There is no clear evidence for a variation of $(B-V)$ as a function of the visual magnitude, and our new photometry obtained after 2015 confirms this conclusion (see Figure 3). This is consistent with Romano's star being hotter (about 30,000 K) than a typical LBV, with the slope of optical spectrum defined by a Raleigh-Jeans power-law tail.

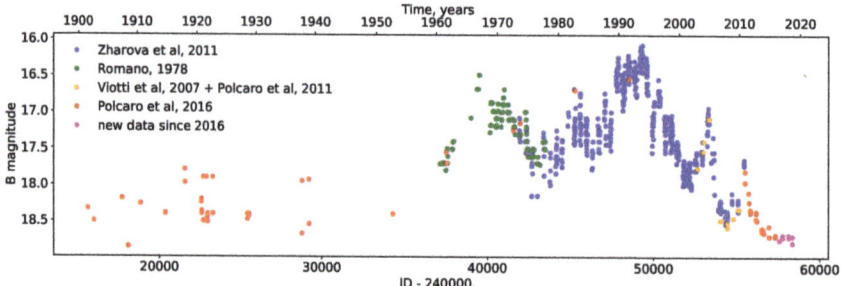

Figure 2. The historical light curve of GR 290 in the B-filter from 1901 to 2019.

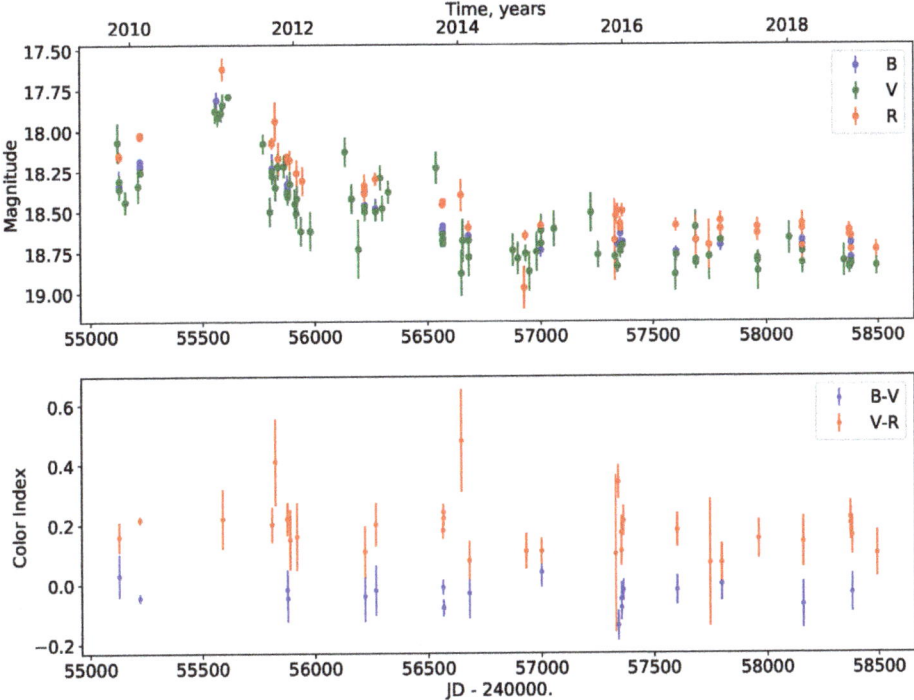

Figure 3. (**top**) Light curve of GR 290 in the B, V and R filters obtained by us between 2010 and 2019 and partially published in Polcaro et al. [10]. (**bottom**) $(B - V)$ and $(V - R)$ colour indices for the same time interval.

In other spectral ranges, the object is much less studied than in optical range. Only a single measurement of its magnitude is available in ultraviolet and infrared ranges, corresponding to different moments of time defined by a mean epoch of the individual survey (Table 4).

Table 4. Stellar magnitudes of GR 290 in ultraviolet and infrared range.

GALEX		2MASS			Spitzer	
Far-UV 1516 Å	Near-UV 2267 Å	J	H	K	3.6 μm	4.5 μm
17.692 ± 0.031	17.438 ± 0.013	16.834 ± 0.128	16.702 ± 0.292	16.657	16.335	15.921
[30]		[31]			[32]	

4. Spectroscopy and Determination of Physical Parameters

The first description of optical spectrum of Romano's star can be found in the article of Humphreys [33]. The spectrum was obtained in August 1978 at Kitt Peak National Observatory, when brightness of the star was $V = 18.00 \pm 0.02$ [33]. Humphreys noted: "Its spectrum shows emission lines of hydrogen and He I. There are no emission lines of Fe II or [Fe II]." and classified the star as a peculiar emission-line object[6] [33].

---

[6] In Humphreys [33], Romano's star is identified as B 601.

In 1992, T. Szeifert obtained a spectrum of Romano's star right before the historical maximum of its brightness. Szeifert [4] described it as "Few metal lines are visible, although a late B spectral type is most likely" (Figure 4). On the other hand, Sholukhova et al. [34] obtained the next spectrum in August 1994 and classified the star as a WN star candidate. Since 1998, regular observations of GR 290 carried out on the Russian 6m [5,35] and spectra published by Sholukhova et al. [35] indicate that the spectrum of GR 290 has not reverted to a B-type spectrum. Thus, Szeifert's [4] spectrum is unique and corresponds to the coldest and brightest state of the star measured so far.

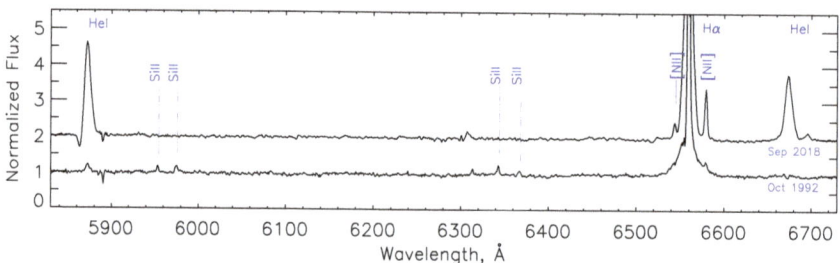

Figure 4. Comparison of normalized optical spectra of GR 290 obtained with Calar Alto/TWIN in October 1992 by Szeifert [4] and with GTC/OSIRIS in September 2018. Spectra are displaced vertically for illustrative purposes.

Studies of GR 290 devoted to its spectral variability show that its spectral type changes between WN11 and WN8 [8,35,36]. Since the beginning of the 2000s, it has made this transition twice [10]. Viotti et al. [37,38] first described an anticorrelation between equivalent width of 4600–4700 Å blend and the brightness. Later, Maryeva and Abolmasov [36] found a correlation of spectral changes and the visual brightness typical for LBVs: the brighter it is, the cooler the spectral type. However, as noted by Humphreys et al. [9], GR 290 does not exhibit S Dor like transitions to the cool state with an optically thick wind, but instead varies between two hot states characterised by WN spectroscopic features. Among all known LBVs, only HD 5980 [39] convincingly shows a hotter spectrum in the minimum of brightness. Other LBV stars showing WN-like spectrum in quiescent "hot" phase usually stop at colder spectral types such as WN11 (for example AG Car [40] and WS 1 [41,42]) or Ofpe/WN9 (for example R 127 [43] and HD 269582 [44]).

As already mentioned, since the autumn of 2013, GR 290 is in a minimum brightness state with $V = 18.7$–18.8 mag. Due to this, it has been challenging to obtain its spectra with good enough quality for wind speeds to be adequately estimated. In summer of 2016, GR 290 was observed with the Optical System for Imaging and low-Intermediate-Resolution Integrated Spectroscopy (OSIRIS) on the *Gran Telescopio Canarias (GTC)* [22]. These observations gave the best spectral resolutions and signal-to-noise ratios ever obtained for this object, and allowed to estimate an average radial velocity (RV) of the object, RV(GR 290) = -163 ± 32 km s^{-1}, which is consistent, within the uncertainties, with the heliocentric velocity -179 ± 3 km s^{-1} of M 33 galaxy.

New spectra of GR 290 were obtained with the OSIRIS spectrograph in September 2018 [23]. Detailed analysis of the spectra obtained in 2016 and 2018 did not reveal any changes (Figure 5). As before, the star displays a WN8h spectrum with forbidden nebular lines.

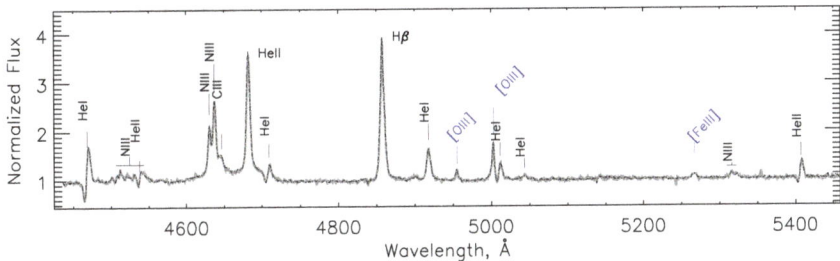

Figure 5. Comparison of normalised optical spectra of GR 290 obtained with GTC/OSIRIS in July 2016 (grey thick line) and September 2018 (black dash-dotted line). Spectra are nearly identical.

The large number of acquired spectra allows tracking the quantitative changes of physical parameters of the star over time. To do it, a numerical modeling of GR 290's atmosphere using CMFGEN code [45] was started by Maryeva and Abolmasov [46], who constructed models for two states—the luminosity maximum of 2005 and the minimum of brightness in 2008. Then, Clark et al. [47] estimated the parameters of GR 290 during the moderate luminosity maximum of 2010. Polcaro et al. [10] built nine models for the most representative spectra acquired between 2002 and 2014. The results of calculations from Polcaro et al. [10], Clark et al. [47] and Maryeva et al. [22] are summarised in Table 5, along with the parameters estimated using the spectrum of September 2018. Comparisons of observed spectra with corresponding models are shown in Figure 6.

Table 5. Derived properties of Romano's star at the moments corresponding to different acquired spectra. H/He indicates the hydrogen number fraction relative to helium, f is the filling factor of the stellar wind. Details of modeling may be found in [10,22,47].

Date	V [mag]	Sp. type	T_{eff} [kK]	$\log T_{eff}$	$R_{2/3}$ [$R_\odot$]	$L_* 10^5$ [$L_\odot$]	$\log L_*$ [$L_\odot$]	$\dot{M}_{cl} 10^{-5}$ [$M_\odot/yr$]	f	v_∞ [km/s]	H/He	Ref.
Oct. 2002	17.98	WN10h	28.1	4.45	31.6	5.6	5.75	1.9	0.15	250 ± 100	1.7	[10]
Feb. 2003	17.70	WN10.5h	28.0	4.45	37	7.5	5.875	2.2	0.15	250 ± 50	1.7	[10]
Jan. 2005	17.24	WN11h	23.6	4.37	54	8.2	5.91	3.5	0.15	250 ± 50	1.7	[10]
Sep. 2006	18.4	WN8h	30.7	4.49	24	4.6	5.66	1.3	0.15	250 ± 100	1.7	[10]
Oct. 2007	18.6	WN8h	33.5	4.53	20	4.5	5.65	1.55	0.15	370 ± 50	1.7	[10]
Dec. 2008	18.31	WN8h	31.6	4.50	23.5	5.0	5.7	1.9	0.15	370 ± 50	1.7	[10]
Oct. 2009	18.36	WN9h	31.6	4.50	23.8	5.1	5.7	1.7	0.15	300 ± 100	1.7	[10]
Sep. 2010 [a]	17.8	WN10h	26	4.41	41.5		5.85	2.18	0.25	265	1.5	[47]
Dec. 2010	17.95	WN10h	26.9	4.43	33	5.3	5.72	2.05	0.15	250 ± 100	1.7	[10]
Aug. 2014	18.74	WN8h	32.8	4.52	19	3.7	5.57	1.4	0.15	400 ± 100	1.7	[10]
Jul. 2016	18.77	WN8h	30.0	4.48	21	3.7	5.57	1.5	0.15	620 ± 50	2.2	[22]
Sep. 2018	18.77	WN8h	30.0	4.48	21	3.7	5.57	1.5	0.15	620 ± 50	2.2	

[a] Clark et al. [47] assumed a distance to M 33 of 964 kpc.

Numerical calculations show that the bolometric luminosity of GR 290 is variable, being higher during the phases of greater optical brightness [10,46]. At the same time, the wind structure of GR 290 also varies in correlation with brightness changes—the slow and dense wind at brightness maxima becomes faster and thinner at minima (Figure 7), and the effective temperature[7] of the star increases from 25 kK (with WN11h spectral type) during the maximum of 2005 year to 31–33 kK (WN8h) during the minima.

---

[7] Effective temperature is defined as a temperature at radius $R_{2/3}$, where the Rosseland optical depth is equal to 2/3.

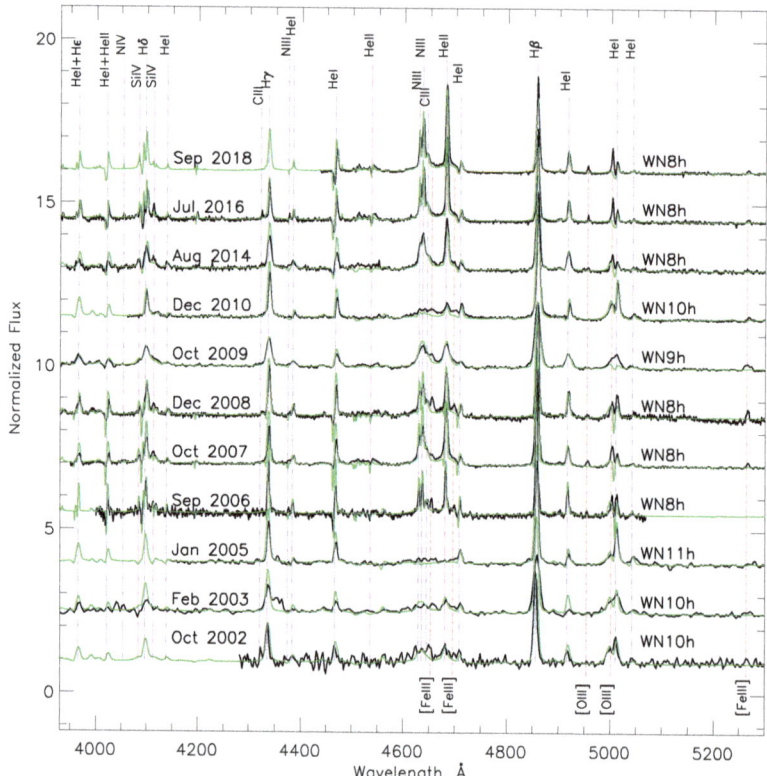

Figure 6. Normalised optical spectra of GR 290 compared with the best-fit CMFGEN models (green line). The model spectra are convolved with a Gaussian instrumental profile. Description of observational data may be found in [10,22,23]. Notice that "September 2006" spectrum was obtained by P. Massey with WIYN 3.5 m telescope [18]. Spectral types are estimated based primarily on relative strengths of N V, N IV, N III, N II and He II λ4686 emission lines [48]. Spectra are displaced vertically for illustrative purposes.

Figure 8 shows the positions of the star in the H-R diagram at different times. The object clearly moves well outside the typical LBV instability strip [49,50], deep inside the region of Wolf-Rayet stars, except for a moment of maximum brightness in 2005. On average, GR 290 lays on the 40–50 $M_\odot$ evolutionary tracks from the Geneva models [51] with rotation. Using CMFGEN, we found that hydrogen mass fraction in the atmosphere of GR 290 is 35% [22], and used this estimation for determination of current stellar mass and age. According to this tracks, the Romano's star should now be 4.5–5.7 Myr old and should have a mass of 27–38 $M_\odot$.

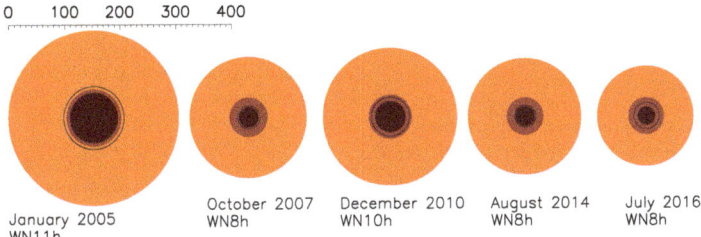

Figure 7. Change of the wind structure and extent over time. The region where $n_e \geq 10^{12}$ cm^{-3} is shown in dark red, $10^{12} \geq n_e \geq 10^{11}$ cm^{-3} in red, and $10^{11} \geq n_e \geq 10^{10}$ cm^{-3} in orange. Solid black line shows the radius where Rosseland optical depth (τ) is 2/3. Scale in units R$_\odot$ is shown at the top.

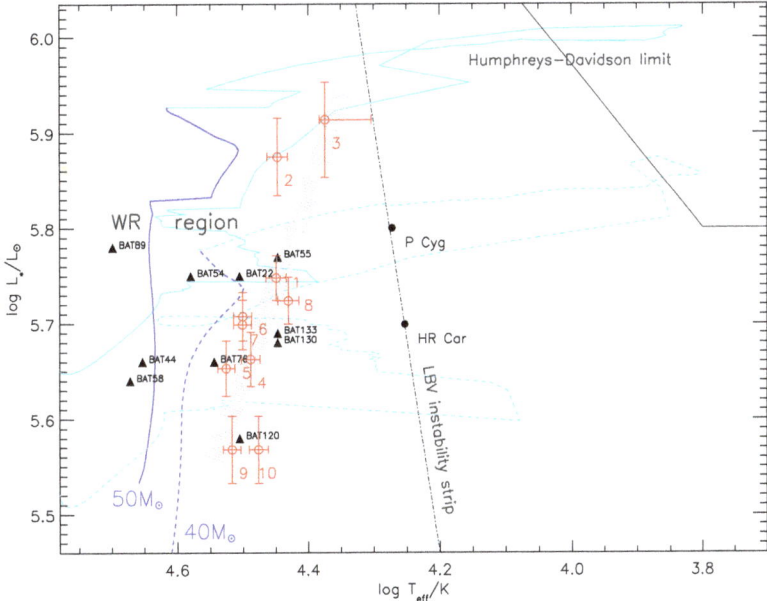

Figure 8. Position of GR 290 in the Hertzsprung-Russell diagram at different times. Numbers correspond to: (1) October 2002; (2) February 2003; (3) January 2005; (4) September 2006; (5) October 2007; (6) December 2008; (7) October 2009; (8) December 2010; (9) August 2014; and (10) July 2016 and September 2018. The hatched strip shows an average line along which GR 290 moved during its recent luminosity cycles. The Geneva tracks [51] for 40 M$_\odot$ (dashed line) and 50 M$_\odot$ (solid line) with rotation are shown by blue lines, with dark blue part corresponding to the hydrogen burning in stellar core. Triangles mark the positions of late-WN stars from Large Magellanic Cloud (LMC), whose data were taken from Hainich et al. [52]. In addition, the positions of LBV stars P Cygni and HR Car are shown with circles. Data for these objects were taken from the works of Najarro [53] and Groh et al. [54]. Grey solid line is Humphreys-Davidson limit [3], grey dash-dotted line is LBV minimum instability strip as defined in [54].

5. Nebula

The presence of forbidden lines [N II] 6548, 6584; [O III] 4959, 5007; [Fe III] 4658, 4701, 5270 and [Ar III] 7136 in the spectrum of GR 290 indicates that it has a nebula, but it is not resolved in direct imaging because of the large distance to M33. In 2005, Fabrika et al. [5] first attempted to detect the nebula and study its spatial structure using the panoramic (3D) spectroscopic data acquired on Russian 6-m telescope, and reported the discovery of an extended structure in the velocity field of Hβ line, with an angular extent of $\sim 9''$ (~ 30 pc) in the NE-SW direction. An excess corresponding to the dust circumstellar envelope around the object has not been detected in the infrared (IR) emission [9].

Maryeva et al. [22] performed a modeling of circumstellar nebula using using CLOUDY photoionisation code [55,56] and the spectrum acquired on GTC/OSIRIS in order to reproduce the observed nebular emission lines that are clearly seen in the spectrum. GR 290 was found to be surrounded by an unresolved compact H II region with a most probable outer radius R = 0.8 pc and a hydrogen density n_H = 160 cm^{-3}, and having chemical abundances that are consistent with those derived from the stellar wind lines. Hence, this compact H II region appears to be largely composed of material ejected from the star.

In addition, the recent analysis of the 2D spectra obtained with GTC/OSIRIS in September 2018 perpendicular to the dispersion at Hα line indicates that the nebula has extended and asymmetric structure [23]. Its size is about 25–30 pc, similar to typical H II regions around O-stars. Based on the similarity of sizes and evolutionary status of GR 290, we speculate that this extended nebula consists of material ejected during O-supergiant phase.

6. Conclusions

GR 290 is is located in the outer spiral arm of the M 33 galaxy at a projected distance of about 4 kpc from the centre. Its spatial location, the proximity to the OB 88 and OB 89 associations, and the similarity of their ages (about 4–5 Myr) as well as a basic concept that a large fraction of all stars, including massive stars, forms in clusters suggest the common origin of GR 290 and OB 89. It is tempting to suggest that GR 290 may have escaped from the association.

The evolution of LBVs during the S Dor cycles seems to occur in most cases roughly at constant bolometric luminosity (see, e.g., [3,57]). However, a decrease of bolometric luminosity from minimum towards the light maximum of the S Dor cycle were observed for several LBVs (e.g., S Dor [58] and AG Car [40]). Lamers [58] interpreted it in terms of the radiative power being partially transformed into mechanical power in order to expand the outer layers of the star from minimum to maximum. In contrast, spectral monitoring of Romano's star during its recent peaks of activity, and the numerical simulation of its stellar atmosphere based on acquired spectra, demonstrated that its bolometric luminosity varies in correlation with its visual brightness, i.e., L_{bol} increases during its visual luminosity maxima [10]. Guzik and Lovekin [59] discussed several mechanisms that could trigger the large outburst activity and variations in bolometric magnitude as observed in GR 290. An interesting possibility is that the interplay between pulsations and rotational mixing lead to an unstable transport of H-rich material to the nuclear burning core. In this context, GR 290 may be the ideal object for testing such theories.

The star is hotter than most other LBVs (Table 6), and lays outside of the LBV instability strip in the H-R diagram. On the other hand, the hydrogen abundance of the envelope appears higher than in late type WN stars, and therefore, from the evolutionary and structural point of view, GR 290 is less evolved than WN8h stars [10]. This suggests that Romano's star may be a post-LBV object, the transition phase between LBVs and Wolf-Rayet stars.

The century long light curve of Romano's star shows that until the 1960s the object was in a long lasting quasi-stationary state, a state to which it has returned in 2013, and since then displaying a WN8h spectrum. While the spectral type during the early "low" state (pre-1960) is unknown, from the observed correlation between the visual magnitude and spectral type, we may suggest that it also was WN8h. The Galactic WN8 stars are known to be significantly more variable than the WRs with hotter spectral types [60]. Thus, it is tempting to speculate on the possibility that, in analogy with GR 290, other WN8s may have just recently passed through the LBV phase. Hence, a systematic investigation of archival data and constructing century long light curves for WN8-WN9 stars using archival photographic plates will probably be able to uncover more objects similar to Romano's star.

Table 6. Comparison of Romano's star with other LBVs and LBV candidates which show WR like spectra.

Star	Sp.type	Ref.	Comments
Wray 15-751	O9.5 I	[61]	LBVc, Milky Way
Sk−69° 279	O9.2 Iaf	[62]	ex-/dormant LBV [63], BSG evolved off the Main Sequence [62]
Hen 3-519	WN11	[64]	ex-/dormant LBV [63], there are no significant changes of brightness [65]
AG Car	WN11	[40]	LBV, Milky Way
WS 1	WN11	[41,42]	LBV, Milky Way
R 127	Ofpe/WN9	[43]	LBV, Large Magellanic Cloud
HD 269582	Ofpe/WN9	[44]	LBV, Large Magellanic Cloud
GR 290	WN8	[10,22]	post-LBV, M33
HD 5980	LBV+WN4+OI	[66,67]	LBV, Small Magellanic Cloud

Author Contributions: R.F.V., spectral analysis; M.C., C.R. and R.G., photometric monitoring and reduction of photometry data; and O.M., numerical modeling of stellar atmosphere and reduced the spectroscopic material and manuscript preparation. G.K. was PI of the 2016 and 2018 GranTeCan observations and performed spectral analysis. All authors discussed the results and commented on the manuscript.

Funding: This research was funded by CONACYT grant 252499, UNAM/PAPIIT grant IN103619, Russian Foundation for Basic Research grant 19-02-00779 and Czech Science Foundation grant GA18-05665S. This project received funding from the European Union's Framework Programme for Research and Innovation Horizon 2020 (2014-2020) under the Marie Skłodowska-Curie Grant Agreement No. 823734.

Acknowledgments: We express our enormous gratitude to V.F. Polcaro, who recently passed away, for having stimulated our interest and studies of this unique object. We thank Roman Zhuchkov, Oleg Egorov and Olga Vozyakova for obtaining the photometric observations on 1.5 m Russian–Turkish telescope and 2.5 m telescope of the Caucasian Mountain Observatory. We thank Thomas Szeifert and Philip Massey for the spectra obtained with Calar Alto/TWIN spectrograph in 1992 and with WIYN 3.5 m telescope in September 2006. We thank the GTC observatory staff for obtaining the spectra and Antonio Cabrera-Lavers for guidance in processing the observations. We thank Guest Editor Prof. Roberta M. Humphreys and our anonymous referees for providing helpful comments and suggestions. In this paper, we use data taken from the public archive of the SAO RAS. The work is partially based on the observation at 2.5-m CMO telescope that is supported by M.V. Lomonosov Moscow State University Program of Development.

Conflicts of Interest: The authors declare no conflict of interest.

Abbreviations

The following abbreviations are used in this manuscript:

CMO	2.5 m telescope of the Caucasian Mountain Observatory of the Sternberg Astronomical Institute of Moscow State University
GTC	Gran Telescopio Canarias
LBV	Luminous blue variable
BSG	Blue supergiant
RSG	Red supergiant
SAO RAS	Special Astrophysical observatory of Russian Academy of Sciences
WR	Wolf-Rayet
WN	Nitrogen-rich Wolf-Rayet stars
WC	Carbon-rich Wolf-Rayet stars
YHG	Yellow hypergiant

Appendix A. Spectra of Stars in Vicinity of GR 290

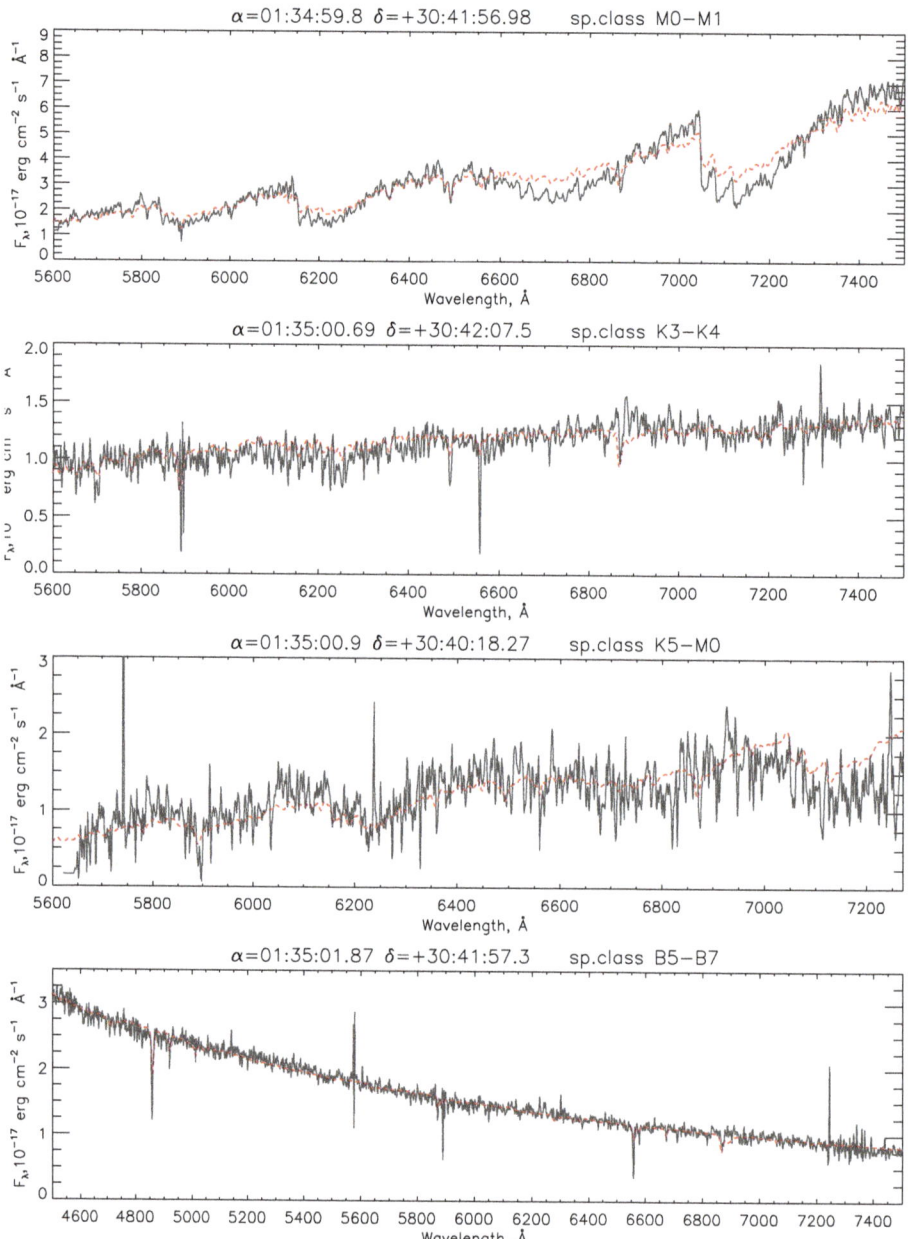

Figure A1. From the top panel downwards, spectra of the stars: J013459.8+304156.98, J013500.69+304207.5, J013500.9+304018.27 and J013501.87+304157.3. For comparison, the reddened spectra of HD 42543 (M1 Ia-ab), HD 154733 (K3 III), HD 146051 (M0.5 III) and HD 164353 (B5 Ib) are shown by red dashed lines.

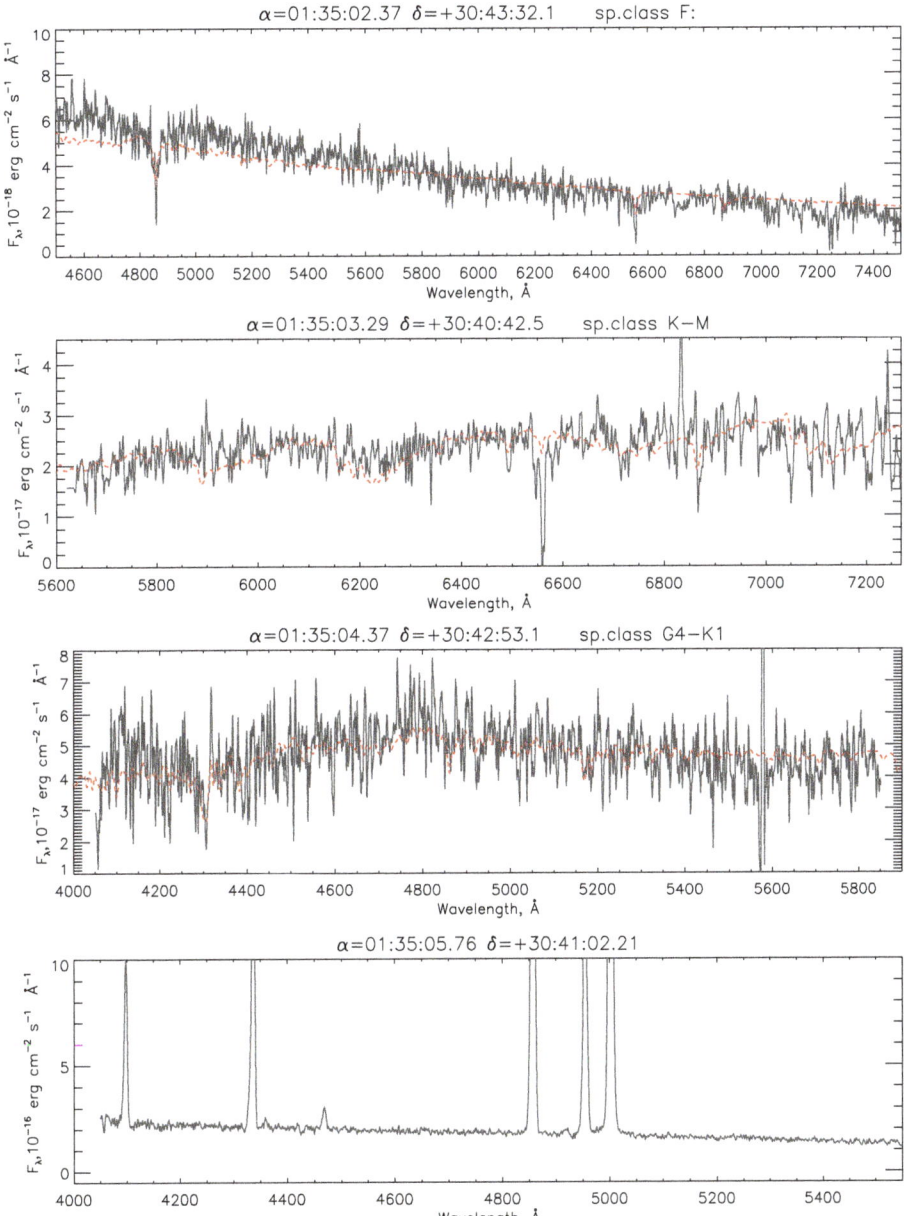

Figure A2. From the top panel downwards, spectra of the stars: J013502.37+304332.1, J013503.29+304042.5, J013504.37+304253.1 and J013505.76+304102.21. For comparison, the reddened spectra of HD 128167 (F2 V), HD 102212 (M1 III) and HD 135722 (G8 III) are shown by red dashed lines.

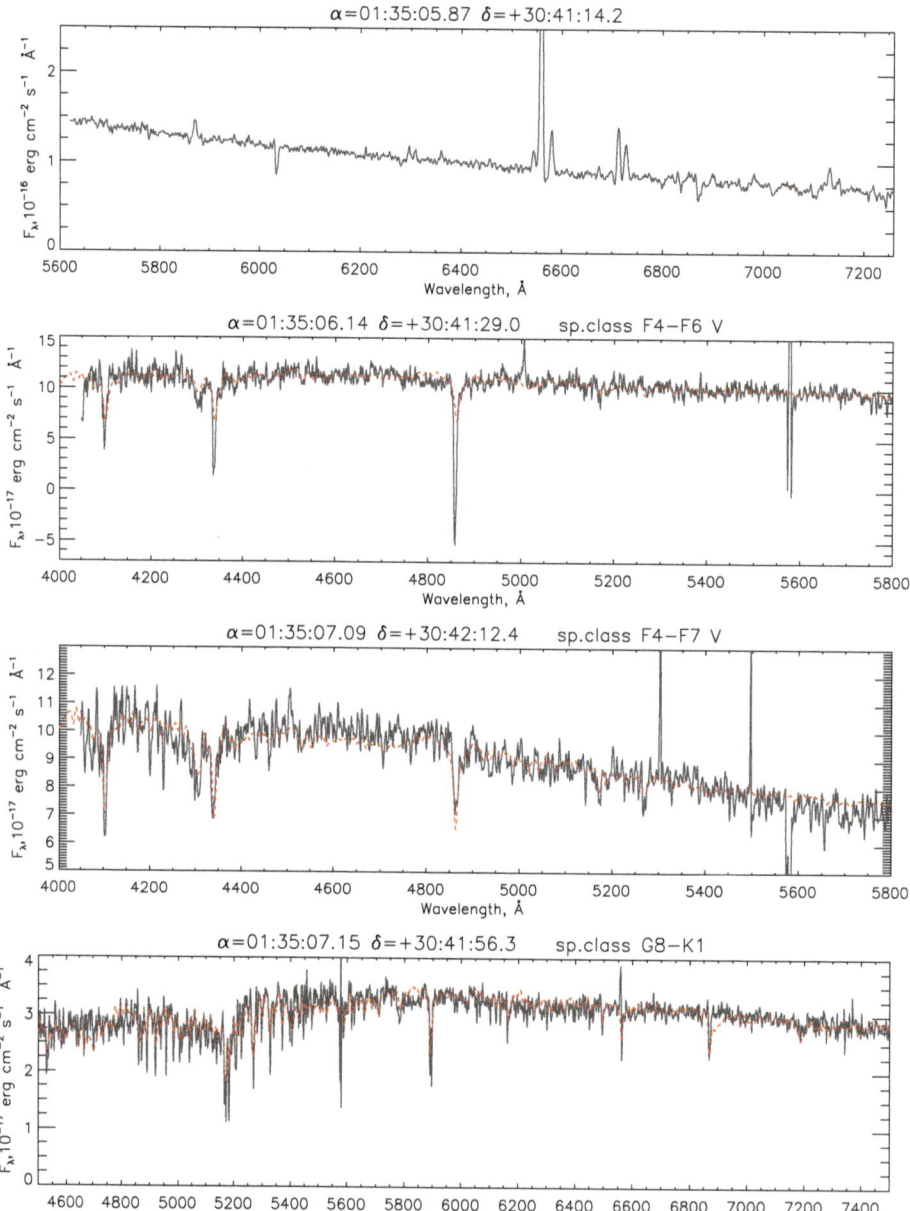

Figure A3. From the top panel downwards, spectra of the stars: J013505.87+304114.2, J013506.14+304129.0, J013507.09+304212.4 and J013507.15+304156.3. For comparison, the reddened spectra of HD 126141 (F5 V), HD 101606 (F4 V) and HD 75532 (G8 V) are shown by red dashed lines.

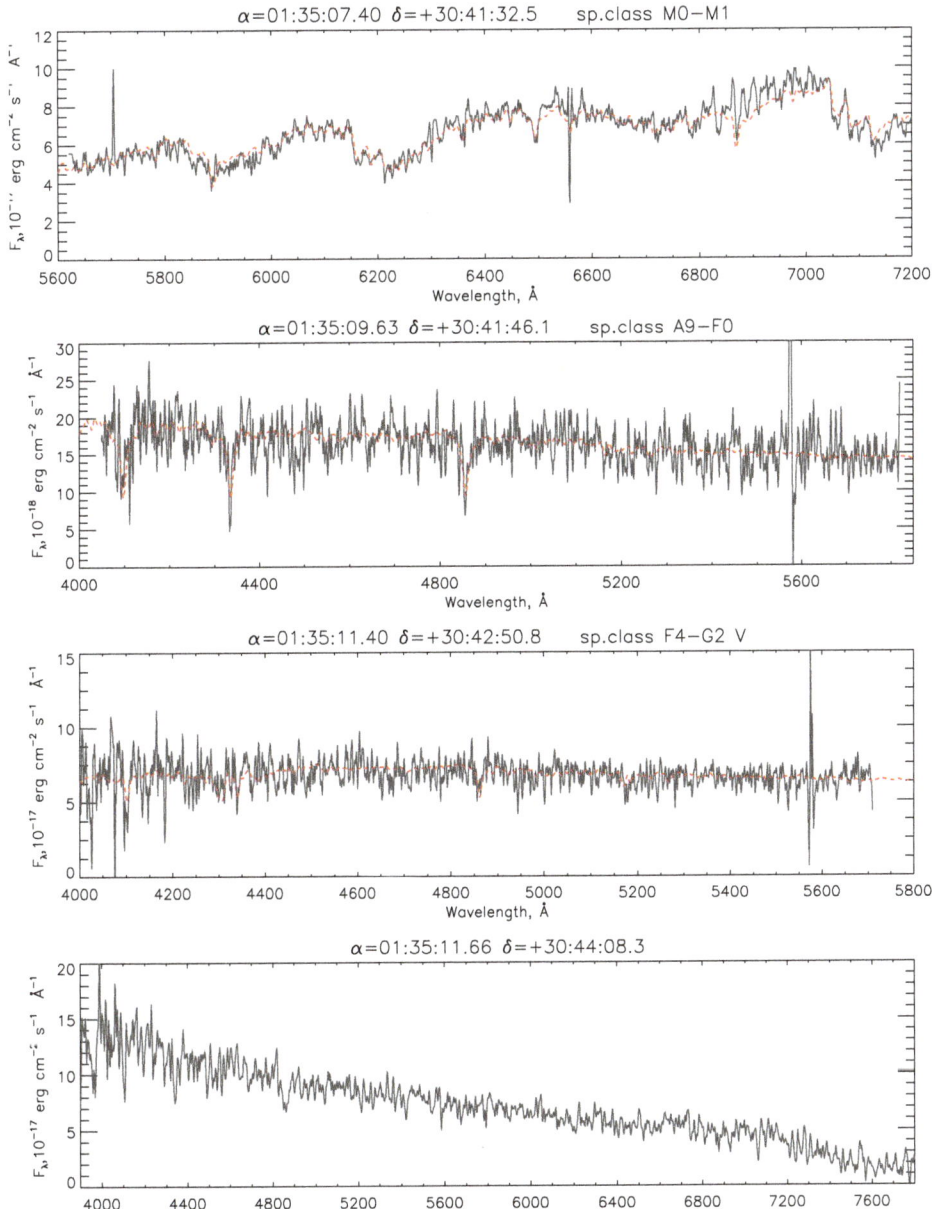

Figure A4. From the top panel downwards, spectra of the stars: J013507.40+304132.5, J013509.63+304146.1, J013511.40+304250.8 and J013511.66+304408.3. For comparison, the reddened spectra of HD 146051 (M0.5 III), HD 50420 (A9 III) and HD 134169 (G1 V) are shown by red dashed lines.

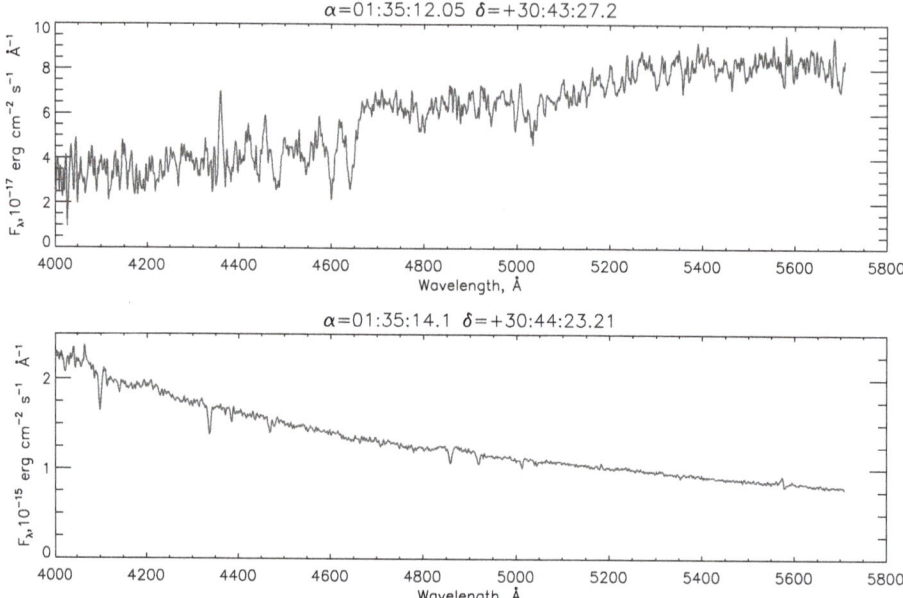

Figure A5. From the top panel downwards, spectra of the stars: J013512.05+304327.2 and J013514.1+304423.21.

References

1. Romano, G. A new variable star in M33. *Astron. Astrophys.* **1978**, *67*, 291–292.
2. Conti, P.S. Basic Observational Constraints on the Evolution of Massive Stars. In *Symposium-International Astronomical Union*; Observational Tests of the Stellar Evolution Theory; Maeder, A., Renzini, A., Eds.; Cambridge University Press: Cambridge, UK, 1984; Volume 105, p. 233.
3. Humphreys, R.M.; Davidson, K. The luminous blue variables: Astrophysical geysers. *Publ. ASP* **1994**, *106*, 1025–1051. [CrossRef]
4. Szeifert, T. LBVs and a late WN-star in M 31 and M 33. In *Liege International Astrophysical Colloquia*; Vreux, J.M., Detal, A., Fraipont-Caro, D., Gosset, E., Rauw, G., Eds.; Université de Liège. Institut d'astrophysique: Liège, Belgium, 1996; Volume 33, p. 459.
5. Fabrika, S.; Sholukhova, O.; Becker, T.; Afanasiev, V.; Roth, M.; Sanchez, S.F. Crowded field 3D spectroscopy of LBV candidates in M 33. *Astron. Astrophys.* **2005**, *437*, 217–226. [CrossRef]
6. Kurtev, R.; Sholukhova, O.; Borissova, J.; Georgiev, L. Romano's Star in M33: LBV Candidate or LBV? *Rev. Mex.* **2001**, *37*, 57–61.
7. Polcaro, V.F.; Gualandi, R.; Norci, L.; Rossi, C.; Viotti, R.F. The LBV nature of Romano's star (GR 290) in M 33. *Astron. Astrophys.* **2003**, *411*, 193–196. [CrossRef]
8. Polcaro, V.F.; Rossi, C.; Viotti, R.F.; Galleti, S.; Gualandi, R.; Norci, L. Optical Spectrophotometric Monitoring of the Extreme Luminous Blue Variable Star GR 290 (Romano's Star) in M 33. *Astron. J.* **2011**, *141*, 18. [CrossRef]
9. Humphreys, R.M.; Weis, K.; Davidson, K.; Bomans, D.J.; Burggraf, B. Luminous and Variable Stars in M31 and M33. II. Luminous Blue Variables, Candidate LBVs, Fe II Emission Line Stars, and Other Supergiants. *Astrophys. J.* **2014**, *790*, 48. [CrossRef]
10. Polcaro, V.F.; Maryeva, O.; Nesci, R.; Calabresi, M.; Chieffi, A.; Galleti, S.; Gualandi, R.; Haver, R.; Mills, O.F.; Osborn, W.H.; Pasquali, A.; Rossi, C.; Vasilyeva, T.; Viotti, R.F. GR 290 (Romano's Star). II. Light History and Evolutionary State. *Astron. J.* **2016**, *151*, 149. [CrossRef]

11. Humphreys, R.M.; Sandage, A. On the stellar content and structure of the spiral Galaxy M33. *Astrophys. J. Suppl. Ser.* **1980**, *44*, 319–381. [CrossRef]
12. Ivanov, G.R.; Freedman, W.L.; Madore, B.F. A catalog of blue and red supergiants in M33. *Astrophys. J. Suppl. Ser.* **1993**, *89*, 85–122. [CrossRef]
13. Massey, P.; Armandroff, T.E.; Pyke, R.; Patel, K.; Wilson, C.D. Hot, Luminous Stars in Selected Regions of NGC 6822, M31, and M33. *Astron. J.* **1995**, *110*, 2715. [CrossRef]
14. Galleti, S.; Bellazzini, M.; Ferraro, F.R. The distance of M 33 and the stellar population in its outskirts. *Astron. Astrophys.* **2004**, *423*, 925–934. [CrossRef]
15. Massey, P.; Neugent, K.F.; Smart, B.M. A Spectroscopic Survey of Massive Stars in M31 and M33. *Astron. J.* **2016**, *152*, 62. [CrossRef]
16. Massey, P.; Olsen, K.A.G.; Hodge, P.W.; Strong, S.B.; Jacoby, G.H.; Schlingman, W.; Smith, R.C. A Survey of Local Group Galaxies Currently Forming Stars. I. UBVRI Photometry of Stars in M31 and M33. *Astron. J.* **2006**, *131*, 2478–2496. [CrossRef]
17. Massey, P.; Johnson, O. Evolved Massive Stars in the Local Group. II. A New Survey for Wolf-Rayet Stars in M33 and Its Implications for Massive Star Evolution: Evidence of the "Conti Scenario" in Action. *Astrophys. J.* **1998**, *505*, 793–827. [CrossRef]
18. Massey, P.; McNeill, R.T.; Olsen, K.A.G.; Hodge, P.W.; Blaha, C.; Jacoby, G.H.; Smith, R.C.; Strong, S.B. A Survey of Local Group Galaxies Currently Forming Stars. III. A Search for Luminous Blue Variables and Other Hα Emission-Line Stars. *Astron. J.* **2007**, *134*, 2474–2503. [CrossRef]
19. Perets, H.B.; Šubr, L. The Properties of Dynamically Ejected Runaway and Hyper-runaway Stars. *Astrophys. J.* **2012**, *751*, 133. [CrossRef]
20. Humphreys, R.M.; Gordon, M.S.; Martin, J.C.; Weis, K.; Hahn, D. Luminous and Variable Stars in M31 and M33. IV. Luminous Blue Variables, Candidate LBVs, B[e] Supergiants, and the Warm Hypergiants: How to Tell Them Apart. *Astrophys. J.* **2017**, *836*, 64. [CrossRef]
21. Afanasiev, V.L.; Moiseev, A.V. The SCORPIO Universal Focal Reducer of the 6-m Telescope. *Astron. Lett.* **2005**, *31*, 194–204. [CrossRef]
22. Maryeva, O.; Koenigsberger, G.; Egorov, O.; Rossi, C.; Polcaro, V.F.; Calabresi, M.; Viotti, R.F. Wind and nebula of the M 33 variable GR 290 (WR/LBV). *Astron. Astrophys.* **2018**, *617*, A51. [CrossRef]
23. Maryeva, O.; Koenigsberger, G.; Karpov, S.; Lozinskaya, T.; Egorov, O.; Rossi, C.; Calabresi, M.; Viotti, R.F. Asymmetrical nebulae of the M 33 variable GR 290 (WR/LBV). 2019, in preparation.
24. Le Borgne, J.F.; Bruzual, G.; Pelló, R.; Lançon, A.; Rocca-Volmerange, B.; Sanahuja, B.; Schaerer, D.; Soubiran, C.; Vílchez-Gómez, R. STELIB: A library of stellar spectra at R~2000. *Astron. Astrophys.* **2003**, *402*, 433–442. [CrossRef]
25. Maryeva, O.V.; Chentsov, E.L.; Goranskij, V.P.; Dyachenko, V.V.; Karpov, S.V.; Malogolovets, E.V.; Rastegaev, D.A. On the nature of high reddening of Cygnus OB2 #12 hypergiant. *Mon. Not. R. Astron. Soc.* **2016**, *458*, 491–507. [CrossRef]
26. Schlegel, D.J.; Finkbeiner, D.P.; Davis, M. Maps of Dust Infrared Emission for Use in Estimation of Reddening and Cosmic Microwave Background Radiation Foregrounds. *Astrophys. J.* **1998**, *500*, 525–553. [CrossRef]
27. Zharova, A.; Goranskij, V.; Sholukhova, O.N.; Fabrika, S.N. V532 in M33. *Peremennye Zvezdy Prilozhenie* **2011**, *11*, 11.
28. Abolmasov, P. Stochastic variability of luminous blue variables. *New Astron.* **2011**, *16*, 421–429. [CrossRef]
29. Calabresi, M.; Rossi, C.; Gualandi, R.; Galeti, S.; Polcaro, V.F.; Viotti, R.; Albanesi, R.; Anzellini, F.; Haver, R.; Caponetto, P.; Gorelli, R. New deep minimum of Romano's Star in M33. *Astron. Telegr.* **2014**, *5846*, 1.
30. Mudd, D.; Stanek, K.Z. GALEX catalogue of UV point sources in M33. *Mon. Not. R. Astron. Soc.* **2015**, *450*, 3811–3821. [CrossRef]
31. Skrutskie, M.F.; Cutri, R.M.; Stiening, R.; Weinberg, M.D.; Schneider, S.; Carpenter, J.M.; Beichman, C.; Capps, R.; Chester, T.; Elias, J.; et al. The Two Micron All Sky Survey (2MASS). *Astron. J.* **2006**, *131*, 1163–1183. [CrossRef]
32. McQuinn, K.B.W.; Woodward, C.E.; Willner, S.P.; Polomski, E.F.; Gehrz, R.D.; Humphreys, R.M.; van Loon, J.T.; Ashby, M.L.N.; Eicher, K.; Fazio, G.G. The M33 Variable Star Population Revealed by Spitzer. *Astrophys. J.* **2007**, *664*, 850–861. [CrossRef]
33. Humphreys, R.M. Studies of luminous stars in nearby galaxies. VI. The brightest supergiants and the distance to M 33. *Astrophys. J.* **1980**, *241*, 587–597. [CrossRef]

34. Sholukhova, O.N.; Fabrika, S.N.; Vlasyuk, V.V.; Burenkov, A.N. Spectroscopy of Hα-emission blue stars in M33. *Astron. Lett.* **1997**, *23*, 458–464. [CrossRef]
35. Sholukhova, O.N.; Fabrika, S.N.; Zharova, A.V.; Valeev, A.F.; Goranskij, V.P. Spectral variability of LBV star V 532 (Romano's star). *Astrophys. Bull.* **2011**, *66*, 123–143. [CrossRef]
36. Maryeva, O.; Abolmasov, P. Spectral Variability of Romano's Star. *Rev. Mex.* **2010**, *46*, 279–290.
37. Viotti, R.F.; Rossi, C.; Polcaro, V.F.; Montagni, F.; Gualandi, R.; Norci, L. The present status of four luminous variables in M 33. *Astron. Astrophys.* **2006**, *458*, 225–234. [CrossRef]
38. Viotti, R.F.; Galleti, S.; Gualandi, R.; Montagni, F.; Polcaro, V.F.; Rossi, C.; Norci, L. The 2006 hot phase of Romano's star (GR 290) in M 33. *Astron. Astrophys.* **2007**, *464*, L53–L55. [CrossRef]
39. Georgiev, L.; Koenigsberger, G.; Hillier, D.J.; Morrell, N.; Barbá, R.; Gamen, R. Wind Structure and Luminosity Variations in the Wolf-Rayet/Luminous Blue Variable HD 5980. *Astron. J.* **2011**, *142*, 191. [CrossRef]
40. Groh, J.H.; Hillier, D.J.; Damineli, A.; Whitelock, P.A.; Marang, F.; Rossi, C. On the Nature of the Prototype Luminous Blue Variable Ag Carinae. I. Fundamental Parameters During Visual Minimum Phases and Changes in the Bolometric Luminosity During the S-Dor Cycle. *Astrophys. J.* **2009**, *698*, 1698–1720. [CrossRef]
41. Kniazev, A.Y.; Gvaramadze, V.V.; Berdnikov, L.N. WS1: one more new Galactic bona fide luminous blue variable. *Mon. Not. R. Astron. Soc.* **2015**, *449*, L60–L64. [CrossRef]
42. Gvaramadze, V.V.; Kniazev, A.Y.; Miroshnichenko, A.S.; Berdnikov, L.N.; Langer, N.; Stringfellow, G.S.; Todt, H.; Hamann, W.R.; Grebel, E.K.; Buckley, D.; et al. Discovery of two new Galactic candidate luminous blue variables with Wide-field Infrared Survey Explorer. *Mon. Not. R. Astron. Soc.* **2012**, *421*, 3325–3337. [CrossRef]
43. Walborn, N.R.; Stahl, O.; Gamen, R.C.; Szeifert, T.; Morrell, N.I.; Smith, N.; Howarth, I.D.; Humphreys, R.M.; Bond, H.E.; Lennon, D.J. A Three-Decade Outburst of the LMC Luminous Blue Variable R127 Draws to a Close. *Astrophys. J. Lett.* **2008**, *683*, L33. [CrossRef]
44. Walborn, N.R.; Gamen, R.C.; Morrell, N.I.; Barbá, R.H.; Fernández Lajús, E.; Angeloni, R. Active Luminous Blue Variables in the Large Magellanic Cloud. *Astron. J.* **2017**, *154*, 15. [CrossRef]
45. Hillier, D.J.; Miller, D.L. The Treatment of Non-LTE Line Blanketing in Spherically Expanding Outflows. *Astrophys. J.* **1998**, *496*, 407–427. [CrossRef]
46. Maryeva, O.; Abolmasov, P. Modelling the optical spectrum of Romano's star. *Mon. Not. R. Astron. Soc.* **2012**, *419*, 1455–1464. [CrossRef]
47. Clark, J.S.; Castro, N.; Garcia, M.; Herrero, A.; Najarro, F.; Negueruela, I.; Ritchie, B.W.; Smith, K.T. On the nature of candidate luminous blue variables in M 33. *Astron. Astrophys.* **2012**, *541*, A146. [CrossRef]
48. Smith, L.J.; Crowther, P.A.; Prinja, R.K. A study of the luminous blue variable candidate He 3-519 and its surrounding nebula. *Astron. Astrophys.* **1994**, *281*, 833–854.
49. Wolf, B. Empirical amplitude-luminosity relation of S Doradus variables and extragalactic distances. *Astron. Astrophys.* **1989**, *217*, 87–91.
50. Clark, J.S.; Larionov, V.M.; Arkharov, A. On the population of galactic Luminous Blue Variables. *Astron. Astrophys.* **2005**, *435*, 239–246. [CrossRef]
51. Ekström, S.; Georgy, C.; Eggenberger, P.; Meynet, G.; Mowlavi, N.; Wyttenbach, A.; Granada, A.; Decressin, T.; Hirschi, R.; Frischknecht, U.; et al. Grids of stellar models with rotation. I. Models from 0.8 to 120 M$_\odot$ at solar metallicity (Z = 0.014). *Astron. Astrophys.* **2012**, *537*, A146. [CrossRef]
52. Hainich, R.; Rühling, U.; Todt, H.; Oskinova, L.M.; Liermann, A.; Gräfener, G.; Foellmi, C.; Schnurr, O.; Hamann, W.R. The Wolf-Rayet stars in the Large Magellanic Cloud. A comprehensive analysis of the WN class. *Astron. Astrophys.* **2014**, *565*, A27. [CrossRef]
53. Najarro, F. Spectroscopy of P Cygni. In *P Cygni 2000: 400 Years of Progress*; Astronomical Society of the Pacific Conference Series; de Groot, M., Sterken, C., Eds.; The Astronomical Society of the Pacific: San Francisco, CA, USA, 2001; Volume 233, p. 133.
54. Groh, J.H.; Damineli, A.; Hillier, D.J.; Barbá, R.; Fernández-Lajús, E.; Gamen, R.C.; Moisés, A.P.; Solivella, G.; Teodoro, M. Bona Fide, Strong-Variable Galactic Luminous Blue Variable Stars are Fast Rotators: Detection of a High Rotational Velocity in HR Carinae. *Astrophys. J. Lett.* **2009**, *705*, L25–L30. [CrossRef]
55. Ferland, G.J.; Korista, K.T.; Verner, D.A.; Ferguson, J.W.; Kingdon, J.B.; Verner, E.M. CLOUDY 90: Numerical Simulation of Plasmas and Their Spectra. *Publ. ASP* **1998**, *110*, 761–778. [CrossRef]

56. Ferland, G.J.; Chatzikos, M.; Guzmán, F.; Lykins, M.L.; van Hoof, P.A.M.; Williams, R.J.R.; Abel, N.P.; Badnell, N.R.; Keenan, F.P.; Porter, R.L.; Stancil, P.C. The 2017 Release Cloudy. *Revista Mexicana de Astronomiá y Astrofiśica* **2017**, *53*, 385–438.
57. De Koter, A.; Lamers, H.J.G.L.M.; Schmutz, W. Variability of luminous blue variables. II. Parameter study of the typical LBV variations. *Astron. Astrophys.* **1996**, *306*, 501.
58. Lamers, H.J.G.L.M. Observations and Interpretation of Luminous Blue Variables. In *International Astronomical Union Colloquium*; Astronomical Society of the Pacific Conference Series; Stobie, R.S., Whitelock, P.A., Eds.; Cambridge University Press: Cambridge, UK, 1995; Volume 83, p. 176.
59. Guzik, J.A.; Lovekin, C.C. Pulsations and Hydrodynamics of Luminous Blue Variable Stars. *Astron. Rev.* **2012**, *7*, 13–47. [CrossRef]
60. Moffat, A.F.J. Wolf-Rayet Stars in the Magellanic Clouds. VII. Spectroscopic Binary Search among the WNL Stars and the WN6/7–WN8/9 Dichotomy. *Astrophys. J.* **1989**, *347*, 373. [CrossRef]
61. Sterken, C.; van Genderen, A.M.; Plummer, A.; Jones, A.F. Wra 751, a luminous blue variable developing an S Doradus cycle. *Astron. Astrophys.* **2008**, *484*, 463–467. [CrossRef]
62. Gvaramadze, V.V.; Kniazev, A.Y.; Maryeva, O.V.; Berdnikov, L.N. Optical spectroscopy of the blue supergiant Sk−69° 279 and its circumstellar shell with SALT. *Mon. Not. R. Astron. Soc.* **2018**, *474*, 1412–1425. [CrossRef]
63. van Genderen, A.M. S Doradus variables in the Galaxy and the Magellanic Clouds. *Astron. Astrophys.* **2001**, *366*, 508–531. [CrossRef]
64. Toalá, J.A.; Guerrero, M.A.; Ramos-Larios, G.; Guzmán, V. WISE morphological study of Wolf-Rayet nebulae. *Astron. Astrophys.* **2015**, *578*, A66. [CrossRef]
65. Davidson, K.; Humphreys, R.M.; Hajian, A.; Terzian, Y. He 3-519—A peculiar post-LBV, pre-WN star? *Astrophys. J.* **1993**, *411*, 336–341. [CrossRef]
66. Foellmi, C.; Koenigsberger, G.; Georgiev, L.; Toledano, O.; Marchenko, S.V.; Massey, P.; Dall, T.H.; Moffat, A.F.J.; Morrell, N.; Corcoran, M.; et al. New insights into the nature of the SMC WR/LBV binary HD 5980. *Rev. Mex.* **2008**, *44*, 3–27.
67. Koenigsberger, G.; Morrell, N.; Hillier, D.J.; Gamen, R.; Schneider, F.R.N.; González-Jiménez, N.; Langer, N.; Barbá, R. The HD 5980 Multiple System: Masses and Evolutionary Status. *Astron. J.* **2014**, *148*, 62. [CrossRef]

© 2019 by the authors. Licensee MDPI, Basel, Switzerland. This article is an open access article distributed under the terms and conditions of the Creative Commons Attribution (CC BY) license (http://creativecommons.org/licenses/by/4.0/).

Review

Radiation-Driven Stellar Eruptions

Kris Davidson

Minnesota Institute for Astrophysics, University of Minnesota, 116 Church St. SE, Minneapolis, MN 55455, USA; kd@umn.edu; Tel.: +1-612-624-5711

Received: 20 December 2019; Accepted: 25 January 2020; Published: 5 February 2020

Abstract: Very massive stars occasionally expel material in colossal eruptions, driven by continuum radiation pressure rather than blast waves. Some of them rival supernovae in total radiative output, and the mass loss is crucial for subsequent evolution. Some are supernova impostors, including SN precursor outbursts, while others are true SN events shrouded by material that was ejected earlier. Luminous Blue Variable stars (LBV's) are traditionally cited in relation with giant eruptions, though this connection is not well established. After four decades of research, *the fundamental causes of giant eruptions and LBV events remain elusive.* This review outlines the basic relevant physics, with a brief summary of essential observational facts. Reasons are described for the spectrum and emergent radiation temperature of an opaque outflow. Proposed mechanisms are noted for instabilities in the star's photosphere, in its iron opacity peak zones, and in its central region. Various remarks and conjectures are mentioned, some of them relatively unfamiliar in the published literature.

Keywords: Eddington Limit; eruption; supernova impostor; LBV; Luminous Blue Variable; stellar outflow; strange modes; iron opacity; bistability jump; inflation

---

1. Super-Eddington Events in Massive Stars

Very massive stars lose much—and possibly most—of their mass in sporadic events driven by continuum radiation. This fact has dire consequences for any attempt to predict the star's evolution. After four decades of research, the instability mechanism has not yet been established; maybe it occurs in the stellar core, or else in a subsurface locale, or conceivably at the base of the photosphere. Without concrete models of this process, massive-star evolution codes can generate only "proof of concept" simulations, not predictive models, because they rely on assumed mass-loss rates adjusted to give plausible results. Even worse, eruptions may illustrate the butterfly effect— the time and strength of each outburst may depend on seemingly minor details, and the total mass loss may differ greatly between two stars that appear identical at birth. And an unexpected sub-topic, involving precursors to supernova events, arose about ten years ago. Altogether, the most luminous stars cannot be understood without a greatly improved theory of radiative mass-loss events. No theorist predicted any of the main observational discoveries in this subject.

Most of the phenomena explored here are either giant eruptions (including supernova impostors, supernova precursors, and shrouded supernovae) or LBV outbursts. They have four attributes in common:

- Their L/M ratios are near or above the Eddington Limit.
- Outflow speeds are usually between 100 and 800 km s^{-1}.
- The eruptive photosphere temperatures range from 6000 to 20,000 K, providing enough free electrons for substantial opacity.
- Observed durations are much longer than relevant dynamical timescales.

Giant eruptions are presumably driven by continuum radiation. They carry far too much kinetic energy to be "line-driven winds." Gas pressure is quite inadequate, blast waves are either absent or inconspicuous, and there is no evidence for sufficient MHD processes. Individual events have peak luminosities ranging from $10^5\,L_\odot$ to more than $10^8\,L_\odot$ while ejecting masses ranging from $10^{-3}\,M_\odot$ to $10\,M_\odot$ or more. Note that driving by continuum radiation (i.e., a super-Eddington flow) is not the root "cause" of an eruption. Logically the cause must be some process or instability that either increases the local radiation flux, or increases its ability to push a mass outflow.

This review does not include eruptions with $L < 10^{5.5}\,L_\odot$, such as "red transients" and nova-like displays. Those lower-luminosity cases generally involve stars with $M < 20\,M_\odot$ or even $M < 10\,M_\odot$, which are vastly more numerous than the very massive stars ($M_{\text{ZAMS}} > 50\,M_\odot$) that most likely produce giant eruptions. The relatively low-luminosity outbursts may be highly abnormal phenomena (e.g., stellar mergers) that occur in only a tiny fraction of the stars. In some cases they might not be above the Eddington Limit, or might not be opaque, or might be accelerated by non-radiative forces. Giant eruptions, by contrast, are highly super-Eddington, tend to look like each other regardless of their causative instabilities, and may occur in a substantial fraction of the most massive stars. Much of Section 4 and part of Section 5 may apply also to the lower-luminosity eruptions, however.

This article is a descriptive review like a textbook chapter, not a survey of publications. It outlines the basic physics and theoretical results with only a minimal account of the observational data. It also includes comments about some of the quoted results, with a few personal conjectures. Some of the generalities sketched here have been unfamiliar to most astronomers, even those who work on supernovae. They are conceptually simple if we refrain from exploring technicalities. One important topic—rotation—is mostly neglected here, because it would greatly lengthen the narrative and there is not yet any strong evidence that it is required for the chief processes. Binary systems are also neglected, except the special case of η Car; see remarks in Section 6.

2. A Checkered History

The origins of this topic are summarized in an appendix in [1]. Since only 2 to 4 giant eruptions or supernova impostors were known before 2000, they were conflated with LBV outbursts. Early discoveries followed two paths, and today we are not yet sure whether those paths really intersect. First, the examples of η Car, P Cyg, SN 1961v, and SN 1954J were known before 1970 [2]. Their eruptions produced supernova-like amounts of radiation, but with longer durations than a supernova and the stars survived. (SN 1961v may have been a true supernova [3,4], but, ironically, that doesn't alter its historical role.) The second path began with the recognition in 1979 of an upper boundary in the empirical HR diagram, the diagonal line in the middle of Figure 1 [5]. Almost no stars are found above and to the right of that line, and the rare exceptions are temporary. Since massive stars evolve almost horizontally across the diagram, this boundary indicates some sort of barrier to the outer-layer evolution of stars with $M > 50\,M_\odot$. The probable explanation involves episodic mass loss as follows.

Note the various zones and boundaries in Figure 1, though in reality they are not so well defined. If a massive star loses a considerable fraction of its mass, then it cannot evolve far toward the right in the HR diagram. Thus a good way to explain the HRD boundary is to suppose that stars above $50\,M_\odot$ lose mass in some process that exceeds their line-driven winds. The S Doradus class of variable stars occurs in the "LBV1" and "LBV2" zones in Figure 1, to the left of the empirical boundary. They are remarkably close to the Eddington Limit (Section 3.1 below), and they exhibit sporadic outbursts which expel more material than their normal winds. If every star above $50\,M_\odot$ behaves in that way after it evolves into the LBV1 zone, then the boundary is an obvious consequence. This scenario was proposed as soon as the empirical limit was recognized [5]. In a variant idea noted in Section 3.1 below, the decisive mass loss occurs just before the stars become LBV's; but both hypotheses invoke eruptions in that part of the HR diagram. No better alternative has appeared in the decades since they were proposed.

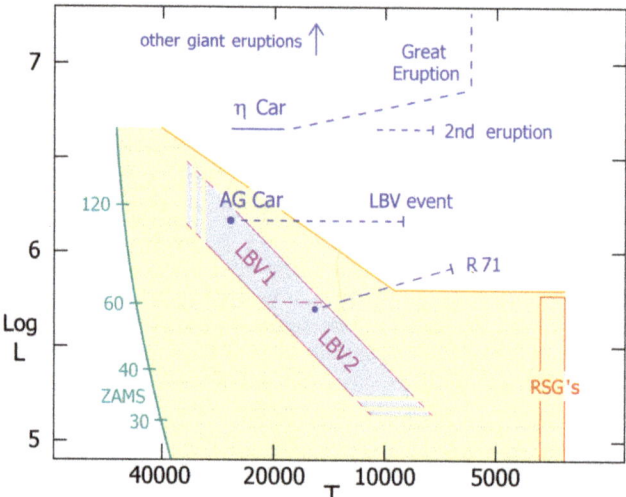

Figure 1. The empirical upper boundary and LBV instability strip in the Hertzsprung-Russell Diagram. In reality they are ill-defined and may depend on rotation and chemical composition. The interval between the LBV strip and the boundary is very uncertain. The zero-age main sequence on the left side shows initial masses, and most of a very massive star's evolution occurs at roughly twice the initial luminosity.

Today, S Dor variables are usually called LBV's, an acronym for "Luminous Blue Variables." Rightly or wrongly, they are frequently mentioned in connection with giant eruptions and supernova impostors. Many of them are easier to observe than eruptions in distant galaxies, and they probably offer hints to the relevant physics, Section 5 below.

Thus the key facts—episodic mass loss, and the existence of giant eruptions—were well recognized before 1995, and credible mechanisms had been noted; see many references in [1]. A decade later, when the mass loss rates of normal line-driven winds were revised downward [6], the same concepts were proposed again as a way to rescue the published evolution tracks (e.g., [7]). Unrelated to that development, extragalactic giant eruptions attracted attention after 2003 and some of them were aptly called "supernova impostors" because their stars survived [8]. Modern SN surveys found many examples [9], often classed among the Type IIn SNae. Some giant eruptions preceded real supernova events, the most notorious being SN 2009ip where the real SN explosion did not occur until 2012 [10,11]. Theoretical explanations continue to be diverse and highly speculative.

3. Categories and Examples

Only a few specific objects are mentioned here, to illustrate the main phenomena.

3.1. LBV's

The term "Luminous Blue Variable" is unfortunate in three respects: Many unrelated luminous blue stars are also variable, LBV's are often not very blue or not strongly variable, and it has caused extraneous objects to be included in lists of LBVs, often without observed outbursts [12,13]. Thus we should regard the trigram "LBV" as an abstract label, not an acronym. Many examples are described and listed in [12,14,15]. The present review includes this phenomenon because it may provide some guidance to the physics of giant eruptions, and LBV's have been observed far more often than giant eruptions. Note, however, that suspected analogies between those two categories have not been proven and may turn out to be illusory.

LBV's are defined by a particular form of variability like AG Car in Figure 1 [1]. Their hot "quiescent" states are located along a strip in the H-R Diagram (HRD) shown in the figure, sometimes called the S Doradus instability strip [16]. Its upper and lower parts, LBV1 and LBV2, represent different stages of evolution—providing a clue for theory, Section 5.1 below.

Most stars in the strip are not LBV's. Spectroscopic analyses consistently show that genuine LBV's have smaller masses than other stars with similar L and $T_{\rm eff}$ [15,17–21]. Consequently, for an LBV the Eddington parameter $\Gamma \equiv \kappa_e L/4\pi cGM = (L/M)/(L/M)_{\rm Edd}$ is close to 0.5 or somewhat larger. This is not surprising for the luminous classical LBV's; a 60 $M_\odot$ star, for example, attains $\Gamma > 0.4$ before the end of central hydrogen burning [12]. But $\Gamma \sim 0.5$ is remarkable in zone LBV2, where most stars have $\Gamma \sim 0.2$. Evidently each lower-luminosity LBV has lost much of its initial mass.

Since the LBV2 stars have luminosities below the upper boundary in Figure 1, they can evolve across the HRD. Hence we can explain their low masses by supposing that they have already passed through a cool supergiant stage where mass loss was very large [1]. After returning to the blue side of the HRD, they now have large L/M ratios which cause them to be LBV's. This surmise is confirmed, or at least very strongly supported, by the fact that LBV1's are generally associated with O-type stars but LBV2's are not [12,13,22]. This fact implies that LBV1's are younger than LBV2's, as expected in the evolved-LBV2 scenario. Classical LBV's (LBV1's) are somewhat more than 3 million years old, near or slightly after the end of core hydrogen burning. LBV2 stars are post-RSG's near or after the end of core helium burning (Figure 1 in [12]).

The above account may seem inconsistent, because LBV's are said to have rapid mass loss but the low masses of LBV2's are ascribed to a different evolutionary stage. This semi-paradox arises because the two types play very different roles in this story. LBV1 outbursts probably cause enough mass loss to shape the appearance of the upper H-R Diagram. LBV2 events do not, but they are pertinent because they suggest a connection between LBV variability and the Eddington parameter Γ as noted above. They also give a strong hint that LBV instability occurs in the outer layers, Section 5.1 below.

Figure 1 shows a well-known classical LBV, AG Carinae. It currently has $L \approx 1.5 \times 10^6 \, L_\odot$ and $M \sim$ 40 to 70 $M_\odot$, with $T_{\rm eff} \sim$ 16000 to 25000 K at times when a major LBV event is not underway [15,23–25]. The initial mass was probably above 85 $M_\odot$ and rotation is non-negligible [25]. In the years 1990–1994, AG Car's photosphere temporarily expanded by an order of magnitude with only a modest change in luminosity [23]. The apparent temperature consequently declined to about 8500 K, shifting much of the luminosity to visual wavelengths. Meanwhile its mass-loss rate increased by a factor of 5 to 10, peaking above $10^{-4} \, M_\odot \, {\rm y}^{-1}$. (The estimated amount depends on assumptions about the wind's inhomogeneity.) Outflow speeds varied in the range 100–300 km s^{-1}. Then, in 1995–1999 the photosphere contracted back to roughly twice its pre-1990 size. The event timescale, about 5 years, was more than 100× longer than the star's dynamical timescale. Perhaps 5 years was a thermal timescale for a particular range of outer layers. AG Car's 1990–1999 event in Figure 1 represents the classic form of high-luminosity LBV event, except that it only partially returned to its pre-1990 state.

Like many other LBV's [26], AG Car has a circumstellar nebula [27–30]. The nebular mass is said to be 5–20 $M_\odot$, ejected thousands of years ago and expanding rather slowly. Either the ejecta from multiple events have piled up there, or the star had one or more giant eruptions larger than any LBV events that have been observed in recent times. (The circumstellar material is almost certainly not due to mass loss in a red supergiant stage of evolution, since AG Car is too luminous to become a RSG—see Figure 1. The same statement applies to various other LBV's that have circumstellar ejecta.)

Figure 1 includes another LBV, R 71, to show that rules can be broken. It had an outburst in the 1970's [31], but a later event starting around 2005 was extraordinary [32,33]. Unlike normal LBV events, the luminosity of R 71 substantially increased while the temperature fell definitely below 7000 K. At minimum temperature it exhibited pulsation on a dynamical timescale (cf. comments in [34]). The mass-loss rate rose well above $10^{-4} \, M_\odot \, {\rm y}^{-1}$, high for its luminosity. Since L is poorly known due to an uncertain amount of interstellar extinction, this object may be either a classical LBV1 or an LBV2.

Note that the empirical limit in Figure 1 does not coincide with the LBV instability strip. The instability strip might extend to the boundary, but this detail should warn us that evolution through the LBV1 stage may involve some unrecognized tricks. Each of the following scenarios would be consistent with available data.

- LBV1 outbursts, described above, may cause enough mass loss to limit the star's later evolution. Given the quoted rates and outburst durations, this idea seems only marginally adequate.
- Or perhaps the crucial mass loss occurs in rare, more extreme LBV eruptions. P Cygni's dramatic brightening about 400 years ago may have been an instance [35], and such an event may have created AG Car's massive ejecta nebula mentioned above.
- Conceivably the most important phenomenon occurs just *before* the LBV1 stage [36–40]. In this scenario, the star first evolves across the LBV strip without incident, and then becomes violently unstable at a stage near or beyond the empirical boundary. A giant eruption occurs, ejecting so much mass that the star moves back to the left in the HR Diagram and becomes an LBV. The pre-LBV evolutionary episode would be too brief for us to have any known examples—though P Cyg and/or η Car might conceivably fill that role (see below). In this view LBV's are results of the boundary, not its cause.

Resourceful theorists can devise other possibilities. This review concerns the nature of mass-loss episodes, not the resulting evolutionary tracks. The latter depend on multiple parameters which are very poorly known.

Concerning LBV photospheres, see Section 4.5 below.

3.2. Giant Eruptions

From a non-specialist's point of view, the observed giant eruptions have too much diversity. Some of them are supernova impostors (i.e., the star survives), while others may be genuine supernovae modified by surrounding material (Section 5.4 below). Both cases are sometimes classified as "Type IIn SNae," which implies narrower emission lines than normal SNae. Major radiation-driven eruptions have several traits:

- The flow is opaque during most of the event—i.e., the continuum photosphere is located in the outflow.
- Photospheric temperatures are usually in the range 6000–20,000 K defined in a particular way, Section 4.1 below.
- Outflow speeds are typically a few hundred km s^{-1}, not thousands, and there are no conspicuous shock waves. Small amounts of material may attain higher speeds at the beginning of the eruption, but they are relatively faint.
- Hα and other bright emission lines have recognizable Thomson-scattered profiles as described in Section 4.2 below. This fact is useful for indicating the nature of the eruption.

An excellent example is SN2011ht [41,42], whose brightness and timescale resembled a supernova (Figure 2). It may have been either a supernova impostor or else a true SN within a dense envelope of prior ejecta (Section 5.4 below); but in either case the observed display was a radiation-driven outflow. Its spectrum (Figure 3) had characteristics explained in Section 4 below, with outward speeds of several hundred km s^{-1} and no hint of a blast wave before the brightness declined. Broad emission line wings were caused by Thomson scattering rather than bulk motion (Section 4.2), and the kinetic energy of visible ejecta was much smaller than in a normal SN [42].

About two months after maximum, the visual-wavelength brightness abruptly decreased by a factor of 60 (Figure 2). Since normal dust formation does not account for this change [42], the simplest interpretation is that most of the trapped radiation escaped through the photosphere just before that time. A normal core-collapse supernova would have remained substantially brighter due to radioactive decays in the ejecta. Some authors *assumed* that 2011ht was a supernova in a discussion of the light curve [43]; but the lack of a radioactive afterglow was decidedly peculiar in that case, and the light curve was reasonable for a non-SN instability (Section 5 below). Spectroscopy gives far more definite information than the shape of a light curve, and strongly implied an opaque continuum-driven outflow far above the Eddington Limit [42,44,45].

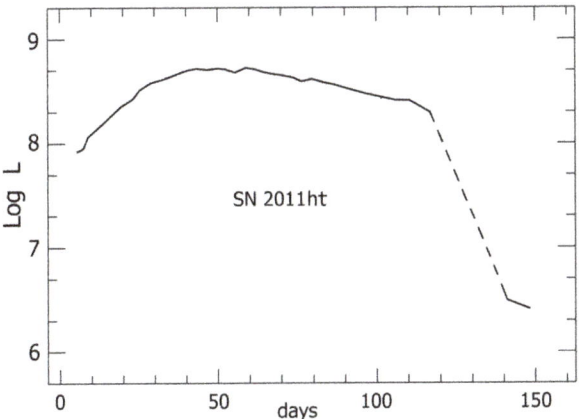

Figure 2. Luminosity record of SN2011ht based on visual-wavelength brightness [42]. The vertical scale, expressed in solar units, neglects variations in the bolometric correction but is adequate for conceptual purposes. The relative faintness after $t \sim 130$ d is highly abnormal if this object was a true supernova.

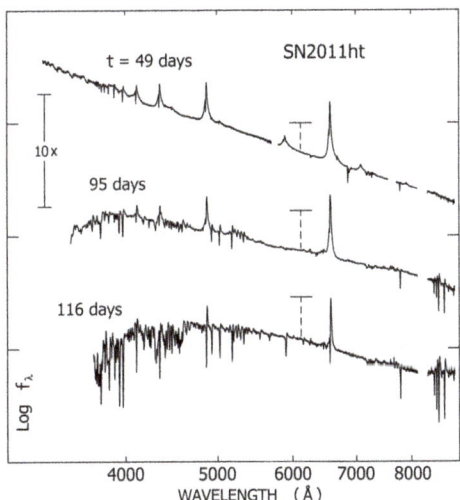

Figure 3. Spectrum of SN2011ht at three different times [42], cf. Figure 2. Both scales are logarithmic, the three tracings have differing vertical offsets, and the marks near 6100 Å indicate $f_\lambda = 10^{-14}$ erg cm^{-2} s^{-1} Å^{-1}. Gaps at $\lambda > 7000$ Å are obscured by terrestrial atmospheric features. Concerning the line profiles, see Section 4.2 and Figure 6.

Historically, the first observed giant eruption was P Cygni about 400 years ago [14,35,46]. Its maximum luminosity was of the order of $10^{6.5}$ $L_\odot$, quite small by giant eruption standards; but, unlike normal LBV events, that amount significantly exceeded the quiescent brightness. P Cyg's outburst (actually two or more episodes) persisted for years so the radiative energy output was probably more than 10^{48} ergs. Today this star is located in the LBV instability strip in the HR Diagram, but it has not exhibited an LBV event. Possibly this is a hint that the interval between episodes is correlated with the strength of the most recent instance, analogous to some forms of relaxation oscillators.

Among the eruptors discovered since 2000, some had luminosities comparable to η Car and/or P Cyg, but behaved very differently. SN2000ch, for instance, has exhibited multiple outbursts considerably brighter than the familiar type of LBV outburst, but not as bright as η Car's Great Eruption [47–49]. Those events were hotter than an LBV outburst, with higher outflow speeds, much shorter durations, and the luminosity may have increased appreciably on each occasion. Altogether its behavior differs from LBV's, η Car, and the brighter giant eruptions noted below. P Cyg might have appeared similar in the years 1600–1650, but this is merely a speculation.

A more extreme object with a different kind of multiplicity was described in [50]. PSN J09132750+7627410, a SN impostor in NGC 2748, attained a luminosity of the order of $10^{7.3}$ $L_\odot$, comparable to η Car's maximum, for several months—even though its quiescent luminosity was probably less than $10^{5.5}$ $L_\odot$. Near maximum its spectrum resembled SN2011ht described above. Its chief peculiarity was the existence of several distinct outflow velocities in each absorption feature: -400, -1100, and -1600 km s^{-1}. These may signify either a series of mass-loss episodes, or structure in the observed episode, or separate ejecta from more than one star. The two larger speeds are much faster than an LBV outflow. Multiple velocities have been seen in a few other eruptive stars—e.g., see [11,51].

Two pre-2000 giant eruptions, SN1954J [52] and SN1961V [2,46], have had enough time to show whether their stars survived. The SN1954J event had a maximum luminosity of the order of 10^7 $L_\odot$ with a duration less than a year [8,9,53]. The surviving star, a.k.a. V12 in NGC 2403, has a likely mass around 20 $M_\odot$ and is seriously obscured by circumstellar dust. Its spectrum includes Thomson-scattered emission line profiles, indicating a present-day opaque outflow (Section 4.2 below). This fact is strong evidence that the observed object really is the survivor of a giant eruption. The star was probably in a post-RSG state when the event occurred [53]. SN1961V, on the other hand, remains doubtful. It achieved a peak luminosity well above 10^8 $L_\odot$ with an overall event duration longer than a normal supernova, but no survivor has been identified with high confidence [3,4].

An unexpected development since 2000 has been the occurrence of precursor eruptions—i.e., giant eruptions that were followed several years later by real supernova events. At first sight this seems unlikely, because the final stages of core evolution have timescales of days, hours, and minutes rather than years. A few years is a likely timescale in the outer layers (Section 5 below), but in the standard view those regions "don't know" the precise state of the core. Hence the precursor events most likely arise in or near the core; but that assessment is too glib to be entirely satisfying, as noted in Section 5 below. The most notorious example of this phenomenon was SN 2009ip, whose blast wave explosion was not observed until 2012 [11]. That object exhibited other events between 2009 and 2012. Evidently some part of the star became unstable a few years before the SN event, but then the observed timescale didn't accelerate with the core evolution. Or perhaps the 2012 shock wave did not represent the real terminal event [10,54]! See comments in Section 5.3 below.

SN 1994W, SN 2009kn, and SN 2011ht probably ejected material months or years before their terminal explosions [42,44,55]. Since those objects became strangely faint at the stage when ^{56}Ni decay normally produces luminosity after a core-collapse SN event, some authors suspect that core collapse did not occur—e.g., [44].

As outlined above, giant eruptions are usually easy to distinguish from LBV events. They have far greater mass loss rates, substantial increases in luminosity, and shorter durations in most cases. A few LBV's, however, have mistakenly been given SN designations. SN 2002kg, for example, is a luminous LBV also known as V37 in NGC 2403 [8,53,56].

3.3. Eta Carinae

The classic example of a supernova impostor, of course, is η Carinae. It merits a separate subsection here, because it has been observed in far more detail than any other relevant object. Following the tradition of classic examples in astronomy, it has abnormal properties. Its event seen in 1830–1860 persisted much longer than other known giant eruptions, and it has a companion star that approaches rather closely at periastron. Many authors reviewed η Car in ref. [57], and later developments have not altered the main facts.

The star's luminosity is roughly $10^{6.6} L_\odot$. Before 1830 its mass was probably in the range 140–200 $M_\odot$, with an apparent temperature close to 20,000–25,000 K [58]. Then its 30-year Great Eruption ejected 10-40 $M_\odot$ at speeds averaging 500 km s^{-1}. It converted more than 10^{50} ergs of energy to roughly equal portions of radiation, kinetic energy of ejecta, and potential energy of escape. The resulting "Homunculus" ejecta-nebula [59] is famously bipolar, indicating a complex role for angular momentum [58]. The ejected material is clearly CNO-processed, with helium mass fraction $Y \sim 0.4$–0.6 [60,61]. The star's subsequent recovery has been unsteady, including a smaller eruption around 1890 and two later disturbances [58,62,63]. Until recently its wind was opaque in the continuum with $\dot{M} \sim 10^{-3} M_\odot$ y^{-1}, and most likely above $10^{-2} M_\odot$ y^{-1} a century ago [62]. That rate was too large for a line-driven stellar wind.

The companion object is most likely an O4-type star with a highly eccentric 5.54-year orbit, see many articles in [57]. The primary star is indeed the eruption survivor, since it has an extremely abnormal wind that has been diminishing since the event. But the hot secondary star alters the situation in several ways.

- It ionizes much of the primary wind, greatly affecting the observed spectrum.
- The periastron separation [64] is so close that it would presumably destabilize the orbit of any third object within 20 AU of the primary star.
- Tidal friction near periastron may transfer orbital angular momentum to the primary star's outer layers. If so, the equilibrium rotation period would be in the range 50–150 days. The star may be highly vulnerable to tidal effects because it is close to the Eddington Limit, but this possibility needs a careful quantitative analysis (see Section 6 below).
- Many authors have noted that the companion star may have triggered the Great Eruption via tidal influence near periastron [58]; but we must not confuse a trigger with the instability mechanism. Perhaps the nearly-unstable primary star gradually expanded until the other star's tidal influence tipped it over the edge. But this idea is not simple, since there is evidence for earlier episodes [26,65] and the present-day orbit eccentricity is too large to be caused entirely by the 19th-century mass loss [58,64]. Incidentally, the present-day orbital period cannot be used to estimate periastron times for the era of the Great Eruption. If we try to extrapolate back to about 1840, the gradual period change due to mass loss causes a phase uncertainty of the order of a year.
- The companion star may be the main reason why η Car's eruption was fainter and more protracted than most supernova impostors [58]. For several years the second star was inside the radius of the eruption photosphere, and near periastron it may have stirred the instability. During the great eruption, and for many years afterward, the secondary star probably accreted some material from the primary's outflow [58,62,66–69]. Possible consequences for the orbit have not yet been examined.
- As noted in Section 6 below, various authors have speculated that η Car was originally a triple system and two of the stars merged. Models of that type have a large number of assumed parameters, they do not agree with each other, and there is no demonstrated need to postulate a third star.

For historical reasons [5], η Car is often called an LBV despite its location in the HR Diagram (Figure 1). The high-luminosity end of the LBV strip in Figure 1 may be misleading, and in principle every star with $L > 10^{6.3} L_\odot$ and $T_{\text{eff}} < 25,000$ K might be an LBV; there are not enough examples to know. But that is only a possibility, and we have no definite reason to classify η Car as an LBV. An LBV-like eruption would be complicated for this object, because the radius of a 10,000 K photosphere would exceed the companion star's periastron distance.

4. The Spectrum of an Opaque Outflow

This section has four main points: (1) Giant stellar eruptions usually have similar colors and spectra even if they're caused by different processes. (2) Certain emission line profiles indicate an opaque outflow. (3) Stellar spectral types are not reliable indicators for outflow temperatures. (4) LBV outflows are not homologous with giant eruptions.

4.1. The Continuum

The apparent radiation temperature of an opaque outflow can be defined in various ways—e.g., based on the photon-energy distribution of the emergent continuum, or its slope at selected wavelengths, or on subsets of absorption features, etc. These alternative T's can differ by 20%, leading to confusion when one attempts to compare values quoted in papers. "Effective temperature" $T_{\rm eff}$ used for stellar atmospheres is not appropriate, because an outflow has no fundamental reference radius that is meaningful for that purpose. (Also note that an optical depth value of 2/3 has no significance in this context. Regarding photon escape probabilities, $\tau_{\rm tot} \sim 1.0$ to 1.3 in a diffuse outflow corresponds to $\tau_{\rm tot} \approx 2/3$ in a plane-parallel atmosphere.)

And we must be careful with the word "photosphere." The region with optical depth $\tau_{\rm tot}(r) \sim 1$ has little effect on an outflow's emergent photon energy distribution, because the dominant opacity is usually Thomson scattering by free electrons. That process has only a weak effect on photon energies. Consider instead a deeper region where absorption and re-emission events are frequent enough to establish $T_{\rm gas} \approx T_{\rm radiation}$. Outside some radius $r_{\rm esc}$, the average photon escapes via multiple scattering before it experiences an absorption event. Evidently the emergent photon energy distribution depends mainly on temperatures that exist just inside radius $r_{\rm esc}$. In this overview "photosphere" means that region.

Classical diffusion theory gives the approximate size of $r_{\rm esc}$ [70,71]. Suppose that local opacities for absorption, scattering, and their sum are $\kappa_{\rm abs}$, $\kappa_{\rm sc}$, and $\kappa_{\rm tot}$, averaged over photon energies in some optimal way. Define a "thermalization opacity"

$$\kappa_{\rm th}(r) \equiv [3\,\kappa_{\rm tot}(r)\,\kappa_{\rm abs}(r)]^{1/2}, \tag{1}$$

with associated optical depth

$$\tau_{\rm th}(r) = \int_r^\infty \rho(r')\,\kappa_{\rm th}(r')\,dr'. \tag{2}$$

Often called thermalization depth or diffusion depth, $\tau_{\rm th}(r)$ is typically of the order of $0.6\,\tau_{\rm tot}(r)$ in a giant eruption or an LBV event photosphere. Calculations show that $r_{\rm esc}$ is approximately the radius where $\tau_{\rm th} = 1$ [70,71], and we can regard the photosphere as the region where $1 < \tau_{\rm th} < 2$. This is not a formal statement, but in practice it applies for any reasonable density law $\rho(r)$ and for large as well as small opacity ratios $\kappa_{\rm abs}/\kappa_{\rm sc}$. The emergent continuum is created mostly at $\tau_{\rm th} \approx 1.5$ to 2.0, while absorption and emission lines are formed mainly at $\tau_{\rm th} < 1$ or perhaps $\tau_{\rm th} < 1.5$. If T_1 and T_2 are the temperatures at $\tau_{\rm th} = 1$ and 2, then we can liken T_1 to the $T_{\rm eff}$ of a star with a similar spectrum, though their values may disagree because they are defined differently. Caveat: In published models of opaque winds, most authors define the photospheric radius by $\tau_{\rm tot} = 1$ or even $\tau_{\rm tot} = 2/3$, rather than $\tau_{\rm th} = 1$. With those choices, a quoted "photosphere temperature" is cooler than the emergent distribution of photon energies.

In a simple model where opacity depends only on ρ and T, the temperature at a given location depends approximately on two quantities, $\tau_{\rm th}$ and $\dot{M}V^{-1}L^{-0.67}$ where V is the local outflow velocity [71,72]. Figure 4 shows examples of T_1 and T_2 in spherical outflows. Corresponding radii are shown in Figure 5. These sketches are intended only for conceptual purposes; they are based on simplified models that ignore some major details (see below).

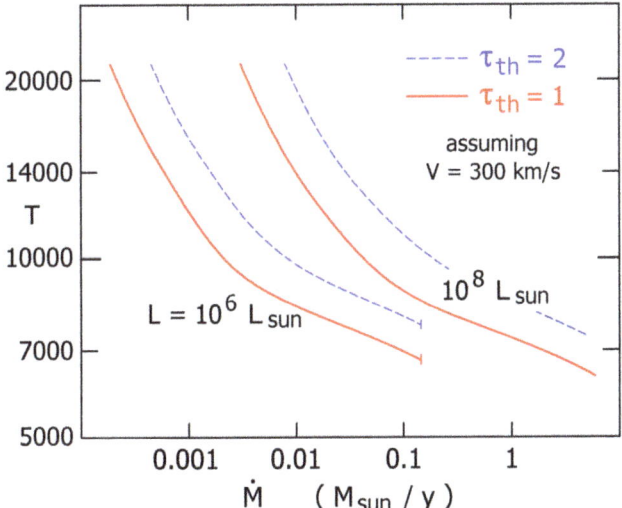

Figure 4. Photosphere temperatures in simplified opaque outflow models with $L = 10^6$ and $10^8 \, L_\odot$ and $V = 300$ km s^{-1}. The curves show temperatures corresponding to thermalization depths of 1 and 2. $T(\tau_{\text{th}})$ depends approximately on the quantity $\dot{M} V^{-1} L^{-0.67}$. Temperature values here are imprecise and very likely overestimated, because the models are highly idealized; see text.

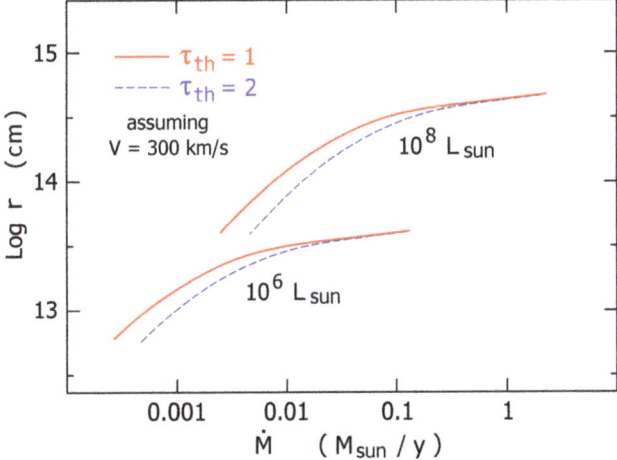

Figure 5. Radii of the photospheric locations shown in Figure 4. For large $\dot{M}$ the photosphere becomes geometrically thin (small $\Delta r/r$) and thus resembles a plane-parallel model.

With Figure 4 in view, imagine a case with constant luminosity L while $\dot{M}/V$ gradually increases. Initially the flow is transparent so our model does not apply. But when $\dot{M}/V$ becomes large enough to be opaque, then it determines the apparent temperature. Increasing $\dot{M}/V$ causes the photosphere to move outward, and $T_1 \propto$ roughly $(\dot{M}/V)^{-0.3}$ as shown in the upper half of Figure 4. Below 9000 K, however, the proportionality changes to $(\dot{M}/V)^{-0.07}$ because the opacities decline rapidly. $T_1 < 7000$ K requires a very large flow density. As noted above, T_1 is a fair indicator of the absorption and emission lines – except for a caveat in Section 4.3 below.

(This paragraph concerns technicalities that don't affect the main concepts.) Each temperature in Figure 4 refers to a location in the flow, which does not represent any specific observable quantity. For comparison with observations, one would need to calculate emergent radiation in a manner resembling [71] and [73]; but a model with wavelength-dependent opacities would be much better. Figure 4 is based on many simplified models with constant luminosities L, mass-loss rates $\dot{M}$, and flow velocities V. It was assumed that $T_{\text{gas}}(r) = T_{\text{rad}}(r)$, with LTE Rosseland mean opacities which are readily available (http://cdsweb.u-strasbg.fr/topbase/, [74], and refs. therein). Hydrogen and helium mass fractions were $X = 0.50$ and $Y = 0.48$. Integrating the spherical radiative transfer equations [73] with those opacities, we can calculate $T(\tau_{\text{th}})$. One recent analysis [72] appears at first sight to favor lower temperatures than Figure 4, but the disagreements involve the distinction between τ_{th} and τ_{tot}, and varying definitions of the observed T. Very likely the true ionization in the outer regions is larger than the LTE values; if so then the temperatures in Figure 4 are overestimates. Errors of that type are probably comparable to the differences between alternative definitions of T for the emergent radiation.

Figure 4 is fairly consistent with observed giant eruptions, such as η Car marked in Figure 1. Exceptionally high flow densities occurred in η Car's 1830–1860 Great Eruption, with $\dot{M} > 1\,M_\odot\,\text{y}^{-1}$ and $L \sim 3 \times 10^7\,L_\odot$ [57]; so it should have had exceptionally low photospheric temperatures. Reflected light echos provide spectra of that event [75–77]. The light-echo researchers deduced a temperature near 5000 K, but that was an informal value, not well-defined and not based on a quantitative analysis. Judging from the published description and reasons noted in Section 4.3 below, the spectrum most likely indicated $T_1 \sim 5500$ to 6500 K in Figure 4 [78]. (Recall that T_1 denotes the temperature at a particular location in the outflow, not an emergent radiation temperature.) If we define "apparent temperature" in a different way, its value may have been as low as 5000 K [72]. A second, lesser eruption of η Car in the 1890's had $T_1 \sim 7500$ K, according to the earliest spectrogram of this object [62]. Several extragalactic giant eruptions have been observed since 2000, usually showing $T_1 > 8000$ K (Section 3.2 above). LBV outbursts have much smaller luminosities, and usually reach 8000–9000 K like AG Car in Figure 1. The assumptions in Figure 4 are not valid for them, because their photospheres are not located in the high-speed part of the flow (Section 4.5 below). However, those observed minimum temperatures of LBV's are very likely determined by the rapid decline of opacity below 9000 K, even though the photospheres resemble static atmospheres.

In summary, there is no known observational reason to doubt the general appearance of Figure 4 for opaque radiation-driven outflows.

4.2. Distinctive Emission Line Profiles

The brightest emission lines from an opaque outflow generally have a certain type of profile, illustrated in Figure 6. Smooth broad line wings extend beyond ± 2000 km s^{-1} even though the wind speed is less than 700 km s^{-1}; and the longer-wavelength side is stronger. These are classic signs of Thomson scattering by free electrons [42,44,79–81]. Since the electrons have r.m.s. thermal speeds of the order of 600 km s^{-1}, some of the photons acquire large Doppler shifts in multiple scattering events before they escape. Meanwhile, expansion of the outflowing material favors shifts toward longer wavelengths. Obviously the resulting profile depends on $\langle \tau_{\text{sc}} \rangle_{\text{em}}$, the line-emitting region's average optical depth for Thomson scattering. The shape in Figure 6 indicates $\langle \tau_{\text{sc}} \rangle_{\text{em}} \sim 0.5$ to 2, and appears to be generic. It specifically represents SN 2011ht [42], but SN 1994w exhibited a similar Hα profile, and so do other giant eruptions and η Car's dense wind [44,63,82].

The moderate size of $\langle \tau_{\text{sc}} \rangle_{\text{em}}$ has a simple explanation. In typical eruptions with $T_1 > 7500$ K, $\kappa_{\text{sc}}/\kappa_{\text{th}} \sim 1$ to 2—a consequence of atomic physics. Hence the continuum photosphere boundary $\tau_{\text{th}} \approx 1$ automatically has $\tau_{\text{sc}} \sim 1$ to 2. Since the emission lines are formed outside the photosphere, we therefore expect $\langle \tau_{\text{sc}} \rangle_{\text{em}} \sim 1$, or perhaps a little smaller, for bright emission lines. The main point here is that an observed profile like Figure 6 is good evidence for an opaque or semi-opaque outflow. It recognizably differs from emission lines seen in normal stellar winds, expanding shells, nebulae, and supernova remnants.

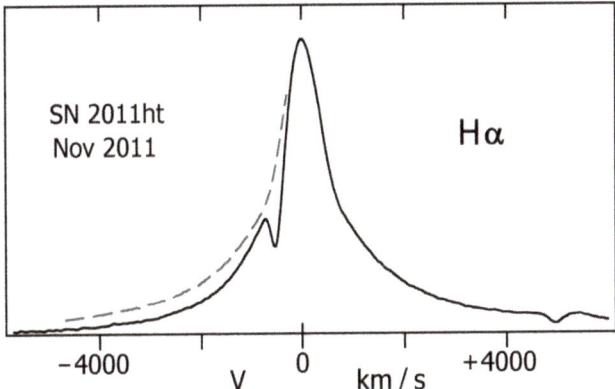

Figure 6. Emission line profile with moderate Thomson scattering. The dashed curve on the left side is a mirror image of the right side. Note that the line wings extend far beyond the velocity indicated by P Cyg absorption. This example is Hα in the radiation-driven outflow of SN 2011ht [42], but other giant eruptions produce similar line shapes.

(Caveat: In some papers a Thomson-scattered line profile is called "Lorentzian," often without recognizing its significance. That usage gives a flatly wrong impression in two respects. First, in physics the word "Lorentzian" has very specific connotations: $1/(1+x^2)$ = the Fourier transform of an exponential decay, the natural shape of an idealized spectral line, closely related to the uncertainty principle. None of these applies to the shape in Figure 6. Second, the wings of a true Lorentz profile are like x^{-2} but the wings of a Thomson-scattered profile are like $e^{-\alpha|x|}$. This difference is fundamental, not just a matter of opinion.)

4.3. Cautionary Remarks about Absorption Features

An opaque outflow also produces absorption lines, but they cannot safely be compared with stellar spectral types. The most dramatic example concerns η Car's Great Eruption. Spectra of that event have been obtained via light echoes, leading to an estimate $T \sim 5000$ K which seemed to contradict an expected value of 7500 K [75–77]. But that conclusion had two disabling flaws. (1) In fact the expected value was far below 7500 K [71,72,78]. (2) More pertinent here, spectral classification standards for stars do not apply to a mass outflow. For instance the light-echo spectra of η Car's Great Eruption showed absorption features of CN, which would indicate $T_{\text{eff}} < 5000$ K in a star—but they may occur in an outflow with $T_1 \sim 6000$ K.

Figure 7 shows why. Each curve represents the column density $\int \rho\, dr$ of material cooler than T. A stellar amosphere with $T_{\text{eff}} \approx 6000$ K has almost no material below 5000 K, but a diffuse outflow with $T_1 \approx 6000$ K can have an appreciable amount of cooler gas at large radii. This difference is a consequence of two facts: (1) the mass distribution in an outflow resembles a power law $\rho(r) \propto r^{-n}$ instead of the exponential $\rho(z) \propto e^{-z/h}$ that roughly describes a stellar atmosphere, and (2) Radiation density in an outflow has a $1/r^2$ "dilution factor." Figure 7 is merely schematic, but it suggests that an eruption with $T_1 \sim 6000$ K can form cool spectral features such as CN in its outer regions. Absorption lines formed at smaller radii may be good indicators of T_1, but they must be chosen carefully. Since LTE is a poor approximation for $T < T_1$, and there are other complications, a realistic model of the absorption line spectrum will be extremely complex (see below). Meanwhile, so far as available information allows us to judge, the light-echo spectra of η Car's eruption appear consistent with standard portrayals of that event.

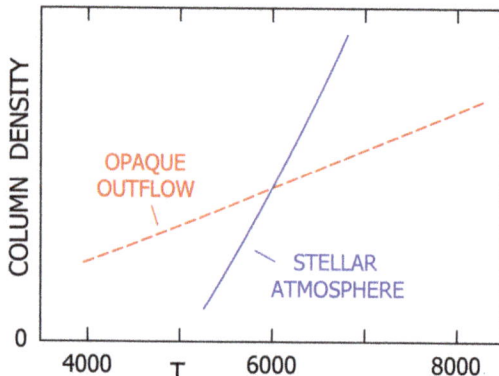

Figure 7. Sketch of the mass column density at temperatures below T, in a stellar atmosphere and in a dense outflow. These curves are based merely on idealized textbook-style models of $\rho(r)$ and $T(r)$, but the difference between them is qualitatively valid.

4.4. Why a Real Outflow Spectrum is Exceedingly Difficult to Calculate

The dependence of T_1 on $\dot{M}$ was first described long ago [71]. Figure 4 and ref. [72] employ modernized opacities, but the resulting quantities are highly imprecise because the models are highly simplified. A truly realistic calculation must include all of the following complications.

1. Real opacities depend on photon energy, and no form of average over $h\nu$ is fully consistent with all of the radiative transfer equations. The model should include $h\nu$-dependences.
2. Standard LTE opacities are very unreliable in the region with $\tau_{th} < 1.5$, because it is far from thermodynamic equilibrium. Existing NLTE codes do not adequately include some effects, e.g., items 4 and 5 below.
3. A realistic model needs a good velocity law $V(r)$, and the functional form used for line-driven stellar winds is probably wrong for an opaque flow. A valid $V(r)$ is surprisingly difficult to calculate, because of item 4 below [83–85]. This fact becomes even worse when we note that opacities due to Fe^+ and other complex species depend on many spectral lines whose interactions depend on item 4 as well as dV/dr!
4. A radiation-driven outflow is obviously unstable—"a light fluid pushing a heavier one"—and breaks up into condensations, greatly complicating the radiative transfer problem [83,84,86]. Modern codes employ a position-dependent "clumping factor" (e.g., [25]) which entails several assumptions and free parameters that are very uncertain. Hence the resulting models are useful guides but there is no reason to assume that they are correct. If each condensation is not transparent, then radiation tends to escape via the easiest paths between condensations. (This definitely happened in η Car's giant eruption [83].) The words "porosity" and "granulation" are often used in this context. Results depend on the condensation sizes, densities, and even shapes. A practical technique, analogous to mixing length theory for convection, is needed for inhomogeneous radiative transfer. Perhaps a recipe can be developed from a large number of specialized three-dimensional simulations. See [85,87].
5. As η Car notoriously shows, spherical symmetry may be a poor approximation. Moreover, since a spherical model maximally entraps the radiation, it represents an extreme case, not typical or average. And the spectrum of a non-spherical outflow depends on the observer's viewing direction.

Omitting any of these complications may cause the results to be almost as inaccurate as the simplified models in Figure 4. Some existing codes employ elaborate radiative transfer with some NLTE effects, but there are reasons for skepticism. Items 3, 4, and 5 have multiple undetermined free parameters, not advertised in most research papers. Items 3 and 4 acting together may invalidate the radiative transfer methods for spectral lines. Most important, the realism of a calculation is very difficult to test. Approximately matching an observed spectrum does not prove correctness, since there are enough free parameters to compensate for omitted effects. In summary, existing codes illustrate most of the chief processes, but they must not be regarded as decisive or authoritative. Their uncertainties may be far worse than most authors assume.

4.5. Are LBV's Relevant?

As noted at the end of Section 3.1 above, the most critical mass-loss episode may conceivably occur just before the LBV stage of evolution. If so, then LBV outbursts may have little in common with continuum-driven giant eruptions. The following remarks and conjectures may be pertinent.

The fast region in any LBV wind (usually 100 to 300 km s^{-1}) is transparent in the visual-wavelength continuum. This fact was noted before 1990 [71], and later motivated the static atmosphere models [25,88–91]. But it might not apply to the inner, slow, denser part of the outflow during a major LBV event. The following remarks concern major LBV1 eruptions with $T_1 < 10,000$ K, not the states above 15,000 K that are emphasized in most analyses since 2010.

Recall the analogy between a stellar wind and flow through a transonic nozzle, with $V = w$, the speed of sound, at radius r_w [92,93]. The subsonic region $r < r_w$ resembles an atmosphere constrained by gravity, while the flow dominates outside r_w. Imagine such a hybrid model of AG Car's major eruption around 1994, when T_1 declined to about 9000 K while $\dot{M}$ rose to about $10^{-4} M_\odot$ y^{-1} [23]. At that time $w \sim 20$ km s^{-1} in the photosphere, and the photospheric outflow speed was of the order of 1 or 2 km s^{-1}. The sonic point was thus two or three scale heights above the photosphere, much closer than in a normal stellar wind. According to Figure 4, a 3× larger value of $\dot{M}$ would have moved the photosphere to the sonic point – so a flow model rather than a static atmosphere would have become appropriate if $\dot{M}$ had grown above that amount. In this order-of-magnitude sense the major eruption was "almost" opaque.

A normal stellar wind has only an indirect relation to the continuum photosphere, with $V < 0.01\,w$ in the photosphere. The much larger V/w in major LBV eruptions, along with proximity to the Eddington Limit, suggests that their outflows are more directly related to their continuum photospheres. This line of thought—or rather, of surmise—motivates a three-step conjecture:

- The rapid decline of opacity below 8500 K encourages the eruptive photosphere to choose a temperature near that value. In other words, for a case with $L \sim 10^6 L_\odot$, the basic parameters of the subphotospheric inflated region would need to become unreasonable in order to push T_1 substantially below 8000 K.
- Due to the role of continuum radiation in the initial acceleration, the sonic point is related to the photosphere. Consequently the outflow speed at $\tau_{\rm th} = 1$ is $V_1 = \xi w$ where ξ is of the order of 0.1.
- Therefore $\dot{M} \approx 4\pi r_1^2 \xi \rho_1 w$.

This is an empirical hypothesis, not a theoretical prediction. There are semi-theoretical methods of predicting LBV mass loss rates, far more elaborate than this reasoning, but their developed versions have been applied mainly to the hotter states above 15,000 K [94]. Anyway, the above expression is consistent with estimated values for major LBV1 eruptions.

If we portray LBV outbursts as quasi-static inflated states [88–91], then unfortunately we de-emphasize the mass outflow. Since the latter is probably more consequential, the "eruption" aspect expresses a broader significance than the inflation. On the other hand, it is conceivable that most of the cumulative LBV mass loss occurs in rare giant eruptions à la P Cygni, not merely major eruptions. In any case the static 1-D models do not answer the main questions, Section 5.1 below.

5. Physical Causes of the Eruptions

The Eddington Limit turns out to be wonderfully subtle and complicated. Relevant instabilities were recognized after 1980 (see many refs. in [1]), but they are difficult to analyze or even to describe.

Practically all of our knowledge of eruption parameters, post-eruptive structure, timescales, and long-term mass loss is still empirical.

Two classic high-mass stellar instabilities were naturally suspected when giant eruptions and LBV events attracted notice in the 1980's: The Eddington ϵ mechanism or Ledoux-Schwarzschild instability energized in the stellar core [95], and radial dynamical instability which occurs if the average adiabatic index falls below 4/3 [96]. But they proved unsuitable [97], and have been replaced by newer ideas which are generally non-adiabatic, involve radiation pressure, and resemble each other. Three regions in the star merit special attention: the photosphere, the "iron opacity bump" region with $T \sim 200,000$ K, and the stellar core, broadly defined.

Details of the instabilities are too lengthy to explore here. Instead, this overview lists a set of essential considerations, including some that have seldom been discussed in research papers. Here rotation does not get the attention that it deserves, because that would greatly lengthen the text. Interacting or merging binary scenarios are omitted because there is no clear need for them at this time, see Section 6.

5.1. The Modified Eddington Limit

Giant eruptions exceed the Eddington Limit. Thus it seems natural to guess that the progenitor stars were very massive, close to the Limit, and vulnerable to the same instabilities as LBV's. This connection between LBV's and giant eruptions may be illusory, but thinking about it leads to some useful ideas.

Observed L/M ratios are clues to the LBV phenomenon in two different ways. First, of course, their proximity to the Eddington Limit (Section 3.1 above) implies a large role for radiation pressure. Second and less obvious, the distinction between two classes of LBV's favors an instability in the outer layers, not the stellar core. Types LBV1 and LBV2 have different central structures because they represent different stages of evolution (Section 3.1 above and [12]). The stars' outer 3% of mass, however, have similar radiation/gas ratios for both classes, and similar variations of opacity.

In the classical Eddington Limit

$$\left(\frac{L}{M}\right)_{\text{Edd}} \equiv \frac{4\pi cG}{\kappa_{\text{sc}}}, \tag{3}$$

the opacity κ_{sc} includes only Thomson scattering by free electrons. Absorption opacity κ_{abs} practically vanishes as a static photosphere approaches the Limit, because density ρ becomes very low. The classical Eddington parameter $\Gamma_{\text{Edd}} \equiv \kappa_{\text{sc}} L / 4\pi cGM$ can thereby approach 1 in an old-fashioned radiative atmosphere model. However, the photospheric κ_{abs} may be appreciable in a model with $\Gamma_{\text{Edd}} < 0.95$, and deeper layers may have larger opacities in any case. During the 1980's this thought inspired the idea of a "Modified Eddington Limit" that takes κ_{abs} into account [1,98–103]. It was an empirical hypothesis, not a theoretical prediction, motivated by η Car and observed LBV behavior.

Strictly speaking, there are two forms of Modified Eddington Limit. (1) It might be a well-defined limit to the allowed values of L/M in a static stellar model, like the classical limit but including realistic convection, incipient porosity, etc. (2) Or, more likely, it may signify an instability that arises when L/M exceeds some value, see Section 5.2 in [1]. In either case the critical L/M depends on opacities in the outer 3% of the star's mass, or maybe the outer 1%.

Rapid rotation reduces the effective gravity mass, thereby altering any form of Modified Eddington Limit. The terms "Ω limit" and/or "ΩL limit" allude to this obvious fact, but the implications are often oversimplified. Two decidedly non-trivial subtleties occur: (1) Rotation causes a star's subsurface temperatures to depend on latitude. Resulting alterations of opacity affect the topics of Section 5.2 and 5.3 below. (2) The specific angular momentum expelled in a giant eruption may be either larger or smaller than in the underlying layers. Consequently the effects of rotation may evolve during the eruption [58].

5.2. The Photosphere, Bistability, and Surface Activity

The easiest place to start is the photosphere. Traditionally, the total energy flow in a stellar interior can exceed $4\pi c G M / \kappa_{tot}$ by inciting convection [104]. But convection becomes inefficient in the photosphere, so radiation must carry nearly all of the energy flux there. Imagine a model wherein κ_{tot} increases inward, and radiative forces are less than gravity outside some radius r_c. Inside r_c, convection carries the excess energy flux. An increase in L/M presumably causes r_c to move outward relative to the stellar material. For some value of L/M, r_c moves into the photosphere; so there may be a practical limit, somewhat smaller than $(L/M)_{Edd}$. By extrapolating normal atmosphere models, one can identify a limiting L/M around $0.9\,(L/M)_{Edd}$ [105–107]. But the reasoning is dangerously subtle, and a different approach suggests that a star may become unstable at a smaller value of L/M, see below.

Now suppose that "Modified Eddington Limit" connotes an instability that arises somewhere in the range $0.5 < \Gamma < 0.9$, rather than a well-defined static limit. At relevant photospheric densities, κ_{tot} has a maximum in the vicinity of $T \sim 13000$ K, involving the ionization ratio Fe^{++}/Fe^{+}. Consequently a high-Γ atmosphere may be very unstable in a particular range of T around that maximum, and might act as a relaxation oscillator jumping back and forth across the unstable range [99,102].

Behavior like that is observed at a somewhat higher temperature [15,18,108,109]. Consider a standard hot line-driven wind model wherein the star gradually becomes cooler. As T_{eff} declines below 20000 K, Fe^{++} and other suitable ion species become numerous enough to drastically increase κ_{tot}; so the wind becomes slower and much denser. The transition occurs across a narrow range of T, hence the term "bistability jump." It was noted around 1990 as a likely cause of LBV events [18,108]. That idea originally meant a difference between two outflow states, but it rapidly evolved into a bistability between two quasi-static states of the star's outer layers [25,88–91]. One state corresponds to a quiescent LBV, the other occurs during an LBV eruption, and intermediate states are more unstable due to the opacity maximum mentioned earlier. In the eruptive state, the outer layers are greatly expanded or "inflated."

But a set of inflated and non-inflated models does not constitute a theory of LBV variability; instead it plays a role more like an existence theorem in mathematics. A proper theory must acknowledge the following questions.

1. What is the state of the outer regions during a major event when $T_1 < 10000$ K? The most elaborate spectral analyses [25,90] focus instead on models with $T_1 > 15000$ K, close to the bistability jump. The cooler state is more difficult but also more consequential. Moreover, all 1-D models disallow some effects that are probably essential [34].
2. Why and when does an LBV eruption end? The star does not merely evolve into an inflated state and remain there until further evolution occurs. Instead it jumps unpredictably back and forth between differing states. Does a major LBV event cease when a critical amount of mass or energy or angular momentum has been lost, or are the reasons chaotic or related to inconspicuous changes in the stellar interior?
3. Is the photospheric opacity behavior sufficient to cause an LBV event? Or is the deeper iron opacity peak (Section 5.2 below) needed?
4. The central LBV problem concerns mass loss, not the star's radius. What factors determine the increased $\dot{M}$? Do they resemble the conjectures in Section 4.5 above? Conventional line-driven wind theory is probably inadequate in this parameter regime (Section 4.4 above). A Monte Carlo radiative transfer technique predicts credible $\dot{M}$ values for LBV's in their hotter phases [94], but it omits many intricate effects seen in a 3-D simulation [34].
5. What determines the event recurrence rate? Is it like a relaxation oscillator wherein the recurrence time depends on details of the preceding event [99,102], or is there some form of periodicity? P Cyg had an extremely large event 400 years ago and has seemed quiet ever since [35].
6. What determines the timescale of a transition to the LBV-event state? Is it a thermal timescale for some relevant set of outer layers?

7. How large is the cumulative amount of mass loss? Does it vary greatly or randomly among LBV's with a given luminosity?
8. How strongly do these answers depend on rotation as well as chemical composition? And how much do the LBV eruptions alter the surface rotation and composition?
9. Do more extreme LBV eruptions occasionally occur, violent enough to substantially increase the luminosity while ejecting far more mass than usual? Observed ejecta nebulae, e.g., around AG Car, may be relics of such events. They might account for most of the cumulative mass loss.

The last item pertains to giant eruptions. In order to expel a mass which greatly exceeds that of the unstable region, the process must be like a geyser: instability begins at the top and moves downward (relative to the material) until some factor stops it. In this way a photospheric instability might even cause a giant eruption. As outer layers depart, a large reservoir of radiative energy is progressively uncovered. At any given time the configuration resembles a steady-state model, since the observed timescale is much longer than the dynamical timescale. Presumably the eruption ends when conditions change at the base of the flow – perhaps when it reaches some particular feature in the pre-eruption interior structure.

Another form of Modified Eddington Limit relates to dynamical processes rather than the temperature dependence of opacity. A static atmosphere dominated by radiation pressure tends to develop inhomogeneities, granulation, and porosity like an outflow; see [84,86] and many refs. therein. Resulting turbulence can engender MHD effects, even though the photosphere is well above the temperatures traditionally associated with stellar activity. These phenomena may influence the outflow rate, and might even determine it. Conceivably, η Car's dense wind a century ago [62] may have involved stellar activity analogous to a red supergiant! [110].

Most of the above possibilities are not mutually exclusive.

5.3. The Iron Opacity Peak

The "iron opacity peak" locale in a star, described below, is probably crucial; but its instabilities are too complex for simple analysis, math expressions, and predictions. A decisive analysis will require numerous 3-D simulations which have not been feasible so far.

Opacity has a dramatic maximum at temperatures around 180,000 K, for reasons concerning ionization stages of iron. In a typical LBV-like very massive star, $\kappa_{tot} > 2\,\kappa_{sc}$ throughout a temperature range such as 100,000 to 300,000 K, though the actual limits depend on mass density. Vigorous convection occurs there because $\kappa_{tot}L$ obviously exceeds $4\pi cGM$, if L signifies the total energy flow. Such a region offers a zoo of instabilities, and dynamically it decouples the outer layers from the stellar interior. Since the associated mass and energy greatly exceed the photosphere, this region is the most promising part of the star for eruption mechanisms. Its usual name, the iron opacity peak zone, might be confused with the iron peak of cosmic abundances; and "iron opacity bump zone" is both inelegant and cumbersome. For convenience, an ugly acronym will be used here: OPR = iron opacity peak region in the star. Although it usually occurs in the outer 1% of the star's mass distribution, its spatial radius may be considerably smaller than the stellar radius R. A second opacity peak will also be mentioned, involving helium at lower temperatures.

The OPR mass and energy are difficult to estimate from observational data. If $\mu(T)$ is the mass column density of layers cooler than T, and radiation pressure dominates, then $\mu \approx P/g \sim aT^4/3g$ so

$$m(T) \sim 4\pi R_T^2\, \mu(T) \sim \frac{4\pi R_T^4 aT^4}{3GM_*}, \qquad (4)$$

where R_T is a radius that has temperature $\approx 0.8T$. But the R_T^4 factor is quite uncertain, because R_T may lie deep within an extended envelope. Consider, for instance, η Car before its giant eruption. If we know only that $M_* \approx 150\,M_\odot$, $L \approx 4 \times 10^6\,L_\odot$, and $T_{\rm eff} \approx 20{,}000$ to $25{,}000$ K [58], then the mass in the temperature range 100,000–300,000 K may have been anywhere in the range 0.002 to 0.1 $M_\odot$.

The thermal timescale for this OPR might have any value ranging from a few days to a few months, depending partly on how we define it. Given these strong dependences, the effects of OPR instabilities may be very sensitive to the evolutionary state and structure of the star—and thus consistent with observed facts about giant eruptions and LBV's.

A standard LBV eruption might expel no more than the OPR mass, and the two amounts may even be related. But a giant eruption rooted in that region must be geyser-like (Section 5.2). Some forms of instability cannot easily function like geysers, for reasons involving timescales – see a remark later below.

Since 1993, almost every stability analysis of very massive stars has emphasized "strange modes" of pulsation [36,40,97,111–113]. Apart from mathematical details, they have the following attributes.

1. Strange modes are essentially dynamical rather than thermal. They resemble accoustic waves, in contrast to thermodynamic Carnot-cycle pulsations driven by the κ mechanism in lower-mass stars.
2. Hence they are fundamentally non-adiabatic. They become especially strong if the local thermal timescale is shorter than the dynamical timescale.
3. They occur if radiation pressure exceeds gas pressure.
4. The density dependence of opacity, $\partial \kappa / \partial \rho$, is critical; but $\partial \kappa / \partial T$ is not.
5. Purely radial strange modes can occur, but non-radial modes may be more important.

These characteristics are almost perfectly suited to the OPR in a star near the Eddington Limit. Item 3 causes the local mass density to be relatively low, thereby enabling item 2. For a very brief account of strange modes, see [112].

Altogether, then, in a star near the Eddington Limit, the OPR forms a queasy sort of cavity between the stellar interior and the outer layers—with strong consequences for pulsation modes. Even if we consider only 1-D radial motions, gas-dynamical simulations reveal phenomena that appear crucial for LBV's and giant eruptions [39,85,113–115]. An essential factor is the time dependence of convection. Normally a massive stellar interior obeys the Eddington Limit by shifting some of the energy flux to convection where necessary [104]. But this assumption fails in a structure that changes rapidly, e.g., in pulsating layers. Convection needs some time to develop, and the dominant convective cells have finite turnover times. Hence the convective energy flux lags behind the total energy flux, especially in the circumstances listed above for strange modes. As explained in the papers cited above, this fact causes the radiative flux to exceed the Eddington Limit at some times and places in a pulsation cycle. No actual runaway outburst occurred in the simulations, but their boundary conditions and lack of non-radial modes may have inhibited such a development.

Three-dimensional simulations show the spatial fluctuations of convection, and reveal some opacity-related phenomena that cannot appear in the 1-D models [34,110,116]. For instance, helium opacity can become large within clumps of gas that have been lifted to regions with $T < 70,000$ K [34]. The result is a second opacity-peak region, indirectly caused by the iron opacity bump. Local regions in and below the photosphere can thus have large radiative accelerations. The outer layers become supersonically turbulent, and local parcels of mass can be ejected in a chaotic way. In this manner we begin to graduate from "pulsations" to "stellar activity" or even "weather"— see Figure 2 in ref. [34]. Unfortunately, the 3-D calculations are so expensive in CPU time that only a few have been attempted.

Given the facts outlined above, the OPR is very likely the root of the LBV phenomenon. It is especially dramatic in stars with LBV-like L/M ratios, and it is rich in phenomena that appear relevant to the questions in Section 5.2 above. Moreover, effects found in numerical simulations can help to accelerate the ejecta. Therefore, contrary to most papers in this topic, we should not assume that LBV outflows are merely line-driven winds—especially during a major outburst (cf. [117]).

But can the OPR incite a giant eruption? No simulation has yet produced an outright eruption. Maybe this is so because the "weather" analogy is apt! A terrestrial atmosphere simulation would usually go for a long time before it produces a typhoon. By analogy, perhaps a stellar eruption results from an infrequent coincidence of several chaotic processes—a Perfect Storm. Note that the inflated

LBV model in [34] was still expanding when the calculations ended after 700 dynamical timescales, only a few percent of a typical event duration.

As mentioned earlier, if a giant eruption can originate in the OPR layers of the star, then it must be a geyser-style process with instability propagating downward through the stellar layers—or rather, the successive layers move outward past the instability zone. The energy budget thereby becomes complicated, because inner regions tend to contract in order to compensate for the lost energy. As noted by [117], the resulting small increase in local temperature can increase nuclear reaction rates; so the overall event may be indirectly powered by hydrogen burning. Nearly all of the mass is close to dynamical equilibrium throughout this process, but thermal equilibrium fails in the outer regions. This story may lend itself to additional instabilities deep within the star.

Unfortunately the geyser analogy may fail for some types of OPR pulsational instability [118,119]. When a pulse of material has been expelled, the driving mechanism needs time to re-establish itself, and that time may be much longer than the dynamical timescale. In that case the instability cannot easily propagate through deeper layers.

At first sight, a supernova precursor eruption (Section 3.2) cannot originate in the OPR, because such events happen only a few years before core collapse, and the outer layers evolve much slower than that. In the outer layers, there is nothing special about the core's last few years. But this view may be too naive, for reasons noted in [120]. During those final years, turbulence in the core can generate unsteady burning and outward waves, which tend to expand the outer layers—"an early warning system for core collapse." The OPR is so sensitive that it may respond violently to even a small change in the outer-layer structure. Thus it seems conceivable that the opacity peak might play a role in every class of eruption from LBV events to pre-SN outbursts.

5.4. Instabilities in and Near the Stellar Core

Some giant eruptions probably originate near the centers of massive stars, rather than in the OPR. But the definite examples concern true supernovae in special circumstances, and the nature of SN impostors (i.e., giant eruptions that are not related to SN events) remains murky.

A supernova can produce a radiation-driven eruption instead of a visible blast wave. Suppose that a star produces an opaque mass outflow in the years preceding its SN explosion. In that case, when the SN blast wave emerges from the star and moves into the surrounding opaque ejecta, photons may diffuse outward faster than the shock speed [121–123]. Radiation thus reaches the $\tau \sim 1$ radius substantially before the shock does; indeed the shock may emerge long after the time of maximum light. The visible event represents "photon breakout" rather than "shock breakout." Maximum luminosity is far above the Eddington Limit.

The photon diffusion rate can be described in terms of a random walk, but the familiar version of that concept doesn't give a unique diffusion speed for comparison with the SN shock speed. Instead, here's a formal example with an constant diffusion speed. Consider pure scattering in a spherical configuration; absorption and re-emission are equivalent to scattering so far as the total energy flux is concerned. Suppose that the scattering coefficient is $k(r) = \zeta/r$, with a constant parameter ζ. (In the notation of Section 4 above, $k = \rho\kappa$.) In this case the time-dependent diffusion equation has a similarity solution that represents an expanding pulse of radiation density:

$$U(r,t) = \left(\frac{E}{8\pi}\right)\left(\frac{3\zeta}{ct}\right)^3 \exp\left(-\frac{3\zeta r}{ct}\right), \tag{5}$$

which has total energy E. Because of the choice $k \propto r^{-1}$, this expression contains a velocity-like ratio r/t. At any given location r, the maximum radiation flux occurs at $t = 0.75\zeta r/c$ when about 24% of the energy has passed. At any given time, half of the radiation is located outside radius $r_{1/2} \approx 0.9ct/\zeta$; so the median diffusion speed is approximately $0.9c/\zeta$. About 10% of the radiation energy moves outward faster than $1.8c/\zeta$. If ζ is small enough for this speed to outrun the SN blast wave, but large enough to make the pre-SN outflow opaque – say $1 < \zeta < 40$—then a radiation-driven eruption rapidly develops.

In a more realistic case with $k(r) \propto r^{-2}$ rather than r^{-1}, the diffusion speed accelerates outward. The light curve can resemble Figure 2, with a sudden decline after most of the radiation has passed through the photosphere. Meanwhile, of course, the radiation accelerates the mass outflow. Later the SN blast wave may emerge after the brightness has declined, with only a modest display. Thus SN 2011ht, for instance, may have been either a true supernova with a hidden shock, or an impostor with no shock [42].

One point about shrouded supernovae is so obvious that it is often underemphasized: *the required circumstellar material was probably ejected in one or more giant eruptions* with $\dot{M} > 10^{-3}$ $M_\odot$ y^{-1}, years or decades before the core-collapse events (Section 3.2 above). Many researchers assume that the pre-SN stars were LBV's, because LBV's are the best-advertised eruptors. But this surmise is not entirely consistent, because the deduced amount of ejecta usually surpasses the familiar type of major LBV eruption [42]. A giant LBV event (Sections 3.1 and 5.2 above) would be needed—i.e., much stronger than any LBV outburst observed in the past few decades. If such large eruptions really do occur as part of the general LBV story, they must be very infrequent. Thus we should be very surprised if several known SN events were closely preceded by random LBV episodes on that scale. It seems far more likely that the pre-SN outbursts were somehow related to the imminent core collapse, i.e., related to the core structure. Hence the deduced pre-SN mass ejection probably had nothing to do with standard LBV behavior. Those stars may have been LBV's, but there is no good reason to assume that they were. The precursor events may have resembled the outbursts of SN 2009ip (Section 3.2 above), but with longer time scales.

Pulsational pair instability attracted attention a decade ago with reference to supernova impostors [124–127], because it can produce repeated eruptions. A star with initial mass around 150 $M_\odot$ eventually becomes a pair-production supernova, wherein core temperatures rise high enough to produce a significant rate of $\gamma + \gamma \rightarrow e^- + e^+$. This transfer of energy to rest mass causes a pressure deficit, while the adiabatic index falls well below 4/3 which implies dynamical instability. Hence the core begins to collapse, raising the temperature so the pair creation accelerates, and runaway nuclear reactions unbind the whole star. But if the star's mass is somewhat smaller, then the central region stabilizes before it is entirely disrupted, and the episode can repeat. This repetition motivates the term "pulsational" instability. It must be very rare because it occurs only in near-terminal stages of very massive stars. The phenomenon seems too indeterminate to be really satisfying; the time interval between events is extremely sensitive to obscure details, and the first such event probably expels all the hydrogen. For the latter reason, supernova impostors such as η Car presumably did not involve this type of event. Apart from having too many syllables, the main fault of pulsational pair instability is the difficulty of making definite statements about it.

Parallel to the computational developments noted in Section 5.3, 3-D simulations have revealed new phenomena in the star's core region. An important fact is that some numerical techniques, especially in 1-D models, entail artificial (i.e., illusory) damping of fluctuations. 3-D convection and turbulence become particularly vigorous during a massive star's final years [120,128], with dynamic effects that cannot be represented in 1-D calculations. Turbulence generates gasdynamic waves, which carry energy outward. Consequently the outer layers, feebly bound because they are close to the Eddington Limit, expand or perhaps even erupt. Mass ejection may occur [120,129,130], while the turbulence also causes the nuclear burning to be unsteady or even explosive. The outer layers are quite vulnerable because their binding energy is much smaller than the nuclear energy being processed in the central region. As mentioned earlier, the opacity-peak region may produce enhanced instabilities because of the waves flowing through it. Given these circumstances, perhaps we should not be surprised that paroxysms occur just before core collapse.

What can we say about core-based eruptions that are *not* related to a SN event? The processes mentioned above would not be suitable. Eta Carinae, for instance, still has considerable hydrogen even after its Great Eruption. Evidently it has not yet evolved far enough to have an exotic core region. It probably has a very capable opacity peak region, but doubts about the geyser process (see above) may require a core-region instability instead. One credible possibility has been suggested in refs. [113,118,119]. In a very massive, moderately evolved star, gravity pulsation modes (like ocean waves rather than pressure waves) may become numerous and strong at the lower boundary of the region that still has some hydrogen. Suppose that they grow enough to mix some hydrogen into the hot dense zones below that boundary. The resulting burst of hydrogen-burning would rapidly lift some material, possibly ejecting a set of outer layers, and then the remaining material would settle down. Events of this type may recur on a thermal timescale, reasonable for an object like η Car. Some remarks in [117], concerning enhanced reaction rates when a star's total energy has been reduced by mass ejection, may be relevant to this idea.

Explorations of core instabilities have naturally concentrated on the final pre-SN state, because the structure is highly complex then and because SN-related processes are most fashionable. With the development of 3-D computation, however, unpredicted phenomena may appear at earlier stages of evolution; anyway that's what we need for giant eruptions if the opacity-peak region turns out to be inadequate.

6. Other Issues

This narrative has omitted stellar rotation even though it is probably important. Rotation would greatly lengthen the narrative, and, more important, would expand the number of free parameters. A traditional exploration strategy makes sense: (1) Begin with simple non-rotating models, (2) learn whether the known processes can account for eruptions without rotation, and then (3) explore the effects of angular momentum. This topic has not yet reached stage 3. In view of the multiple parameters required for a distribution of angular momentum, this approach is particularly justified for expensive 3-D simulations (Section 5.3 above). Apart from η Car as noted below [58] and the morphology of LBV ejecta-nebulae [26], there is little observational evidence concerning angular momentum in radiation-driven eruptions.

The same attitude is even more justified for eruption scenarios that require interactions of binary or multiple stars, particularly merger events. As noted many years ago, speculations in that vein allow theorists to "ascend into free-parameter heaven" [131]. Generically they require either small orbits or unusual orbit parameters. Such models are credible for lower-luminosity events that are not discussed in this review (e.g., red transients), because moderate-luminosity star systems are very numerous. The observed lower-luminosity outbursts can be explained by supposing that a tiny fraction of stars experience mergers and other exotic interactions. Stars with $L > 10^{5.5} L_\odot$, however, are scarce; so we should not see the observed number of LBV's and giant eruptors if unusual circumstances are required. It is true that most massive stars have companions, but only a small fraction of them are close enough for major interactions [132]. Equally important, *there is no evident need* for eruption models of that type. The HRD upper limit in Figure 1 applies to practically all stars above 50 $M_\odot$, not just those with close companions. The LBV instability strip becomes much harder to explain if we suppose that it depends on multi-parameter interacting binaries [1,13]. And, perhaps most important, the single-star processes in Section 5 appear sufficiently promising until proven otherwise. In summary: Binary and multiple-system phenomena certainly deserve attention, but they have not yet earned a well-defined place in the giant eruption puzzle.

Binarity does play a role for our best-observed supernova impostor, η Car, but it probably did not provide the basic instability mechanism. This object merits additional paragraphs here because so much is known about it, especially regarding some potentially instructive abnormalities. For instance, consider the hot secondary star's high orbital eccentricity, $\epsilon \approx 0.85$, with a periastron distance only about 3× or 4× larger than the primary star's radius [64]. Tidal effects are significant during about 3% of the 5.5-year orbital period, and may have triggered the Great Eruption as noted in Section 3.3. But this is not a straightforward idea! When we take the Eddington factor Γ into account, the companion star's maximum tidal effect is of the order of 10% as strong as effective gravity at the star's surface [58]. The iron opacity peak region is less perturbed because it has a smaller radius, and the core region is practically unaffected. Hence the periastron tidal-trigger conjecture requires an instability that began fairly near the surface—the geyser concept again. Moreover, the eruption did not begin suddenly; instead the star's brightness began to rise and fluctuate years earlier [63,133]. Later the mass outflow persisted long after tidal forces became negligible. Since the tidal maximum at periastron had a duration comparable to the star's dynamical timescale, it was neither an adiabatic nor an impulsive perturbation. Nonetheless the trigger concept has undeniable appeal. One can easily imagine a star expanding due to evolution, until it encountered a radius limit enforced by its companion. This differs from a familiar Roche lobe story in two respects: it was close to the Eddington Limit, and the tidal force made itself felt only for a few weeks near each periastron.

Two other points should be noted about η Car's periastron passages. First, after a sufficiently long time, tidal friction should cause the star's outer layers to rotate synchronously with the orbital rate at periastron, like the planet Mercury. The surface rotation period would then be roughly 90 days. In fact the X-rays show a quasi-period of that length [134]. Second, why is the orbit so eccentric? Its period would be only about 130 days if it were circular with $r=$ the present-day periastron distance. If the orbit was circular a few thousand years ago, then the simplest explanation for large ϵ has two or three parts: (1) Most of the eruptive mass loss must have occurred near periastron, in order to eccentrify the orbit. (2) Several giant eruptions like 1830–1860 were necessary in order to attain $\epsilon \approx 0.85$. (3) However, since that value is very high, some additional factor was probably needed—e.g., asymmetric mass flows. See [58] and references therein.

Another of η Car's oddities concerns its equatorial skirt of ejecta. It is manifestly not a rotating disk, but instead appears to consist of radial spikes of ejecta [57,135]. Velocities and proper motions indicate that they formed at about the same time as the Homunculus lobes.

Various authors have speculated that η Car's giant eruption was a merger event, entailing a former third star [136–140]. Their scenarios employ at least 8 adjustable parameters, plus qualitative assumptions that are not emphasized, in order to account for 5 or fewer observed quantities. There is no evident need to postulate a third object; the primary star appears well suited to the single-star ideas listed in Section 5 above. (For instance, it is near the Eddington Limit without any reference to companion objects, and probably has a substantial iron opacity peak region.) The most detailed merger model [139] predicted too low a helium abundance, its stated quiescent brightness was far too low, and it was vague about the ejecta morphology. Exotic models can be interesting, but there is no reason to guess that they are necessary for this object. The single-star processes, modified by the known companion star, intuitively seem very promising for η Car and have not yet been analyzed in sufficient detail.

High-velocity material associated with η Car has been interpreted as evidence for either a blast wave or an merger event [77,140]. Some outlying ejecta have Doppler velocities of 1000–3000 km s^{-1} [26], and light-echo spectra of the Great Eruption may show velocities as fast as 10,000 km s^{-1} [77]. However, other interpretations appear more likely according to the "maximum simplicity" criterion. Judging from Hα images of the outer ejecta, the high-speed mass and kinetic energy are probably less than $10^{-5}\ M_\odot$ and 10^{44} ergs, and possibly much less. These amounts are substantially smaller than the mass and thermal energy of the star's opacity peak region, for instance. If an eruptive instability begins suddenly, a small amount of leading material may be ejected to very high speeds, analogous to the acceleration of a SN blast wave as it moves through a negative density gradient. Indeed an acceleration feature like that can be seen in Figure 2 of [34]. The standard super-Eddington flow becomes established after the initial transient burst. This explanation may be wrong, but it as well-developed as the exotic interpretations, and more credible because it fits the other characteristics of η Car's ejecta [57]. Moreover, the very-high-velocity line wings in the light echo spectra are so faint that they may be either instrumental artifacts or features caused by Thomson scattering in dense locales of the outflow.

As emphasized in Section 4 above, the brightness and spectrum of a radiation-driven eruption do not tell us much about the star and its structure. However, the post-eruption behavior may give some useful information. At any given time during the event, the entire configuration is close to dynamical equilibrium (including flow processes) but far from thermal and rotational equilibrium. This remains true after the event subsides, leaving a star with a peculiar thermal structure. It should then recover—i.e., find a new equilibrium state—in a few thermal timescales. This process has been observed in η Car, and the record is interesting in two respects: it has taken longer than the expected 50 years, and it has been quite unsteady [58]. Major changes occurred at 50-year intervals [58,63], and the spectrum has evolved more rapidly during the past 20 years [141,142]. This temporal structure surely depends on the star's thermal and rotational structure. A preliminary assessment of the recovery problem was reported in [143], but multiple 3-D simulations are needed.

As mentioned near the end of Section 5.2, stellar activity and turbulent MHD may occur in the outer layers of LBV's and/or related stars. This would not be terribly surprising, since one can write the Schwarzschild criterion in a form that looks much like the Eddington Limit. The point is that MHD waves, or similar processes, may assist the outward acceleration forces, and might even produce violent instabilities.

Finally, a point in Section 4.4 merits repetition because it affects this entire topic: a radiation-driven outflow is difficult to calculate. If one writes 1-D analytic equations for radiative transfer and acceleration, they give nonsensical results because a real outflow automatically becomes inhomogeneous. Acceleration and radiation leakage depend on the sizes, spacing, and even the shapes of the granules. These effects are too intricate to calculate ab initio for every model or sub-model. Therefore it might be valuable, and certainly would be interesting, to have some sort of general prescription based on many specialized 3-D simulations. As a first step, those simulations could include only Thomson scattering. What factors determine the characteristic size scales and time scales and density distributions? Cf. [83,84,117].

Funding: This research received no external funding, and was supported primarily by photons.

Acknowledgments: I am grateful to R.M. Humphreys, J. Guzik, I. Appenzeller, C. de Jager, M. Schwarzschild, E.E. Salpeter, and A.S. Eddington for indicating good points of view for this topic.

Conflicts of Interest: The author declares no conflicts of interest.

References

1. Humphreys, R.M.; Davidson, K. Luminous Blue Variables: Astrophysical Geysers. *Publ. Astron. Soc. Pac.* **1994**, *106*, 1025–1051. [CrossRef]
2. Zwicky, F. Supernovae. In *Stellar Structure, Stars and Stellar Systems vol. VIII*; Aller, L., McLaughlin, D.B., Eds.; University of Chicago Press: Chicago, IL, USA, 1965; pp. 367–424.
3. Kochanek, C.S.; Szczygiel, D.M.; Stanek, K.Z. The Supernova Impostor SN 1961V: Spitzer Shows That Zwicky Was Right (Again). *Astrophys. J.* **2011**, *737*, 76. [CrossRef]
4. van Dyk, S.D.; Matheson, T. It's Alive! The Supernova Impostor 1961V. *Astrophys. J.* **2012**, *746*, 179. [CrossRef]
5. Humphreys, R.M.; Davidson, K. Studies of Luminous Stars in Nearby Galaxies. III. The Evolution of the Most Massive Stars in the Milky Way and the Large Magellanic Cloud. *Astrophys. J.* **1979**, *232*, 409–420. [CrossRef]
6. Fullerton, A.W.; Massa, D.L.; Prinja, R.K. The Discordance of Mass-Loss Estimates for Galactic O-type Stars. *Astrophys. J.* **2006**, *637*, 1025–1039. [CrossRef]
7. Smith, N.; Owocki, S.P. On the Role of Continuum-driven Eruptions in the Evolution of Very Massive Stars. *Astrophys. J.* **2006**, *645*, L45–L48. [CrossRef]
8. Van Dyk, S.D. The η Carinae Analogs. In *The Fate of the Most Massive Stars, A.S.P.Conf. 332*; Humphreys, R.M., Stanek, K.Z., Eds.; Astronomical Society of the Pacific: San Francisco, CA, USA, 2005; pp. 47–57, ISBN 1-58381-195-8.
9. Van Dyk, S.D.; Matheson, T. The Supernova Impostors. In *Eta Carinae and the Supernova Impostors, ASSL 384*; Davidson, K., Humphreys, R.M., Eds.; Springer: New York, NY, USA, 2012; pp. 249–274, ISBN 978-1-4614-2274-7.

10. Pastorello, A.; Cappellaro, E.; Inserra, C.; Smartt, S.J.; Pignata, G.; Benetti, S.; Valenti, S.; Fraser, M.; Takats, K.; Benitez, S.; et al. Interacting Supernovae and Supernova Impostors: SN 2009ip, is this the End? *Astrophys. J.* **2013**, *767*. [CrossRef]
11. Margutti, R.; Milisavljevic, D.; Soderberg, A.M.; Chornock, R.; Zauderer, B.A.; Murase, K.; Guidorzi, C.; Sanders, N.E.; Kuin, P.; Fransson, C.; et al. A Panchromatic View of the Restless SN 2009ip Reveals the Explosive Ejection of a Massive Star Envelope. *Astrophys. J.* **2014**, *780*, 21. [CrossRef]
12. Humphreys, R.M.; Weis, K.; Davidson, K.; Gordon, M.S. On the Social Traits of Luminous Blue Variables. *Astrophys. J.* **2016**, *825*, 64. [CrossRef]
13. Davidson, K.; Humphreys, R.M.; Weis, K. LBVs and Statistical Inference. *arXiv* **2016**, arXiv:1608.02007.
14. Weis, K.; Bomans, D. Luminous Blue Variables. *Galaxies* **2019**, submitted. [CrossRef]
15. Vink, J.S. Eta Carinae and the Luminous Blue Variables. In *Eta Carinae and the Supernova Impostors, ASSL 384*; Davidson, K., Humphreys, R.M., Eds.; Springer: New York, NY, USA, 2012; pp. 221–247, ISBN 978-1-4614-2274-7.
16. Wolf, B. Empirical Amplitude-luminosity Relation of S Doradus Variables and Extragalactic Distances. *Astron. Astrophys.* **1989**, *217*, 87–91.
17. Kudritzki, R.P.; Gabler, A.; Gabler, R.; Groth, H.G.; Pauldrach, A.W.A.; Puls, J. *Model Atmospheres and Quantitative Spectroscopy of Luminous Blue Stars Physics of Luminous Blue Variables, ASSP 157*; Davidson, K., Moffat, A., Lamers, H., Eds.; Kluwer: Dordrecht, The Netherlands, 1989; pp. 67–82. [CrossRef]
18. Pauldrach, A.W.A.; Puls, J. Radiation-driven Winds of Hot Luminous Stars. VIII. The Bistable Wind of P Cygni. *Astron. Astrophys.* **1990**, *237*, 409–424.
19. Stahl, O.; Wolf, B.; Kare, G.; Juettner, A.; Cassatella, A. Observations of the New Luminous Blue Variable R 110. *Astron. Astrophys.* **1990**, *228*, 379–386.
20. Sterken, C.; Gosset, E.; Juttner, A.; Stahl, O.; Wolf, B.; Axer, M. HD 160529: A New Galactic Luminous Blue Variable. *Astron. Astrophys.* **1991**, *247*, 383–392.
21. Vink, J.S.; de Koter, A. Predictions of Variable Mass Loss for Luminous Blue Variables. *Astron. Astrophys.* **2002**, *393*, 543–553. [CrossRef]
22. Aadland, E.; Massey, P.; Neugent, K.F.; Drout, M.R. Shedding Light on the Isolation of Luminous Blue Variables. *Astron. J.* **2016**, *156*, 294. [CrossRef]
23. Stahl, O.; Jankovics, I.; Ková, J.; Wolf, B.; Schmutz, W.; Kaufer, A.; Rivinius, Th.; Szeifert, Th. Long-term Spectroscopic Monitoring of the Luminous Blue Variable AG Carinae. *Astron. Astrophys.* **2001**, *375*, 54–69. [CrossRef]
24. Groh, J.H.; Hillier, D.J.; Damineli, A.; Whitelock, P.A.; Marang, F.; Rossi, C. On the Nature of the Prototype Luminous Blue Variable AG Carinae. *Astrophys. J.* **2009**, *698*, 1698–1720. [CrossRef]
25. Groh, J.H.; Hillier, D.J.; Damineli, A. On the Nature of the Prototype Luminous Blue Variable AG Carinae. II. Witnessing a Massive Star Evolving Close to the Eddington and Bistability Limits. *Astrophys. J.* **2011**, *736*, 46. [CrossRef]
26. Weis, K. The Outer Ejecta. In *Eta Carinae and the Supernova Impostors, ASSL 384*; Davidson, K., Humphreys, R.M., Eds.; Springer: New York, NY, USA, 2012; pp. 171–194, ISBN 978-1-4614-2274-7.
27. Thackeray, A.D. Some Southern Stars Involved in Nebulosity. *Mon. Not. R. Astron. Soc.* **1950**, *110*, 524–530. [CrossRef]
28. Nota, A.; Livio, M.; Clampin, M.; Schulte-Ladbeck, R. Nebulae around Luminous Blue Variables: A Unified Picture. *Astrophys. J.* **1995**, *448*, 788–796. [CrossRef]
29. Smith, L.J.; Stroud, M.P.; Esteban, C.; Vichez, J.M. The AG Carinae Nebula: Abundant Evidence for a Red Supergiant Progenitor? *Mon. Not. R. Astron. Soc.* **1997**, *290*, 265–275. [CrossRef]
30. Vamvatira-Nakou, C.; Hutsemékers, D.; Royer, P.; Cox, N.L.J.; Nazé, Y.; Rauw, G.; Waelkens, C.; Groenewegen, M.A.T. The Herschel View of the Nebula around AG Carinae. *Astron. Astrophys.* **2015**, *578*, A108. [CrossRef]
31. Wolf, B.; Appenzeller, I.; Stahl, O. IUE and Ground-based Spectroscopic Observations of the S Dor-type LMCvariable R71 During Minimum State. *Astron. Astrophys.* **1981**, *103*, 94–102.
32. Mehner, A.; Baade, D.; Rivinius, T.; Lennon, D.J.; Martayan, C.; Stahl, O.; Stefl, S. Broad-band Spectroscopy of the Ongoing Large Eruption of the Luminous Blue Variable R71. *Astron. Astrophys.* **2013**, *555*, A116. [CrossRef]
33. yyy Mehner, A.; Baade, D.; Groh, J.H.; Rivinius, T.; Hambsch, F.-J.; Bartlett, E.S.; Asmus, D.; Agliozzo, C.; Szeifert, T.; Stahl, O. Spectroscopic and Photometric Oscillatory Envelope Variability during the S Doradus Outburst of the Luminous Blue Variable R71. *Astron. Astrophys.* **2017**, *608*, A124. [CrossRef]

34. Jiang, Y.-F.; Cantiello, M.; Bildsten, L.; Quataert, E.; Blaes, O.; Stone, J. Outbursts of Luminous Blue Variable Stars from Variations in the Helium Opacity. *Nature* **2018**, *561*, 498–501. [CrossRef]
35. de Groot, M.; Sterken, C. (Eds.) *P Cygni 2000: 400 Years of Progress, ASP Conf. 233*; Astronomical Society of the Pacific: San Francisco, CA, USA, 2001; ISBN 1-58381-070-6.
36. Stothers, R.B.; Chin, C.-W. Dynamical Instability as the Cause of Massive Outbursts in Eta Carinae and Other Luminous Blue Variables. *Astrophys. J.* **1993**, *408*, L85–L88. [CrossRef]
37. Stothers, R.B.; Chin, C.-W. Luminous Blue Variables at Quiescence: The Zone of Avoidance in the H-R Diagram. *Astrophys. J.* **1994**, *426*, L43–L46. [CrossRef]
38. Stothers, R.B.; Chin, C.-W. The Brightest Supergiants Predicted by Theory. *Astrophys. J.* **1999**, *522*, 960–964. [CrossRef]
39. Lovekin, C.C.; Guzik, J.A. Pulsations as a Driver for LBV Variability. *Mon. Not. R. Astron. Soc.* **2014**, *445*, 1766–1773. [CrossRef]
40. Glatzel, W.; Kiriakidis, M. Stability of Massive Stars and the Humphreys/Davidson Limit. *Mon. Not. R. Astron. Soc.* **1993**, *263*, 375–384. [CrossRef]
41. Roming, P.W.A.; Pritchard, T.A.; Prieto, J.L.; Kochanek, C.S.; Fryer, C.L.; Davidson, K.; Humphreys, R.M.; Bayless, A.J.; Beacom, J.F.; Brown, P.J. The Unusual Temporal and Spectral Evolution of the Type IIn Supernova 2011ht. *Astrophys. J.* **2012**, *751*, 92. [CrossRef]
42. Humphreys, R.M.; Davidson, K.; Jones, T.J.; Pogge, R.W.; Grammer, S.H. The Unusual Temporal and Spectral Evolution of SN2011ht. II. Peculiar Type IIn or Impostor? *Astrophys. J.* **2012**, *760*, 93. [CrossRef]
43. Mauerhan, J.C.; Smith, N.; Silverman, J.M.; Filippenko, A.V.; Morgan, A.N.; Cenko, S.B.; Ganeshalingam, M.; Clubb, K.I.; Bloom, J.S.; et al. SN 2011ht: Confirming a class of interacting supernovae with plateau light curves (Type IIn-P). **2013**, *431*, 2599–2611. [CrossRef]
44. Dessart, L.; Hillier, D.J.; Gezari, S.; Basa, S.; Matheson, T. SN 1994W: An Interacting Supernova or Two Interacting Shells? *Mon. Not. R. Astron. Soc.* **2009**, *394*, 21–37. [CrossRef]
45. Kankare, E.; Ergon, M.; Bufano, F.; Spyromilio, J.; Mattila, S.; Chugai, N.N.; Lundqvist, P.; Pastorello, A.; Kotak, R.; Benetti, S.; et al. SN 2009kn - the twin of the Type IIn supernova 1994W. *Mon. Not. R. Astron. Soc.* **2012**, *424*, 855–873. [CrossRef]
46. Humphreys, R.M.; Davidson, K.; Smith, N. Eta Carinae's Second Eruption and the Light Curves of the Eta Car Variables. *Publ. Astron. Soc. Pac.* **1999**, *111*, 1124–1131. [CrossRef]
47. Wagner, R.M.; Vrba, F.J.; Henden, A.A.; Canzian, B.; Luginbuhl, C.B.; Filippenko, A.V.; Chornock, R.; Li, W.; Coil, A.L.; Schmidt, G.D.; et al. Discovery and Evolution of an Unusual Luminous Variable Star in NGC 3432 (SN 2000ch). *Publ. Astron. Soc. Pac.* **2004**, *116*, 326–336. [CrossRef]
48. Pastorello, A.; Botticella, M.T.; Trundle, C.; Taubenberger, S.; Mattila, S.; Kankare, E.; Elias-Rosa, N.; Benetti, S.; Duszanowicz, G.; Hermansson, L.; et al. Multiple major outbursts from a restless luminous blue variable in NGC 3432. *Mon. Not. R. Astron. Soc.* **2010**, *408*, 181–198. [CrossRef]
49. Van Dyk, S.D.; Cenko, S.B.; Clubb, K.I.; Fox, O.D.; Zheng, W.; Kelly, P.L.; Filippenko, A.V.; Smith, N. PSN J10524126+3640086 is the Continued Outburst of SN 2000ch. *Astron. Telegr.* **2013**, *4891*.
50. Humphreys, R.M.; Martin, J.C.; Gordon, M.S.; Jones, T.J. Multiple Outflows in the Giant Eruption of a Massive Star. *Astrophys. J.* **2016**, *826*, 191. [CrossRef]
51. Mauerhan, J.C.; Van Dyk, S.D.; Graham, M.L.; Zheng, W.; Clubb, K.I.; Filippenko, A.V.; Valenti, S.; Brown, P.; Smith, N.; Howell, D.A.; et al. SN Hunt 248: A super-Eddington outburst from a massive cool hypergiant. *Mon. Not. R. Astron. Soc.* **2015**, *447*, 1922–1934. [CrossRef]
52. Tammann, G.A.; Sandage, A. The Stellar Content and Distance of the Galaxy NGC 2403 in the M81 Group. *Astrophys. J.* **1968**, *151*, 825–860. [CrossRef]
53. Humphreys, R.M.; Davidson, K.; Van Dyk, S.D.; Gordon, M.S. A Tale of Two Impostors: SN2002kg and SN1954J in NGC 2403. *Astrophys. J.* **2017**, *848*, 86. [CrossRef]
54. Fraser, M.; Inserra, C.; Jerkstrand, A.; Kotak, R.; Pignata, G.; Benetti, S.; Botticella, M.-T.; Bufano, F.; Childress, M.; Mattila, S.; et al. SN 2009ip à la PESSTO: No evidence for core collapse yet. *Mon. Not. R. Astron. Soc.* **2013**, *433*, 1312–1337. [CrossRef]
55. Fraser, M.; Magee, M.; Kotak, R.; Smartt, S.J.; Smith, K.W.; Polshaw, J.; Drake, A.J.; Boles, T.; Lee, C.-H.; Burgett, W.S.; et al. Detection of an Outburst One Year Prior to the Explosion of SN 2011ht. *Astrophys. J. Lett.* **2013**, *779*, L8. [CrossRef]

56. Weis, K.; Bomans, D. SN 2002kg—The brightening of LBV V37 in NGC 2403. *Astron. Astrophys.* **2005**, *429*, L13–L16. [CrossRef]
57. Davidson, K.; Humphreys, R.M. (Eds.) *Eta Carinae and the Supernova Impostors, ASSL 384*; Springer: New York, NY, USA, 2012; ISBN 978-1-4614-2274-7.
58. Davidson, K. The Central Star: Instability and Recovery. In *Eta Carinae and the Supernova Impostors, ASSL 384*; Davidson, K., Humphreys, R.M., Eds.; Springer: New York, NY, USA, 2012; pp. 43–65, ISBN 978-1-4614-2274-7.
59. Smith, N. All Things Homunculus. In *Eta Carinae and the Supernova Impostors, ASSL 384*; Davidson, K., Humphreys, R.M., Eds.; Springer: New York, NY, USA, 2012; pp. 145–169, ISBN 978-1-4614-2274-7.
60. Davidson, K.; Dufour, R.J. Walborn, N.R.; Gull, T.R. Ultraviolet and Visual Wavelength Spectroscopy of Gas around Eta Carinae. *Astrophys. J.* **1986**, *305*, 867–879. [CrossRef]
61. Dufour, R.J.; Glover, T.W.; Hester, J.J.; Currie, D.G.; van Orsow, D.; Walter, D.K. HST-FOS UV-optical Spectra of Ejecta from η Carinae. In *Eta Carinae at the Millenium, ASP Conf. 179*; Morse, J.A., Humphreys, R.M., Damineli, A., Eds.; Astronomical Society of the Pacific: San Francisco, CA, USA, 2005; pp. 134–143, ISBN 1-58381-003-X.
62. Humphreys, R.M.; Davidson, K.; Koppelman, M. The Early Spectra of Eta Carinae 1892 to 1941 and the Onset of its High Excitation Emission Spectrum. *Astron. J.* **2008**, *135*, 1249–1263. [CrossRef]
63. Humphreys, R.M.; Martin, J.C. Eta Carinae from 1600 to the Present. In *Eta Carinae and the Supernova Impostors, ASSL 384*; Davidson, K., Humphreys, R.M., Eds.; Springer: New York, NY, USA, 2012; pp. 1–24, ISBN 978-1-4614-2274-7.
64. Davidson, K.; Ishibashi, K.; Martin, J.C. Concerning the Orbit of Eta Carinae. *New Astron.* **2017**, *1*, 6. [CrossRef]
65. Davidson, K.; Humphreys, R.M. Eta Carinae and Its Environment. *Annu. Rev. Astron. Astrophys.* **1997**, *35*, 1–32. [CrossRef]
66. Soker, N. Accretion by the Secondary in η Carinae During the Spectroscopic Event. *Astrophys. J.* **2005**, *635*, 540–546. [CrossRef]
67. Soker, N. Accretion onto the Companion of η Carine During the Spectroscopic Event. *Astrophys. J.* **2007**, *661*, 482–489. [CrossRef]
68. Kashi, A.; Soker, N. Possible Implications of Mass Accretion in Eta Carinae. *New Astron.* **2009**, *14*, 11–24. [CrossRef]
69. Davidson, K.; Mehner, A.; Humphreys, R.M.; Martin, J.C.; Ishibashi, K. Eta Carinae's 2014.6 Spectroscopic Event. *Astrophys. J.* **2015**, *801*, L15. [CrossRef]
70. Rybicki, G.B.; Lightman, A.P. *Radiative Processes in Astrophysics*; Wiley: New York, NY, USA, 1979; pp. 33–45, ISBN 0-471-04815-1.
71. Davidson, K. The Relation between Apparent Temperature and Mass-Loss Rate in Hypergiant Eruptions. *Astrophys. J.* **1987**, *317*, 760–764. [CrossRef]
72. Owocki, S.P.; Shaviv, N.J. The Spectral Temperature of Optically Thick Outflows with Application to Light Echo Spectra from Eta Carinae's Giant Eruption. *Mon. Not. R. Astron. Soc.* **2016**, *462*, 345–351. [CrossRef]
73. Hummer, D.G.; Rybicki, G.B. Radiative Transfer in Spherically Symmetric Systems. *Mon. Not. R. Astron. Soc.* **1971**, *152*, 1–19. [CrossRef]
74. Badnell, N.R.; Bautista, M.A.; Butler, K.; Delahaye, F.; Mendoza, C.; Palmeri, P.; Zeippen, C.J.; Seaton, M.J. Updated Opacities from the Opacity Project. *Mon. Not. R. Astron. Soc.* **2005**, *360*, 458–464. [CrossRef]
75. Rest, A.; Prieto, J.L.; Walborn, N.R.; Smith, N.; Bianco, F.B.; Chornock, R.; Welch, D.L.; Howell, D.A.; Huber, M.E.; Foley, R.J.; et al. Light Echoes Reveal An Unexpectedly Cool Eta Carinae During Its Great Eruption. *Nature* **2012**, *482*, 375–378. [CrossRef] [PubMed]
76. Prieto, J.L.; Rest, A.; Bianco, F.B.; Matheson, T.; Smith, N.; Walborn, N.R.; Hsiao, E.Y.; Chornock, R.; Paredes Alvarez, L.; Campillay, A.; et al. Light Echoes from Eta Carinae's Great Eruption. *Astrophys. J.* **2014**, *787*, L8. [CrossRef]
77. Smith, N.; Rest, A.; Andrews, J.E.; Matheson, T.; Bianco, F.B.; Prieto, J.L.; James, D.J.; Smith, R.C.; Strampelli, G.M.; Zenteno, A. Exceptionally Fast Ejecta Seen in Light Echos of Eta Carinae's Great Eruption. *Mon. Not. R. Astron. Soc.* **2018**, *480*, 1457–1465. [CrossRef]
78. Davidson, K.; Humphreys, R.M. The Great Eruption of Eta Carinae. *Nature* **2012**, *486*, E1. [CrossRef]
79. Weymann, R.J. Electron-Scattering Line Profiles in Nuclei of Seyfert Galaxies. *Astrophys. J.* **1970**, *160*, 31–41. [CrossRef]

80. Davidson, K.; Ebbets, D.; Weigelt, G.; Humphreys, R.M.; Hajian, A.R.; Walborn, N.R.; Rosa, M. HST/FOS Spectroscopy of Eta Carinae. *Astron. J.* **1995**, *109*, 1784–1796. [CrossRef]
81. Chugai, N.N. Broad Emission Lines from the Opaque Electron-Scattering Environment of SN 1998S. *Mon. Not. R. Astron. Soc.* **2001**, *326*, 1448–1454. [CrossRef]
82. Mehner, A.; Davidson, K.; Humphreys, R.M.; Walter, F.M.; Baade, D.; de Wit, W.J.; Martin, J.; Ishibashi, K.; Rivinius, T.; Martayan, C.; et al. Eta Carinae's 2014.6 Spectroscopic Event. *Astron. Astrophys.* **2015**, *578*, A122. [CrossRef]
83. Shaviv, N.J. The Porous Atmosphere of Eta Carinae. *Astrophys. J.* **2000**, *532*, L137–L140. [CrossRef]
84. Owocki, S.P.; Shaviv, N.J. Instability and Mass Loss Near the Eddington Limit. In *Eta Carinae and the Supernova Impostors, ASSL 384*; Davidson, K., Humphreys, R.M., Eds.; Springer: New York, NY, USA, 2012; pp. 275–299, ISBN 978-1-4614-2274-7.
85. Guzik, J.A.; Fryer, C.; Urbatsch, T.J.; Owocki, S.P. Radiation Transport Through Super-Eddington Winds. In *Third BRITE Science Conference*; Wade, G.A., Baade, D., Guzik, J.A., Smolec, R., Eds.; Polish Astronomical Society: Warsaw, Poland 2018; pp. 33–36, ISBN 978-83-540430-1-7.
86. Shaviv, N.J. The Nature of the Radiative Hydrodynamic Instabilities in Radiatively Supported Thomson Atmospheres. *Astrophys. J.* **2001**, *549*, 1093–1110. [CrossRef]
87. Owocki, s.P.; Hirai, R.; Podsiadlowski, P.; Schneider, F.R.N. Hydrodynamic Simulations and Similarity Relations for Eruptive Mass Loss from Massive Stars. *Mon. Not. R. Astron. Soc.* **2019**, *485*, 988–1000. [CrossRef]
88. Leitherer, C.; Schmutz, W.; Abbott, D.C.; Hamann, W.-R.; Wessolowski, U. Atmospheric Models for Luminous Blue Variables. *Astrophys. J.* **1989**, *346*, 919–931. [CrossRef]
89. de Koter, A.; Lamers, H.J.G.L.M.; Schmutz, W. Variability of Luminous Blue Variables II. Parameter Study. *Astron. Astrophys.* **1996**, *306*, 501–518.
90. Gräfener, G.; Owocki, S.P.; Vink, J.S. Stellar Envelope Inflation near the Eddington Limit. *Astron. Astrophys.* **2012**, *538*, A40. [CrossRef]
91. Sanyal, D.; Grassitelli, L.; Langer, N.; Bestenlehner, J.M. Massive Main-sequence Stars Evolving at the Eddington Limit. *Astron. Astrophys.* **2015**, *580*, A20. [CrossRef]
92. Shore, S.N. *An Introduction to Astrophysical Hydrodynamics*; Academic Press: San Diego, CA, USA, 1992; pp. 260–273, ISBN 0-12-640670-7.
93. Lamers, H.J.G.L.M.; Cassinelli, J.P. *Introduction to Stellar Winds*; Cambridge University Press: Cambridge, UK, 1999. ISBN 0-521-59398-0.
94. Vink, J.S. Fast & Slow Winds from Supergiants and Luminous Blue Variables. *Astron. Astrophys.* **2018**, *619*, A54. [CrossRef]
95. Schwarzschild, M.; Härm, R. On the Maximum Mass of Stable Stars. *Astrophys. J.* **1959**, *129*, 637–646. [CrossRef]
96. Stothers, R.B. A Semiempirical Test for Dynamical Instability in Luminous Blue Variables. *Astrophys. J.* **1999**, *516*, 366–368. [CrossRef]
97. Glatzel, W. Instabilities in the Most Massive Evolved Stars. In *The Fate of the Most Massive Stars, ASP Conf. 332*; Humphreys, R., Stanek, K., Eds.; Astronomical Society of the Pacific: San Francisco, CA, USA, 2005; pp. 22–32, ISBN 1-58381-195-8.
98. Humphreys, R.M.; Davidson, K. The Most Luminous Stars. *Science* **1984**, *223*, 243–249. [CrossRef]
99. Appenzeller, I. Instability in Massive Stars - An Overview. In *Luminous Stars and Associations in Galaxies, IAU Symp. 116*; De Loore, C.W.H., Willis, A.J., Laskarides, P., Eds.; Reidel: Dordrecht, The Netherlands, 1986; pp. 139–149.
100. Lamers, H.J.G.L.M. P Cygni Type Stars—Evolution and Physical Processes. In *Luminous Stars and Associations in Galaxies, IAU Symp. 116*; De Loore, C.W.H., Willis, A.J., Laskarides, P., Eds.; Reidel: Dordrecht, The Netherlands, 1986; pp. 157–178.
101. Davidson, K. Giant Outbursts of the Eta Carinae—P Cygni Type. In *Instabilities in Luminous Early Type Stars, ASSL 136*; Lamers, H.J.G.L.M., de Loore, W.H., Eds.; Reidel: Dordrecht, The Netherlands, 1987; pp. 127–141. [CrossRef]
102. Appenzeller, I. The Role of Radiation Pressure in LBV Atmospheres. In *Physics of Luminous Blue Variables, IAU Colloq. 113*; Davidson, K., Moffat, A.F.J., Lamers, H.J.G.L.M., Eds.; Kluwer: Dordrecht, The Netherlands, 1989; pp. 195–204, ISBN 0-7923-0443-8.
103. Asplund, M. The Stability of Late-type Stars Close to the Eddington Limit. *Astron. Astrophys.* **1998**, *330*, 641–650.

104. Joss, P.C.; Salpeter, E.E.; Ostriker, J.P. On the Critical Luminosity in Stellar Interiors. *Astrophys. J.* **1973**, *181*, 429–438. [CrossRef]
105. Lamers, H.J.G.L.M.; Fitzpatrick, E.L. The Relationship between the Eddington Limit, the Observed Upper Luminosity Limit for Massive Stars, and the Luminous Blue Variables. *Astrophys. J.* **1988**, *324*, 279–287. [CrossRef]
106. Gustafsson, B.; Plez, B. Can Classical Model Atmospheres be of Any Use for the Study of Hypergiants? In *Instabilities and Evolved Supergiants and Hypergiants*; de Jager, C., Nieuwenhuijzen, Eds.; North-Holland: Amsterdam, The Netherlands, 1992; pp. 86–97.
107. Ulmer, A.; Fitzpatrick, E.L. Revisiting the Modified Eddington Limit for Massive Stars. *Astrophys. J.* **1998**, *504*, 200–206. [CrossRef]
108. Lamers, H.J.G.L.M.; Snow, T.P.; Lindholm, D.M. Terminal Velocities and the Bistability of Stellar Winds. *Astrophys. J.* **1995**, *455*, 269–285. [CrossRef]
109. Vink, J.S.; de Koter, A.; Lamers, H.J.G.L.M. On the Nature of the Bi-stability Jump in the Winds of Early-type Supergiants. *Astr. Astrophys* **1999**, *350*, 181–196.
110. Jiang, Y.-F.; Cantiello, M.; Bildsten, L.; Quataert, E.; Blaes, O. Local Radiation Hydrodynamic Simulations of Massive Star Envelopes at the Iron Opacity Peak. *Astrophys. J.* **2015**, *813*, 74. [CrossRef]
111. Gautschy, A.; Glatzel, W. On Highly Non-Adiabatic Stellar Pulsations and the Origin of Strange Modes. *Mon. Not. R. Astron. Soc.* **245**, *597*, 597–613.
112. Saio, H. Strange Modes. In *Communications in Asteroseismology, Vol. 158, Proceedings of 38th Liege International Astrophysical Colloquium*; Noels, A., Aerts, C., Montalban, J., Miglio, A., Briquet, M., Eds.; Springer: Berlin, Germany, 2009; pp. 245–250.
113. Guzik, J.A.; Lovekin, C.C. Pulsations and Hydrodynamics of Luminous Blue Variable Stars. *arXiv* **2014**, arXiv:1402.0257.
114. Onifer, A.J.; Guzik, J.A. Pulsation-Initiated Mass Loss in Luminous Blue Variables: A Parameter Study. In *Massive Stars as Cosmic Engines, IAU Symp. 250*; International Astronomical Union: Paris, France, 2005; pp. 83–88. [CrossRef]
115. Lovekin, C.C.; Guzik, J.A. Behaviour of Pulsations in Hydrodynamic Models of Massive Stars. In *New Windows on Massive Stars, IAU Symp. 307*; International Astronomical Union: Paris, France, 2014; pp. 176–181. [CrossRef]
116. Jiang, Y.-F.; Cantiello, M.; Bildsten, L.; Quataert, E.; Blaes, O.; Stone, J. Three Dimensional Radiation Hydrodynamic Simulations of Massive Star Envelopes. *arXiv* **2018**, arXiv:1809.10187.
117. Quataert, E.; Fernández, R.; Kasen, D.; Klion, H.; Paxton, W. Super-Eddington Stellar Winds Driven by Near-Surface Energy Deposition. *Mon. Not. R. Astron. Soc.* **2016**, *458*, 1214–1233. [CrossRef]
118. Guzik, J.A. Instability Considerations for Massive Star Eruptions. In *The Fate of the Most Massive Stars, A.S.P.Conf. 332*; Humphreys, R.M., Stanek, K.Z., Eds.; Astronomical Society of the Pacific: San Francisco, CA, USA, 2005; pp. 204–210, ISBN 1-58381-195-8.
119. Guzik, J.A.; Cox, A.N.; Despain, K.M. Pulsation-driven Outflows in Luminous Blue Variables. In *Eta Carinae at the Millenium, ASP Conf. 179*; Morse, J.A., Humphreys, R.M., Damineli, A., Eds.; Astronomical Society of the Pacific: San Francisco, CA, USA, 2005; pp. 347–353, ISBN 1-58381-003-X.
120. Arnett, D.W.; Smith, N. Preparing for an Explosion: Hydrodynamic Instabilities and Turbulence in Presupernovae. *Astrophys. J.* **2014**, *785*, 82. [CrossRef]
121. Chugai, N.N.; Blinnikov, S.I.; Cumming, R.J.; Lundqvist, P.; Bragaglia, A.; Filippenko, A.V.; Leonard, D.C.; Matheson, T.; Sollerman, J. The Type IIn Supernova 1994W: Evidence for the Explosive Ejection of a Circumstellar Envelope. *Mon. Not. R. Astron. Soc.* **2004**, *352*, 1213–1231. [CrossRef]
122. Chevalier, R.A.; Irwin, C.M. Shock Breakout in Dense Mass Loss: Luminous Supernovae. *Astrophys. J. Lett.* **2011**, *729*, L6. [CrossRef]
123. Moriya, T.J.; Tominaga, N. Diversity of Luminous Supernovae from Non-steady Mass Loss. *Astrophys. J.* **2012**, *747*, 113. [CrossRef]
124. Woosley, S.E.; Blinnikov, S.; Heger, A. Pulsational Pair Instability as an Explanation for the Most Luminous Supernovae. *Nature* **2007**, *450*, 390–392. [CrossRef]
125. Heger, A. The Final Stages of Massive Star Evolution and Their Supernovae. In *Eta Carinae and the Supernova Impostors, ASSL 384*; Davidson, K., Humphreys, R.M., Eds.; Springer: New York, NY, USA, 2012; pp. 299–326, ISBN 978-1-4614-2274-7.

126. Chatzopoulos, E.; Wheeler, J.C. Hydrogen-poor Circumstellar Shells from Pulsational Pair-instability Supernovae with Rapidly Rotating Progenitors. *Astrophys. J.* **2012**, *760*, 154. [CrossRef]
127. Leung, S.-C.; Nomoto, K.; Blinnikov, S. Pulsational Pair-instability Supernovae. I. Pre-collapse Evolution and Pulsational Mass Ejection. *Astrophys. J.* **2019**, *887*, 72. [CrossRef]
128. Arnett, D.W.; Meakin, C. Toward Realistic Progenitors of Core-Collapse Supernovae. *Astrophys. J.* **2011**, *733*, 78. [CrossRef]
129. Quataert, E.; Shiode, J.H. Wave-driven Mass Loss in the Last Year of Stellar Evolution: Setting the Stage for the Most Luminous Core-Collapse Supernovae. *Mon. Not. R. Astron. Soc.* **2012**, *423*, L92–L96. [CrossRef]
130. Shiode, J.H.; Quataert, E. Setting the Stage for Circumstellar Interaction in Core-Collapse Supernovae. II. Wave-driven Mass Loss in Supernova Progenitors. *Astrophys. J.* **2014**, *780*, 96. [CrossRef]
131. Gallagher, J.S. Close Binary Models for Luminous Blue Variable Stars. In *Physics of Luminous Blue Variables, ASSP 157*; Davidson, K., Moffat, A., Lamers, H., Eds.; Kluwer: Dordrecht, The Netherlands, 1989; pp. 185–194. [CrossRef]
132. Neugent, K.; Massey, P. The Wolf-Rayet Content of the Galaxies of the Local Group and Beyond. *Galaxies* **2019**, *7*, 74. [CrossRef]
133. Frew, D.J. The Historical Record of η Carinae I. The Visual Light Curve, 1595–2000. *J. Astron. Data* **2004**, *10*.
134. Davidson, K.; Ishibashi, K.; Corcoran, M.F. The Relationship between Two Periodicities Observed in Eta Carinae. *New Astron.* **1998**, *3*, 241–245. [CrossRef]
135. Artigau, E.; Martin, J.C.; Humphreys, R.M.; Davidson, K.; Chesneau, O.; Smith, N. Penetrating the Homunculus—Near-IR Adaptive Optics Images of Eta Carinae *Astrophys. J.* **2011**, *141*, 202. [CrossRef]
136. Livio, M.; Pringle, J.E. Can Eta Carinae be a Triple System? *Mon. Not. R. Astron. Soc.* **1998**, *295*, L59–L60. [CrossRef]
137. Lamers, H.J.G.L.M.; Livio, M.; Panagia, N.; Walborn, N. On the Multiplicity of η Carinae. *Astrophys. J.* **1998**, *505*, L131–L133. [CrossRef]
138. Kundt, W.; Hillemanns, C. Eta Carinae—An Evolved Triple System? *Chin. J. Astr. Astrophys.* **2003**, *3*, 349–360. [CrossRef]
139. Portegies Zwart, S.F.; van den Heuvel, E.P.J. Was the 19th Century Giant Eruption of Eta Car a Merger Event in a Triple System? *Mon. Not. R. Astron. Soc.* **2016**, *456*, 3401–3412. [CrossRef]
140. Smith, N. A Blast Wave from the 1843 Eruption of Eta Carinae *Nature* **2008**, 201–203. [CrossRef]
141. Mehner, A.; Davidson, K.; Humphreys, R.M.; Martin, J.C.; Ishibashi, K.; Ferland, G.J.; Walborn, N.R. A Sea Change in Eta Carinae. *Astrophys. J. Lett.* **2010**, *717*, L22–L25. [CrossRef]
142. Davidson, K.; Ishibashi, K.; Martin, J.C.; Humphreys, R.M. Eta Carinae's Declining Outflow Seen in the UV, 2002–2015. *Astrophys. J.* **2018**, *858*, 109. [CrossRef]
143. Kashi, A.; Davidson, K.; Humphreys, R.M. Recovery from Giant Eruptions n Very Massive Stars. *Astrophys. J.* **2016**, *817*, 66. [CrossRef]

© 2020 by the author. Licensee MDPI, Basel, Switzerland. This article is an open access article distributed under the terms and conditions of the Creative Commons Attribution (CC BY) license (http://creativecommons.org/licenses/by/4.0/).

Review

The Wolf–Rayet Content of the Galaxies of the Local Group and Beyond

Kathryn Neugent [1,2,*] and Philip Massey [2,3]

1   Astronomy Department, Box 351580, University of Washington, Seattle, WA 98195, USA
2   Lowell Observatory, 1400 W Mars Hill Road, Flagstaff, AZ 86001, USA
3   Department of Physics and Astronomy, Northern Arizona University, Flagstaff, AZ 86011-6010, USA
*   Correspondence: kneugent@lowell.edu

Received: 12 July 2019; Accepted: 17 August 2019; Published: 21 August 2019

Abstract: Wolf–Rayet stars (WRs) represent the end of a massive star's life as it is about to turn into a supernova. Obtaining complete samples of such stars across a large range of metallicities poses observational challenges, but presents us with an exacting way to test current stellar evolutionary theories. A technique we have developed and refined involves interference filter imaging combined with image subtraction and crowded-field photometry. This helps us address one of the most controversial topics in current massive star research: the relative importance of binarity in the evolution of massive stars and formation of WRs. Here, we discuss the current state of the field, including how the observed WR populations match with the predictions of both single and binary star evolutionary models. We end with what we believe are the most important next steps in WR research.

Keywords: massive stars; Wolf–Rayet stars; local group galaxies; stellar evolution

1. Wolf–Rayet Star Primer

Wolf–Rayet (WR) stars are hot, luminous stars whose spectra are dominated by strong emission lines, either of helium and nitrogen (WN-type) or helium, carbon, and oxygen (WC and WO type). It is generally accepted that these are the He-burning bare stellar cores of evolved massive stars [1]. Mass loss (whether from binary interactions or stellar winds) first strips away the outer layers of a massive star to reveal the products of CNO hydrogen-burning, nitrogen and helium, creating a nitrogen-rich WR (WN-type). If enough subsequent mass loss occurs, these layers are then stripped away, revealing the triple-α helium-burning products, carbon and oxygen, creating a WC star. Further evolution and mass loss may result in a rare-type oxygen-rich WR (WO-type).

The mass loss that shapes the evolution of these stars can occur through two main channels: binary and single-star evolution. The relative importance of each method is still one of the most important questions facing massive star evolution today. In a binary system, the more massive star will expand first and be stripped by the companion star, revealing the bare stellar core of a WR. In single star evolution, the star will follow the Conti scenario [2,3]. In the Conti Scenario, stars with initial masses greater than $\sim 30 M_\odot$ will form on the main-sequence as massive O-type stars. As they evolve, the stellar winds will continue to strip more and more material from their surfaces until they first turn into WNs, and then (depending on the strength of the stellar winds), WCs and possibly WOs. Stars with initial masses greater than $85 M_\odot$ will also briefly pass through the turbulent Luminous Blue Variable (LBV) phase, shedding material that way.

Single-star evolution is highly dependent on the strength of the stellar-wind mass-loss rates, which are in turn dependent on the metallicity of the birth environment. Since this mass-loss is driven by radiation pressure on highly ionized metal atoms, a massive star born in a higher metallicity environment will have a higher mass-loss rate, and thus the mass limit for becoming a WR would be lower in a higher metallicity environment. If stellar winds dominate the mass-loss mechanism (as opposed to binary evolution), it follows that WC stars will be more common relative to WN stars in high metallicity galaxies while low metallicity galaxies will have few or even no WCs. It also follows that, assuming only single-star evolution, WOs will be rare in all except the highest-metallicity galaxies. Thus, the presence of WOs in a low-metallicity environments (as we discuss later) suggests that binary-evolution plays an important role in the creation and evolution of WRs in at least some cases [4,5], or, as J. J. Eldridge and collaborators have put it [5], "Single-star stellar winds are not strong enough to create every WR star we see in the sky."

Determining the relative number of WC-type and WN-type WRs (the WC to WN ratio) allows us to test stellar evolutionary models by comparing what we see observationally to what the models predict as they scale with the metallicity of the environment. Reliable evolutionary tracks affect not only the studies of massive stars, but the usefulness of population synthesis codes such as STARBURST99 [6], used to interpret the spectra of distant galaxies. For example, the inferred properties of the host galaxies of gamma-ray bursts depend upon exactly which set of stellar evolutionary models are included [7]. It is also important for improving our knowledge of the impact of massive stars on nucleosynthesis and hence the chemical enrichment of galaxies [8]. Thus, determining an accurate ratio of WC to WN stars in a galaxy turns out to have its uses far beyond the massive star community [9]. Additional diagnostics include the relative number of red supergiants (RSGs) to WRs, and the relative number of O-type stars to WRs.

The galaxies of the Local Group provide an excellent test-bed for such comparisons between the observations and models because they allow us to determine a *complete* population of different types of stars. In all except the most crowded of regions (such as 30 Doradus in the Large Magellanic Cloud), stars can be individually resolved by ground-based telescopes and instruments. Such photometric studies have been done previously (such as the Local Group Galaxy Survey [LGGS] [10]), but photometry alone can't be used to detect Wolf–Rayet stars. Thus, as we will discuss in this article, other methods such as interference filter imaging and image subtraction must be employed. The WR-containing galaxies of the Local Group span a range in metallicity from $0.25\times$ solar in the Small Magellanic Cloud (SMC) [11] to $1.7\times$ solar in M31 [12]. This allows us to compare the observations against the model predictions across a large range of metallicities, which is important given the strong dependence on stellar evolution to mass-loss rate. Thus, here we focus our discussions on WRs in the galaxies of the Local Group.

In this review paper, we will first discuss how WRs were found in the past as well as current methods. We'll review the current WR content of the Local Group Galaxies and Beyond while discussing a few important and surprising findings made along the way. Next, we'll discuss the important issue of binarity and how it influences the evolution of WRs. Finally, we'll describe how to obtain the physical parameters of such stars using spectral modeling programs before ending with a discussion of how the evolutionary models compare to our observed number of WRs.

2. Surveys for Wolf–Rayet Stars

2.1. The Milky Way

The first survey for Wolf–Rayet stars (inadvertently) began in 1867 when Charles Wolf and Georges Rayet were examining spectra of stars in Cygnus using a visual spectrometer on the 40-cm Foucault telescope at the Paris Observatory. They came across three very unusual stars. While the spectra of most stars are dominated by absorption lines, these stars had mysterious strong, broad emission lines. (These stars were later designated and classified as HD 191765, WN5; HD 192103, WC8; and HD 192641, WC7.)

The correct identification of the spectral features was lacking for nearly 60 years after their discovery: it was Carlyle Beals, a Canadian astronomer, who correctly identified the lines as due to ionized helium, nitrogen, and carbon [13]. The width of these lines were understood as being due to

Doppler broadening of thousands of km s^{-1}, a result of the outflow rates of the strong stellar winds in the formation region of these lines [14–16]. Example spectra are shown in Figure 1.

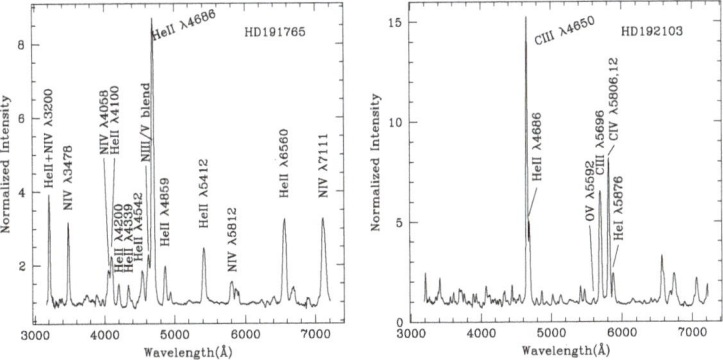

Figure 1. The spectra of two of the first discovered WR stars. Left: HD 191765 is a WN star, with unusually broad and strong lines. Its classification is a "WN5" subtype. Right: HD 192103 is a WC star, with a "WC8" subtype.

WN-type Wolf–Rayet stars are further classified primarily by the relative strengths of N III λ4634,42, N IV λ4058, and N V λ4603,19, while the classification of WC-type WRs is based upon the relative strengths of O V λ5592, C III λ5696, and C IV λ5806,12. The system was first proposed by Lindsey Smith [17], although some extension to earlier and later types of WNs have been made by others [18,19]; a classification scheme for WO stars was developed by Paul Crowther and collaborators [20]. As with normal stars, a lower number is indicative of higher excitation, i.e., WN2 (hotter) vs. WN9 (cooler), WC4 (hotter) vs. WC9 (cooler), WO1 (hotter) vs. WO4 (cooler).

The late-type WNs are morphologically similar to O-type supergiants, known as "Of-type" type stars, in that the latter show N III λ4634,42 and He II λ4686 emission, also the result of stellar winds. The late-type WNs are more extreme, however, with stronger lines. In general, WNs (and WRs in general) do not show absorption lines; rather, all of the lines are formed in the stellar winds. There are, however, exceptions, such as HD 92740, a singled-lined WR binary in which the emission and absorption move together in phase [21]. It was the similarity between Of-type and WNs that led in part to the Conti scenario [2].

As summarized in [22], a total of 52 similar stars were discovered by Copeland, Fleming, Pickering, and Respighi in the 25 years that followed Wolf and Rayet's discovery. These findings, and early visual work by Vogel in 1885, and photographic studies of their spectra by Pickering in 1890, are discussed in the contemporary review by Julius Scheiner and Edwin Frost in their 1894 publication *A Treatise on Astronomical Spectroscopy* [23]. William Campbell (who served as director of Lick Observatory 1901–1930) published the first catalog of these 55 Galactic WRs in 1894 [24]. Additional WRs were discovered as by Williamina Fleming, Annie J. Cannon and coworkers as part of the Henry Draper catalog project, and accidental discoveries continued to be made over the years. The first modern catalog of Galactic Wolf–Rayet stars compiled by Karel van der Hucht and collaborators in 1981 [18]. Titled "The VIth Catalog" (Campbell's was considered the first), the work included extensive bibliographies and references to earlier studies. This catalog contained 168 WRs. The next edition, in 2001 [19], listed 227 spectroscopically confirmed Galactic WRs, with an "annex" in 2006 [25] bringing the number known to 298. The most-up-to-date catalog of Milky Way WRs is maintained online by Paul Crowther[1], which contained 661 entries as of of this writing, June 2019.

---

[1] http://www.pacrowther.staff.shef.ac.uk/WRcat/

Systematic searches for WRs in the Milky Way are stymied by the vast angular extent that needs to be examined (the entire sky!), and variable and sometimes high reddening. The Henry Draper catalog is probably complete down to an apparent magnitude of 10th or 11th, except in regions of crowding. Spectroscopic surveys of young clusters or OB associations reveal additional WR finds yearly; possibly the most extreme example is that of Westerlund 1 and various open clusters near the Galactic Center; see [25] and references therein. However, the large increase in the number of WR stars known in the Galaxy in the past 15 years has has come about primarily from the use near- and mid-IR colors to identify WR candidates [26–30], a method first pioneered by Schuyler van Dyk and Pat Morris, plus the use of narrow-band IR imaging in the K-band [31,32], pioneered by Mike Shara. Optical or near-IR spectroscopy is then used to confirm the color-selected candidates.

With the advent of *Gaia*, it is now possible for the first time to actually derive distances to many of these Wolf–Rayet stars. However, difficulties of constructing meaningful volume-limited samples remain for Galactic studies. As discussed later, WN-type WRs are harder to find than WC-type due to their weaker lines; at the same time, WC stars may be dustier (and thus fainter) than WN stars in the same location. They also cover a limited range in metallicity compared to what can be achieved by using the non-MW members of the Local Group. Finally, observations of Galactic WRs may be more difficult due to reddening than those in much further, but less reddened, regions. Thus, Galactic studies still have limited value for testing models of stellar evolution theory. Thus, for the rest of this review, we will focus on the WR content of galaxies outside our own.

2.2. Early Searches for Extra-Galactic WRs

2.2.1. Large Magellanic Cloud

As part of the Harvard spectral surveys, Anne J. Cannon and Cecilia Payne (later Payne–Gaposchkin) identified 50 Wolf–Rayet stars in the Large Magellanic Cloud (LMC) according to Bengt Westerlund and Alexander Rodgers (1959) [33] quoting an early review article on the stellar content of the LMC by Gerard de Vaucouleurs and collaborators [34]. Westerlund and Rodgers carried out their own search of the LMC, the first systematic search for WR stars in another galaxy, using slitless (objective prism) spectroscopy to identify 50 WRs, 36 of which were in common with the Harvard studies [33]. They note that nine Harvard O-type stars in the 30 Doradus region had been recently reclassified as WN by Michael Feast and coworkers [35] in the previous year. Two decades later, Marc Azzopardi and Jacques Breysacher (1979) completed an even more powerful objective prism survey using an interference filter to further reduce the effects of crowding [36]. This increased the number of known WRs in the the LMC to 100. Accurate spectral types of these 100 LMC WRs were subsequently published by Breysacher in 1981 [37]. In that paper, Breysacher estimated that the LMC likely contained a total of 144 ± 20 LMC WRs, with 44 left to be discovered. He further speculated that the majority of these undiscovered WRs would be found deep within the cores of dense H II regions where slitless spectroscopy often fails. (Indeed, the "final census" catalogue of LMC WRs, discussed below, lists 154 separate WRs [38], well within Breysacher's estimate of 144 ± 20.) These early studies culminated in Breysacher's et al.'s "Fourth Catalog" of LMC WRs [39] (hereafter BAT99), which listed 134 LMC WRs.

The R136 cluster merits separate attention, as investigations of its stellar content led to the recognition that not all luminous stars with WR-like spectra are evolved objects. R136 is of course the central object at the heart of the 30 Doradus nebula in the LMC. Once thought to house a supermassive star, early *Hubble Space Telescope* (*HST*) images showed it was even more interesting, the core of a super star cluster, with over 3500 stars (120 of which are blue and more luminous than $M_V \sim -4$) most of which lie within 8" (2pc) of the semistellar R136 cluster [40]. Using ground-based spectroscopy in 1985, Jorge Melnick had identified 12 WR stars in or near the central cluster [41]. When Deidre Hunter and collaborators analyzed the first *HST* images of the cluster in 1995, this created a conundrum: the isochrones indicated that the lower mass stars had ages of only 1–2 Myr, but the presence of WR stars implied ages for the massive stars of 3–4 Myr [40]. Why had the formation of high mass stars, with their strong stellar winds, not stopped star formation in the cluster? Melnick had also found early-type O stars in the cluster, possibly as early as O3, although the presence of strong nebulosity made this classification uncertain, and this also seemed to conflict with the ages of the WR stars, as the O3 phase lasts for only a million years. Massey and Hunter obtained *HST* spectroscopy of 65 of the hottest, bluest stars in the cluster, and discovered two amazing facts: (1) the vast majority of these stars were of O3, and that (2) the WR stars were not common, garden-variety WNs [42]. Rather, they were 10× more luminous in the V-band than normal WRs, and their spectra were still rich in hydrogen. Massey and Hunter argued that a similar situation existed in the Galactic giant H II region NGC 3603, where both O3 stars and WRs were known [43]; they examined the archival spectra and concluded that those WR stars were like the H-rich super-bright WR in R136. The obvious conclusion was that these were young (1–2 Myr) objects still burning hydrogen whose high luminosities simply resulted in WR-like emission features, in essence, Of-type stars on steroids [42]. This interpretation built on the important result the previous year by Alex de Koter and collaborators who found that one of the over-luminous, hydrogen-rich WR stars in the core of the R136 cluster had a normal hydrogen abundance, and who had originally suggested that this and similar were still in the hydrogen-burning phase [44].

2.2.2. The Small Magellanic Cloud

The identification of WRs in the SMC followed a similar pattern, but, thanks to its smaller angular size compared to the LMC, a complete census became possible earlier than for the galaxies discussed above. As summarized in an earlier review [45], four WRs had been found by general spectroscopic studies [46] when Azzopardi and Breysacher used the same technique of objective prism and interference filter photography to find four additional WRs, bringing the total up to eight [47]. A ninth WR was found by spectroscopy from objective prism photography [48]. In 2001, Massey and summer student Alaine Duffy carried out the first CCD survey for WRs in the SMC [49]. They used an on-band, off-band interference filter imaging campaign with the wide-field CCD camera on the CTIO Curtis Schmidt to cover most of the SMC. Photometry of 1.6 million stellar images helped identify a number of candidates, including all of the known SMC WRs, at high significance levels. Two new WNs were then confirmed by follow-up spectroscopy, bringing the total to 11. The survey also found a number of Of-type stars, demonstrating that the survey was sensitive to even the weakest-lined WNs. However, shortly after this, a 12th WR star was discovered in the SMC [50]. This star had been too crowded to have been found in the Massey and Duffy survey. Of these 12 WRs, 11 are of WN-type and only 1 is of WC-type. (Actually, the strength of O VI lines qualifies this as a WO-type star [20].) This low WC/WN ratio is consistent with our expectations based upon the SMC's low metallicity.

Quantitative studies of the strength of He II $\lambda 4686$ emission in SMC WN stars by Peter Conti and collaborators [51] showed that the line was weaker than in WNs of similar types in the Milky Way or LMC, also consistent with the expectation that stellar winds would be weaker in lower-metallicity environments.

2.2.3. Beyond the Magellanic Clouds

The first WR stars to be discovered beyond the Magellanic Clouds were in the nearby spiral galaxy M33. In 1972, James Wray and George Corso pioneered the interference-filter method of searching for WRs by comparing images of M33 taken through an interference filter centered on the C III $\lambda 4650$ and He II $\lambda 4686$ emission complex with that of a continuum image [52]. WR candidates would stand out by being brighter in the on-band compared to non-WR stars in the field. Their paper contained spectroscopic confirmation of two of their 25 candidates (thanks to Roger Lynds); both stars were of WC-type, although Lindsey Smith is quoted as saying that the spectra were "not quite like any I have seen from either the Galaxy or the Magellanic Clouds." (This was probably more due to the poor quality of these early spectroscopic efforts on these faint objects, which pushed the limits of photographic spectroscopy at that time.) Spectroscopy of three other candidates followed five years later by Alex Boksenberg, Allan Willis, and Leonard Searle using one of the first digital photon-counting systems [53]. A search using photographic "grism" imaging on the Kitt Peak 4-meter (a technique similar to objective prism survey but using a grating prism and a much larger telescope) carried out by Bohannan, Conti, and Massey revealed a host of H II regions in M33, but only five more WRs [54]. Spectroscopy of the stars in M33's H II regions by Conti and Massey in 1981 was more effective, identifying 14 more WRs [55]; some were in common with the nearly contemporaneous study of the stellar content of NGC 604, the largest H II region in M33, by Mike Rosa and Sandro D'Odorico [56,57]. The properties of some of these stars were highly unusual, with higher luminosities and more hydrogen than normal WR stars, similar to what would be eventually noted in the R136 cluster as mentioned above. A photographic search with the 3.6-meter Canada–France–Hawaii telescope with followup spectroscopy on the Kitt Peak 4-meter provided the first galaxy-wide survey, including 41 newly found WRs [58]. This 1983 Massey and Conti catalog included all previous known WRs, for a total of 79 WRs, and revealed a trend in the relative number of WCs to WNs as a function of galactocentric distance within M33. Quantitative analysis of the lines (measurements of line strengths and widths) and absolute magnitudes showed no gross differences between the M33 WRs and those of the Milky Way or Magellanic Clouds [58,59], refuting the Smith's first impression from the Lynds' earlier spectroscopy.

The first use of CCDs to survey for WRs was carried out by Taft Armandroff and Massey in 1985 using the newly implemented prime-focus CCD camera on the Cerro Tololo Blanco 4-meter telescope [60]. They had refined the interference-filter method to include a three-filter system, with one centered on C III $\lambda 4650$, another on He II $\lambda 4686$, and a third on neighboring continuum, and used these with a CCD to search for WRs in the dwarf galaxies IC 1613 and NGC 6822, as well as two M33 test fields. One WR star had been previously identified in IC 1613, a WC star (now considered a WO) discovered in an H II region by D'Odorico and Rosa in 1982 [61], and subsequently studied by Kris Davidson and Tom Kinman [62]. Similarly a WN-type WR had previously been found in NGC 6822 by Westerlund and coworkers using an objective prism [63]. These early CCDs were incredibly tiny compared to what are in use today, and multiple fields were needed to cover even these relatively small galaxies. These CCDs were also incredibly noisy (with read-noise of 100 e- compared to typically 3 e- today). Armandroff and Massey found 12 "statistically significant" WR candidates in NGC 6822 and 8 in IC 1613. However, only four of the NGC 6822 WR candidates proved to be real (including the one that was previously known), and the only IC 1613 WR candidate that checked out was the one already known [64].

A search for WR stars in the dwarf galaxy IC 10 proved the most surprising of any of these early studies. Despite its small size, 16 WR candidates were initially found by Massey, Armandroff, and Conti [65], 15 of which were quickly confirmed [66], causing the authors to recognize this as the nearest starburst galaxy. Despite the galaxy's low metallicity, the relative proportion of WC stars was very large. Was this suggestive of a top-heavy initial mass function as has been historically suggested for other starbursts [67], or is indicative that an even larger number of WRs (predominantly WN) remained to be discovered, as suggested by [68]? This issue is still not settled. The current count is 29 spectroscopically confirmed WRs [69], with additional candidates still under investigation.

The situation for M31 was probably the worst. Interference photography by Tony Moffat and Mike Shara identified a few of the strongest-lined WRs [70,71]; CCD imaging through interference filters by Massey and collaborators went much deeper but covered only a small portion of the galaxy [64,72].

These early studies culminated in the 1998 paper by Massey and Olivia Johnson [73], who identified additional M33 WR stars found using a larger format (and less noisy) CCD, and provided a catalog of all of the known extragalactic WR stars beyond the Magellanic Clouds. For the purposes of this review, we will consider that the end of the "early era" of WR searches. Although completeness indeed would prove to be a problem, the following facts had emerged:

- The WC/WN ratio appeared to be strongly correlated with metallicity, with the exception of the starburst galaxy IC 10.

- Late-type WC stars (WC7-9) were found only in regions of high metallicity, while WCs in low-metallicity regions were invariably of early type (WC4s).
- The spectral properties of a given WR type were generally similar regardless of the environment, although weaker emission is found in the WNs of lower metallicity, indicative of smaller mass-loss rates.
- Giant H II regions (NGC 604, 30 Dor, NGC 3603) contained very luminous stars whose spectra showed WR-like features, but which were hydrogen-rich. These stars were basically "super Of-type stars," stars that are so massive and luminous that their atmospheres are extended creating WR-like features but which are likely still hydrogen-burning objects.

2.3. Motivation for New Studies

As of the early 2000s, our knowledge of the LMC's WR population was thought to be relatively complete thanks to the work of Breysacher's BAT99 catalog [39]. However, other galaxies of the Local Group, namely M31 and M33, still lacked galaxy-wide surveys. Figure 2 shows the observed WC/WN ratio compared to the 2005 Geneva Evolutionary Group's model predictions [1]. (These were the first complete set of models at different metallicities which included the important effect of rotation.) Notice first that the observed relative number of WCs to WNs increases with metallicity. This is exactly what we would expect given single-star evolution because higher metallicity environments will allow more WCs to form. This increase in ratio vs. metallicity is additionally what the models predict. However, a comparison between the models and the observations show that the relative number of predicted WRs is not consistent between the two. Additionally, the models do a particularly poor job of predicting the WC to WN ratio at higher metallicities, such as in M31 and M33.

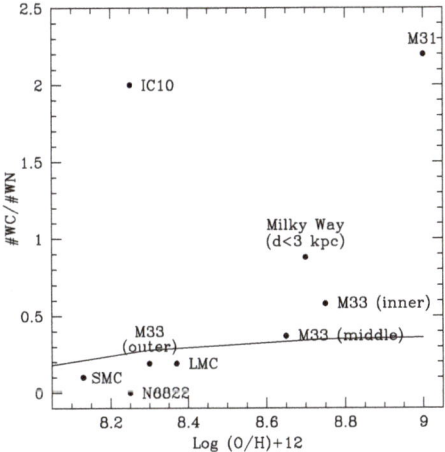

Figure 2. The state of our knowledge of the WC/WN ratio vs metallicity in the mid 2000s. The points are from the 1998 Massey and Johnson summary [73]. The solid curve shows the predictions based upon the 2005 Geneva evolutionary models that included rotation for the first time [1]. Note that, while both show an increase in the WC/WN ratio with metallicity, there is a large discrepancy between the observed results and model predictions at higher metallicity values. Recall that NGC 6822 contains only four WRs (all of WN-type) and the SMC only 12 WRs (one of which is a WC/WO), thus deviations from the models for these two galaxies are not significant.

Clearly, a problem existed, but was it a failing of the models or observations (or both)? Given the complexities of modeling the physics at the end of a massive star's life, it made sense that there could be some deficiencies in the models. However, there were a few reasons that suggested that the observations were actually at fault. For one, as discussed above, there was still no galaxy-wide targeted survey of WRs in the LMC, M31 or M33; only the SMC had been well covered by the Massey and Duffy survey. The vast majority of WRs that had been discovered within those galaxies had been discovered either by accident or as part of a survey of a limited portion of the galaxy. Additionally, crowding of tight OB associations (where we expect to find the vast majority of WRs) makes finding even bright, strong-lined WRs difficult. Thus, telescopes with more resolving power could help disentangle the tightly-packed regions. Finally, and perhaps more importantly, there is a strong observational bias towards detecting WC-type stars over WNs.

The basis for this observational bias is shown in Figure 3. The strongest emission feature in WCs is nearly 4× stronger than the strongest line in WNs, making WNs much more difficult to detect than WCs of similar brightness [74]. (More accurately, this is an issue of line fluxes; see treatments in [73] and [38].) Thus, while a galaxy (or catalog such as BAT99) might be complete for WC-type stars, there might be a number of missing WNs since their emission lines are so much weaker. The exclusion of these stars would bias the WC to WN ratio to higher values, much like we see when we compare the relative number of WRs observed to that predicted by the Geneva Evolutionary models. Indeed, this was particularly a problem for M31. The ratio of 2.2 shown in Figure 2 is the galaxy-wide average for M31, including the older photographic work; if, instead, one used only the eight CCD fields, this value would drop to 0.9 [73], giving strong credence to selection effects being responsible for the problem.

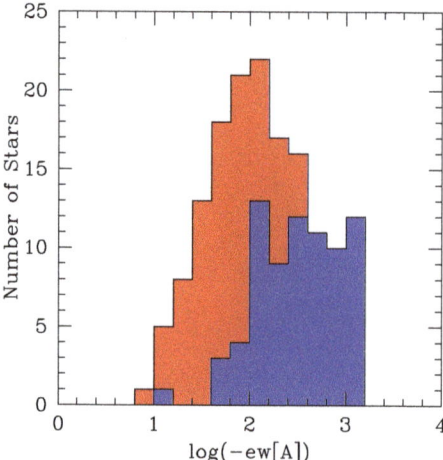

Figure 3. Line strengths of galactic and LMC WRs. The red histogram shows the line strengths (measured as the log of the equivalent width) for a WN's strongest emission line, HII $\lambda 4686$. The blue histogram shows the same for the WC's strongest emission line, CIII/IV $\lambda 4650$. The WC's strongest line is up to 4× stronger than the WN's strongest line making WC stars much easier to detect.

The lack of a galaxy-wide survey for M31 or M33 as well as the possibility of crowding and a strong observational bias against WN stars led us to conduct our own survey for WRs in M31 and M33.

3. New Era of Discoveries

As discussed above, as of 2005, the observed WC/WN ratio was quite poorly aligned with the theoretical predictions at higher metallicities. Thus, M31 and M33 were two ideal regions to study. M31 has the highest metallicity of the Local Group galaxies at $\log(O/H) + 12 = 8.9$ [12,75]. M33 has a strong metallicity gradient going from $\log(O/H) + 12 = 8.3$ in the outer regions up to $\log(O/H)+12 = 8.7$ in the inner regions [76]. Thus, these two galaxies presented the perfect opportunity to re-examine the differences between theory and observations.

In 1985, Massey and Armandroff had pioneered the use of interference filter imaging with CCDs to identify WR candidates [60]. However, the small size of the CCDs available at that time limited the area that could be covered and the large read-noise limited the sensitivity. An equally large problem, however, was the use of photometry to identify candidates. This method was far superior to "blinking by eye," as had been used in the photographic studies by [52,70,71], and allowed "statistically significant" candidates to be identified. However, the fraction of false positives was overwhelming, simply given the large number of stars involved.

In the mid-2000s, large format CCD Mosaic cameras came along, such as those implemented on the Kitt Peak and Cerro Tololo 4-meter telescopes. CCDs now have read-noises of 3 e- rather than 100 e-, and these mosaic cameras made it practical to cover all of M31 and M33 in a finite number of fields. Equally importantly, supernova and transient searches had required the development of the powerful technique of image subtraction, where the the PSFs were matched between two images, and one image subtracted from another to identify images. We took advantage of both of these improvements in conducting our own searches.

3.1. Identification of Candidate WRs

Searching for candidate WRs was done using the same method in both galaxies as is detailed in [77,78]. Overall, the method combines photometric observations using an interference filter system with image subtraction and photometry for candidate detection.

Thanks to the WR's strong emission lines, they're relatively simple to detect using the appropriately designed interference filters. Taft Armandroff and Massey used spectrophotometry for WR and non-WR to design a three-filter system that was optimized identifying WRs in the optical [60]. All three filters have $\sim$50 Å wide bandpasses, with one centered on the strongest optical line in a WC's spectrum, CIII/IV λ4650 ("WC" filter), another centered on the strongest optical line in a WC's spectrum, HeII λ4686 ("WN" filter) and a third on the neighboring continuum at λ4750 ("CT" filter). (Placement of the continuum filter to the red of the emissionline filters is crucial; otherwise, red stars show up as candidates.) The bandpasses are shown placed atop the spectrum of both an LMC WC- and WN-type WR in Figure 4. This filter set was used by [60] to search for WRs in the Local Group galaxy dwarfs NGC 6822 and IC 1613, as well as two small test regions of M33. Such work was then extended to selected regions of M33 [73] and M31 [72], and for the galaxy-wide survey of the SMC [49] discussed above. With these interference filter images in hand, there are two main methods of determining stars that are brighter in the on-band filters (WC and WN) vs. in the continuum (CT). The first is using image subtraction and the second is using photometry.

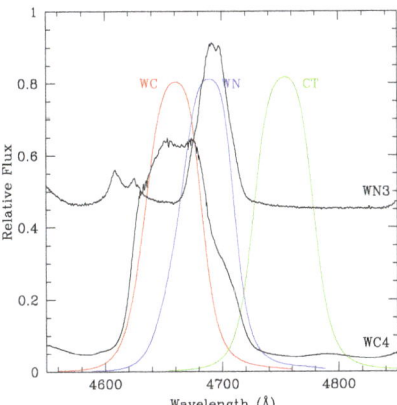

Figure 4. Filter bandpasses of WN, WC and CT filters. The WN and WC filters are centered on the strongest lines of the WC and WN-type WRs while the CT is centered on the neighboring continuum; this figure was adapted from [79].

As mentioned above, image subtraction has been used with great success to detect small brightness changes between on and off band photometry by the supernovae community [80]. Simply subtracting the CT from the WC filter should yield candidate WCs while subtracting the CT from the WN filter should yield candidate WNs. However, seeing variability and small changes in pixel scales across the images turn this simple idea into a complex problem and thus cross-convolution methods and point-spread fitting techniques must be used. Example programs include the Astronomical Image Subtraction by Cross-Convolution program [81] and High Order Transform of PSF and Template Subtraction (HOTPANTS) [82]. An example resulting image is shown in Figure 5 where the background stars have been subtracted out and the candidate WRs are left behind.

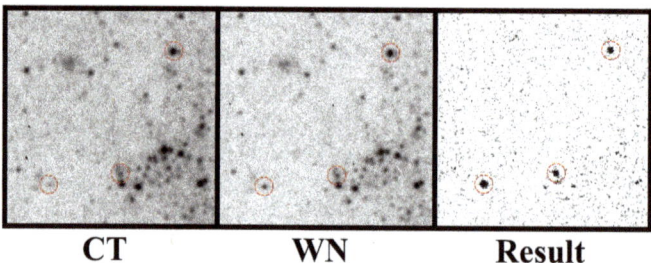

Figure 5. WR-detection through image subtraction. Three known WRs are outlined in red dashed circles. After subtracting the continuum filter from the WN filter, the resulting image shows three WRs as black stars. This method was used to search for candidate WRs; this figure is from [78].

As discussed above, most WRs are formed in dense OB associations (in fact, Neugent and Massey found that 80% of the WRs in M33 were found in OB associations [78] with only 2% being truly isolated). This dictates the need for crowded field photometry to determine the magnitude differences between the WC-CT and WN-CT filters. Armandroff and Massey had adopted Peter Stetson's DAOPHOT crowded field photometry software [83], with subsequent modifications and porting to IRAF [84]. Careful matching in crowded regions must be performed by eye. Photometry is obtained for all the stars on each on-band exposure (WC, WN), and then matched with the photometry for the same stars on the CT exposure. A zero-point adjustment is then made so that the average difference was zero, and then stars that were more than 3σ brighter on either the WC or WN filter exposure when compared to the continuum exposure can be identified.

3.2. M33

Neugent et al. completed the first galaxy-wide survey for WRs using a combination of the image subtraction and photometric method as discussed above [78]. Overall, they discovered 54 new WRs bringing the total number of confirmed WRs in M33 up to 206, a number they believe is complete to ~5%. A majority of these new discoveries were WNs suggesting that the previous WC/WN ratio had been biased towards the easier to find WCs. The locations of the known WRs across the disk of the galaxy are shown in Figure 6. Notice that the galaxy has been divided up into three regions representing the strong metallicity gradient with the inner region having a higher metallicity than the outer region.

As discussed in the Introduction, the formation of WRs is highly dependent on mass-loss rates, which is, in turn, dependent on the metallicity of the environment. In higher metallicity environments, the mass-loss rates will be higher leading to the creation of more WCs. Thus, we expect the WC/WN ratio to be higher in regions of high metallicity, such as in the center of M33. Indeed, this is what we find. While the full comparison of WC/WN ratios vs. metallicity will be discussed later, Table 1 shows the WC/WN ratio vs. metallicity for the inner, middle, and outer regions of M33. (The cut-offs for these regions are a little different than had been used in the earlier study by [73] shown in Figure 2).

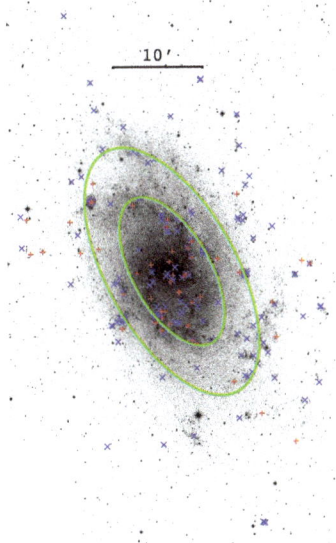

Figure 6. Location of known WC and WN stars in M33. WN stars are represented as blue ×s while WC stars are represented as red +s. The green ovals represent distances of $\rho = 0.25$ (1.9 kpc) and $\rho = 0.50$ (3.8 kpc) within the plane of M33. The metallicity gradient extends outward with higher metallicity in the middle and lower in the outer regions; this figure is from [78].

Table 1. WC/WN ratio vs. metallicity for the inner, middle, and outer regions of M33.

Region	$\bar{\rho}$	log(O/H) + 12	# WCs	# WNs	WC/WN
$\rho < 0.25$	0.16	8.72	26	45	0.58 ± 0.09
$0.25 \geq \rho < 0.50$	0.38	8.41	15	54	0.28 ± 0.07
$\rho \geq 0.50$	0.69	8.29	12	54	0.22 ± 0.06

The metallicity gradient of M33 also allows us to probe the relative number of early and late type WCs vs. metallicity. Smith first discovered that nearly all of the late-type WCs are found in higher metallicity environments than the early-type WCs [85]. Additionally, late-type WCs have C IV $\lambda 5806$ lines that both have smaller equivalent widths and smaller full width half max values than early-type WCs. Thus, plotting these two values against each other vs. metallicity shows that the spectral type becomes earlier as metallicity decreases. This is shown in Figure 7. This proves, independent of any direct metallicity measurements, that the metallicity of M33 increases towards the center of the galaxy.

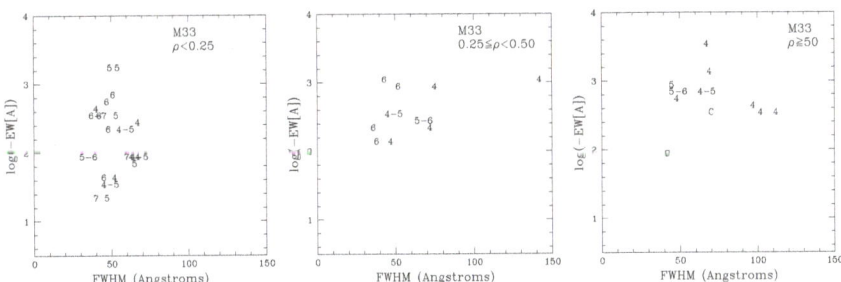

Figure 7. WC line strength vs. line width. The line strength of the C IV $\lambda 5806$ line is plotted against its line width with the spectral subtypes indicated (with "C" used if the subtype has not been well established). Notice how the FWHM increases and subtype decreases as the metallicity decreases (larger values of ρ); this figure is from [78].

With this new data discussed in Neugent and Massey, the WC/WN ratio was determined for three regions of medium to high metallicity [78] and the number of WRs was thought to be complete to 5%.

3.3. M31

The next study was done in M31 by Neugent et al. [77] which has an even higher metallicity than that of the inner region of M33. By using the same detection methods of interference filter imaging, image subtraction and photometry, they discovered 107 new WRs (79 WNs and 28 WCs) bringing the total number of WRs in M31 up to 154, a number they argue is good to within 5%. They additionally found that 86% of the observed WRs were within known OB associations as determined by van den Bergh [86]. The locations of the WRs are shown in Figure 8. Due to the addition of the new WNs, the WC/WN ratio dropped from 2.2 down to 0.67. While this helped bring the observations closer to that of the theoretical model predictions, the full story will be told in Section 7.

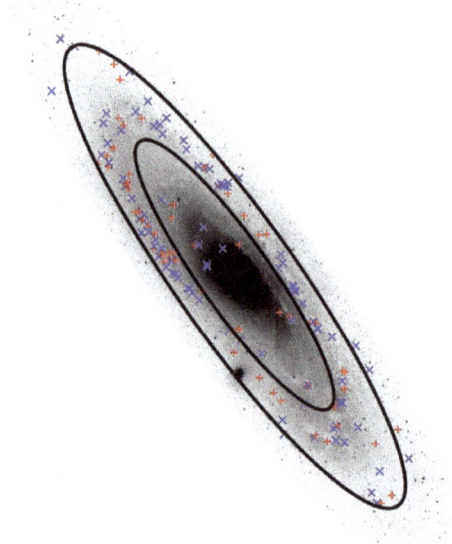

Figure 8. Locations of all known WR stars in M31. The blue ×s represent WN stars while the red +s represent WC stars. The inner black ellipse is at 9 kpc ($\rho = 0.43$) within the plane of M31, and the outer one is at 15 kpc ($\rho = 0.71$), which represents the location of the majority of OB forming associations within M31; this figure is from [77].

Subsequent to the this study, Mike Shara and collaborators discovered an additional WR star in M31, a WN/C star [87]. Such objects have WN-like spectra but strong C IV λ5806,12 line. The star is located in strong nebulosity and is described as heavily reddened (although no specific values are given), and the authors speculate based on this one object that there might be a large population of unfound WRs lying on the "far side" M31's disk, i.e., that only lightly reddened specimens have been found so far. Is this reasonable? First, we note that the width of the "blue plume" (denote OB stars) in the color magnitude diagram of M31 has a similar width to that of the LMC; compare Figures 10 and 12 in [88]. If there were a huge population of highly reddened stars, we would expect the blue plume to be high asymmetric, with a large tail extending to redder magnitudes. Secondly, we can do a crude estimate of what we might expect. We note that the total extinction through the MW's disk is ~0.4 mag in B [89]. If M31 is similar, then, at an inclination of 77° to the line of sight, we expect the total extinction in B from one side to the other to be about 1.8 mag, or in V, about 1.4 mag. This is only 0.6 mag greater than the 0.8 mag in A_V found for OB stars in some of the handful of well-studied

OB associations [72], and is unsurprising. Thus, although a handful of heavily reddened WRs may certainly have been missed (consistent with the ten that Shara et al. estimate), it seems unlikely that there is an opaque wall obscuring WRs on the far side of M31.

3.4. Magellanic Clouds

Thanks to previous surveys, such as the BAT99 catalog, the population of WRs in the MCs was thought to be complete. However, over the years, a few unexpected discoveries were made. Perhaps the most surprising of which was of a rare strong-lined WO discovered in the LMC in the rich OB association of Lucke–Hodge 41 [90]. Since the BAT99 catalog, six new WRs were discovered before the addition of this new WO suggesting that perhaps our knowledge of the WR content of the LMC was still not complete. Thus, a new search for WRs in the MCs was launched [79,91,92]. A summary of the results can be found in [38].

The overall process of this survey was similar to finding WRs in M31 and M33. The entire optical disks of both the LMC and SMC were observed using the 1-m Swope telescope on Las Campanas, with the three-filter interference system and then a combination of image subtraction and photometry was used to detect candidate WRs before they were spectroscopically confirmed.

In the SMC, no new WRs were discovered. However, this isn't too surprising given that there are only 12 known WRs in the entire galaxy [49], and that the Massey and Duffy survey had covered the entire galaxy. All of them are of WN type except one binary WO. Further characteristics, such as their physical properties and binary status, are discussed later.

The LMC, however, held many surprises. Overall, the new study found 15 new WRs bringing the total number of WRs in the LMC up to 152. Five of them were normal WNs that had been missed due to crowded fields and faint emission lines. However, ten of them were unlike any WR we had seen before.

The spectra of these stars contain absorption lines like that of a O3 star with emission lines like that of a WN3, thus leading to a designation of WN3/O3s [93]. A spectrum of one such star showing both the narrow absorption lines and broad emission lines is shown in Figure 9. While their spectra initially suggests binarity, these stars are simply too faint to be WN3 + O3V binaries. The absolute magnitude of an O3V by itself is $M_V \sim -5.5$ while the absolute magnitudes of these WN3/O3s are around $M_V \sim -2.5$. Thus, they could not be in systems with even brighter O3Vs. For this, and other reasons detailed in [93], these stars are single in nature. A further description of their physical parameters and hypothesized place in massive star evolution is discussed in Section 6.

In Figure 10, we now show the effect that the recent work of ourselves and others have made in our knowledge of the WC/WN ratio as a function of metallicity. Clearly, the biggest improvements have come about for M31 and IC10. However, even for IC 10, the results are still very uncertain, with [68] finding many additional candidates that have not yet been certified by spectroscopy, and [69] finding a small number that have also not yet been observed. For the Milky Way (MW), we took the current 661 in Paul Crowther's online catalog, and selected only those with Gaia distances <3 kpc using the (model-dependent) catalog of Bailer-Jones et al. [94]. This found 99 WRs. Despite the vast improvement in the distances available since the estimate of the MW's WC/WN by Massey and Johnson [73], the value for the WC/WN ratio is essentially unchanged. Still, as emphasized earlier, construction of a volume-limited sample for the MW is fraught with difficulties.

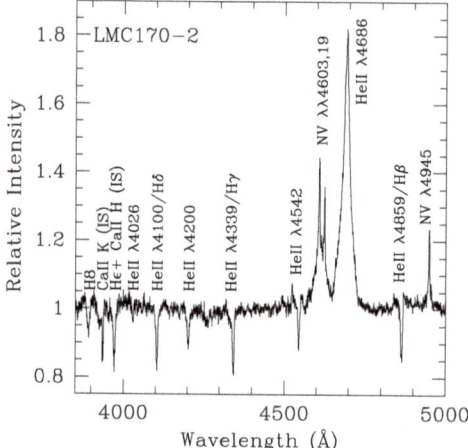

Figure 9. Spectrum of LMC170-2, one of our newly discovered WN3/O3 type stars. The WN3 classification comes from the star's N V emission ($\lambda\lambda$ 4603,19 and λ4945), but lack of N IV. The O3 classification comes from the strong He II absorption lines but lack of He I; this figure is from [93].

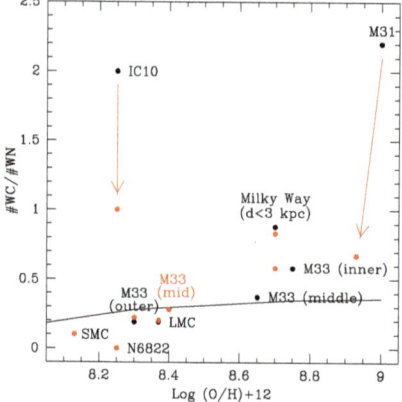

Figure 10. Updated comparison of WC/WN ratio of observed results vs. Geneva Evolutionary models. Notice the drastic changes between the old and new values for IC10 and M31. However, the lack of agreement between the models and observations at high metallicities remains.

4. Wolf–Rayets beyond the Local Group

4.1. Individual WR Populations

WR stars have been found in a number of more distant galaxies. NGC 300 is a spiral galaxy in the Sculptor Group (1.9 Mpc) [95], the nearest galaxy group outside the Local Group. Broad WR features were found in the spectra of several of NGC300s H II regions in the 1980s [57,96]. Eighteen individual WRs were subsequently identified in the early 1990s by interference imaging and follow-up spectroscopy by Testor, Schild, and Breysacher [97–99], with a nineteenth one discovered by accident by Fabio Bresolin and collaborators [100]. A concerted survey with the 8-meter VLT by Schild and collaborators brought this total up to 60, a value which they state is close to complete [101]. Subsequently, an additional nine WRs were found by Crowther and collaborators [102], bringing the total to 69.

How complete do we expect such surveys to be? The distance to NGC 300 is 2.4× larger than the distance to M33, and, with similar reddenings, WR stars will be nearly 6× fainter; crowding will be also be 2.4× larger. Thus, given what was involved in obtaining a (nearly) complete sample of WRs in M33 by Neugent et al. using imaging on a 4-m telescope, one may question how well completeness can be achieved by a telescope only twice as large in aperture. M33 has 206 WRs. What would we expect the population to be scaling by the integrated Hα luminosities? The integrated Hα luminosity is considered to be one of the "gold standards" of recent star formation activity, and (corrected for reddening and distance) is about 2.1× greater in M33 than in NGC300 [103]. Thus, one would naively expect NGC300s WR population to number about 100.

The most interesting discovery to come out of the NGC 300 studies was Crowther et al.'s discovery that one of the WR stars is coincident with a bright, hard X-ray source [102]. Prior to this, only the Milky Way's Cyg X-3 and IC10-X1 were known as a WR+compact companion (neutron star or black hole) system; see, e.g., discussion and references in [102]. Analysis by Crowther and his team led to a mass of 37$M_\odot$ for the WR star, and >10$M_\odot$ for the compact companion, placing it firmly in the black hole camp.

Other surveys have been carried out for WR stars in even more distant systems with the 8-m VLT by Hadfield, Bibby and Crowther: IC 4662 (2.3 Mpc) [104], NGC 7793 (3.4 Mpc) [105], NGC 1313 (4.1 Mpc) [106], M83 (4.5 Mpc) [107], NGC 5068 (5–7 Mpc) [108], NGC 6744 (7–11 Mpc) [109]. Most interesting, perhaps, has been their *HST* study of M101, a large spiral located at a (relatively speaking) modest 6.7 Mpc distance [110,111], with followup spectroscopy with the Gemini 8-m [112]. To these, we note the more recent study of the WR content of NGC 625 (3.9 Mpc) [113] by integral field spectroscopy on the VLT by Ana Monreal-Ibero and collaborators.

Although these systems are all too far for completeness to be reached to determine the WC/WN ratio reliability, or provide other diagnostics for testing evolutionary models, they are potentially very useful were one of these stars to become a Type Ibc supernova sometime in the near future [114–116]. Thus, patience may be required to achieve the scientific benefits of these studies of more distant systems. It is also worth noting that no supernova progenitor has yet to be identified as a WR star [117,118].

4.2. Integrated WR Populations

Distant starburst galaxies (by "distant" we means not resolved into stars) often display a WR "bump" in their optical spectra at rest wavelengths of 4650–4670 Å, due to a mixture of WN and WC stars in the integrated spectrum. The first such system was identified in the compact dwarf He 2-10 [119]; quantitative analysis in theory allows one to derive the relative number of WR and O stars [120]; for a more on this subject, see [121] and other papers in their series.

Sokal and collaborators detected the WR bump in an emerging "super star clusters," massive clusters which are just now clearing out their natal material, demonstrating that the time to clear out such material is comparable to the time it takes for massive stars to evolve to the WR phase (~3 Myr) [122,123].

5. Binarity

One of the most heavily debated questions in massive star research is the issue of binarity. Observations have shown that a significant but still contested fraction of massive stars are found in binary systems. Studies of un-evolved massive stars typically find an observed binary fraction of 30–35% for O-type stars in relatively short period (less than ~100 days) systems [124,125]. When long-period systems are included, this percentage approaches 70% or higher [126,127]. This question of binarity also extends to WRs. Methods range from light curve analysis, searching for spectral signatures (such as radial velocity variations), and the presence of X-ray emission. As discussed earlier, the galaxies of the Local Group provide an excellent test-bed for such studies as we are able to determine a complete sample of WRs with which to study the binary fraction.

Over the decades, many papers have attempted to tackle the issue of binarity head-on. In 1981, Massey and Conti found that the fraction of Galactic WR stars that were close WR+O star systems was ~25%, and thus the total fraction must be <50% when the issue of compact companions were included [128]. In 2001, van der Hucht compiled an updated list of WRs in the Galaxy bringing the total up to 227 [19]. They found that the binary fraction of observed and probable binaries was around 40%. Foellmi et al. published papers in 2003 looking at the Magellanic Clouds finding close binary fractions of 40% in the SMC and 30% in the LMC [129,130]. More recently, in 2014, Neugent et al. obtained multi-epoch spectra of nearly all of the WRs in M31 and M33 and searched for short period binary systems by observing radial velocity variations within the prominent emission and hydrogen absorption lines. Such hydrogen lines tend to suggest the presence of an O-type star companion (with the notable exceptions being the WN3/O3s, and some hydrogen-rich WRs found in the Galaxy and in the SMC) [131]. This study found that ~30% of the WRs within M31 and M33 were in short-period binary systems. They additionally found that there was no correlation between binarity and metallicity. Thus, overall, the close binary fraction of WRs appears to be around 30–40% within all metallicity cases, similar to what is observed for O-type stars. (The exact definition of "close" is a debatable one, but we use here a "spectroscopist's definition", corresponding to detection of orbital motions on the order of several 10s of km s^{-1}, corresponding to periods of order 100 days or less for massive stars.)

One further way of searching for WR binaries is through the presence of hard X-ray emission. Most single WRs show soft X-ray emission produced by the winds of the single stars. However, in WR binaries, harder, more luminous X-ray emission forms due to the macroscopic shock interactions between the winds in a binary bound system [132,133]. Such X-ray signatures have been found in a few known binary WRs. One of the most extreme such examples is Mk 34 located in the rich OB association of 30 Doradus in the LMC. It has been classified as a WN5ha and is thought to have a (disputably) high mass of 380$M_\odot$ as derived through spectroscopic analysis [134], but see also [135]. Garofali et al. additionally found a candidate colliding wind binary (WC + O star) in M31 that is located in the dense HII region NGC 604. It is not nearly as bright as Mk 34, but it still shows X-ray emission as discovered by *Chandra* [136]. While searching for X-ray emission is not the most prominent way of detecting WR binaries, it is more frequently being used as a method of determining binarity.

As one of our good friend and colleague often reminds us, "One can never prove any star is *not* a binary." That said, another colleague has noted that the presence of a companion star often makes itself known in the spectrum, albeit in subtle ways.

In single star evolution, the type of WR is heavily influenced by the metallicity of the gas out of which the star formed. As discussed in the introduction, WN stars that show the hydrogen burning byproducts will appear before WC stars which show the helium burning byproducts. Thus, in a low metallicity environment, one expects to find fewer WCs than in a high metallicity environment. However, once binary evolution is considered, this metallicity dependence decreases because the stripping is being done by Roche-lobe overflow instead of metal-driven stellar winds. Thus, one test of binarity is to look for an excess of WCs in an environment—or, even more compelling, is to identify the even more evolved WOs (oxygen-rich WRs) in low metallicity environments. There are two prime examples of such stars that were most likely created through binary evolution. The first is the WO star in the SMC. As discussed earlier, there are only 12 known WRs in the SMC (a low metallicity environment of 0.25× solar) and 11 of them are WNs, as expected. However, the 12th one is a WO that should only form in a high metallicity environment [4,137]. There is an additional example of a WO forming in the low metallicity environment of IC1613 [138], which has a metallicity of ~0.15× solar [139]. Although evolution to the WO stage is not expected by even the most massive single stars in low metallicity environments, models that include binary evolution do predict WOs in low metallicity environments [5]. These two stars are thus examples of WRs likely forming through binary evolution; undoubtedly, there are many more.

While many studies have shown the close binary fraction to be around 30–40%, the actual value is still hotly debated. Proponents of binary evolution argue that the currently single WR stars were once multiple, but their companions have merged. There is little evidence, however, to support this conjecture. There is additionally the question of whether the WRs that formed from binary evolution began with initial masses great enough to suggest that they would have become WRs anyway and the binary mechanism simply sped up the process. Thus, it is possible that the importance of binary evolution may be somewhat overstated, even if the fraction of WRs in binary systems is higher than currently observed.

6. Physical Parameters

As is characteristic of stars approaching the Eddington Limit, a WR's spectrum is heavily influenced by strong stellar winds and high mass-loss rates [140]. Keeping the model's luminosity near, but below, the Eddington limit can make modeling WRs quite a challenge. Additionally, the stars' high surface temperatures mean that the assumption of local thermodynamic equilibrium (LTE) is no longer valid. Instead, the high degree of ionization (and correspondingly decreased opacity) causes the radiation field to decouple from the local thermal field. Furthermore, WR atmospheres are significantly extended when compared to their radius. Thus, plane-parallel geometry cannot be used, and instead spherical geometry must be included. The emission lines that characterize WR spectra are produced in the outflowing winds, with mass-loss rates of order $10^{-5} M_\odot$ yr^{-1}. Finally, WR models must be fully blanketed and include the effects of thousands of overlapping metal lines, which occur at the (unobservable) short wavelengths (<1000 Å) where most of the flux of the star is produced. Two codes are currently capable of including these complexities: the Potsdam Wolf–Rayet Models, or PoWR [141], and the CoMoving Frame GENeral spectrum analysis code, CMFGEN [142]. For a much more detailed description of the physics and complexities involved in modeling a WR, see, e.g., [143–145].

There have been few modeling campaigns of complete samples of WRs in galaxies other than the Magellanic Clouds. In M31, for example, 17 late-type WNs were modeled using PoWR in an attempt to learn more about the wind laws of such stars in different metallicity environments [146]. One limitation of this study was the lack of UV spectroscopy. Nevertheless, they were able to place luminosity constraints on the modeled WRs for values between 10^5 and $10^6 L_\odot$ and suggest that WRs in M31 form from initial mass ranges between 20 and $60 M_\odot$. This is similar to that found in both the Galaxy and Magellanic Clouds. However, no modeling has taken place for the WC stars in a high metallicity environment like M31.

Conversely, much modeling has been done of WRs in the Magellanic Clouds. Over the past few years, surveys of single and binary WNs in both the SMC and LMC, and the WN3/O3 stars in the LMC have all been performed. In 2014, Hainich et al. determined physical parameters of over 100 WNs in the LMC using grids of PoWR [147] models. They concluded that the bulk (∼88%) of the WRs analyzed had progressed through the RSG before becoming WRs, thus implying that they evolved from 20–40$M_\odot$ progenitors. They also found that these results were well aligned with studies of Galactic WRs suggesting that there is no metallicity dependence on the range of main sequence masses that evolve into WRs. This research in the LMC was extended to the WR binaries by Shenar et al. in 2019 [148], who looked at the 44 binary candidates and found that 28 of them have composite spectra and five of them show periodically moving WR primaries. They conclude that while 45 ± 30% of the WNs in the LMC have most likely interacted with a companion via mass-transfer, many of these WRs would have evolved to become WRs through single star evolution.

Both the binary and single WNs in the SMC have also been modeled using the PoWR code [149,150]. As discussed earlier, many of the WNs in the SMC have absorption lines that, if not due to a companion, could simply be photospheric lines that are inherent to the stars because of their weak stellar winds. Thus, studying them for photometric and radial velocity variability is necessary to determine their binarity. Based on modeling with the PoWR code, it was concluded again that, while some of these stars are binaries now, they still would have become WRs through single-star evolution given their high initial main-sequence masses.

As discussed above, there has been additional modeling of the LMC WN3/O3s using CMFGEN. All ten of these stars show strong absorption and emission lines as is shown in Figure 9 for one of the newly discovered stars. CMFGEN spectral line fitting was used to determine the physical parameters of these ten stars. Table 2 shows the range of values for the 10 WN3/O3s compared to typical values for an O3V and WN3 star in the LMC (WN3 parameters from [147]. O3V parameters from [151].). While the temperature is a bit on the high side for what we would expect for a LMC WN, the majority of the parameters are within the expected ranges. The one exception is the mass-loss rate which is more similar to that of an O3V than of a normal LMC WN.

Table 2. Physical parameters of WN3/O3s, WNs, and O3Vs in the LMC.

	WN3/O3s	WN3	O3V
$T_{\rm off}$ (K)	100,000–105,000	80,000	48,000
$\log \frac{L}{L_\odot}$	5.6	5.7	5.6
$\log \dot{M}$	−6.1 − −5.7	−4.5	−5.9
He/H (by #)	0.8–1.5	1.0–1.4	0.1
N (by mass)	5–10× solar	5–10× solar	0.5× solar
M_V	−2.5	−4.5	−5.5

Although other WN stars with intrinsic absorption lines are known, the WN3/O3s appear to be unique [93,148], and their place in the evolution of massive stars still unknown. Neugent's study [93] considered the possibility that these stars were the products of homogenous evolution, a situation that can occur if the star is rotating so rapidly that mixing keeps the composition nearly uniform within the star (see, e.g., [152]). However, they ruled this out based upon the stars' low rotational velocities combined with low mass-loss rates, as the latter implies that the high angular momentum could not have been carried off by stellar winds. Based on their absolute magnitudes they are not WN + O3V binaries, though they could be hiding a less-massive companion. It is additionally possible that binarity influenced their previous evolution. However, it is currently thought that instead these stars represent an intermediate stage between O stars an WNs. More research is ongoing in an attempt to answer this question.

7. Comparisons to Evolutionary Models

As discussed in the Introduction, comparing the observed WC/WN ratio with evolutionary model predictions is one of the most important reasons to search for WRs. Currently, we have complete samples of the WR populations for the Magellanic Clouds, M31, and M33. The galaxy's metallicities and WC/WN ratios are shown in Table 3. We have included the Milky Way, although here the data are far less certain that the statistical uncertainties would indicate. As is expected, the WC/WN ratio increases with increasing metallicity due to the strength of the stellar winds. We can now compare these observational results to those of the evolutionary models.

Table 3. WC/WN ratio vs. metallicity for the Local Group Galaxies.

Region	log(O/H) + 12	# WCs and WOs	# WNs	WC/WN
SMC	8.13	1	11	0.09 ± 0.09
M33 outer	8.29	12	54	0.22 ± 0.06
LMC	8.37	28	124	0.23 ± 0.01
M33 middle	8.41	15	54	0.28 ± 0.07
Milky Way	8.70	46	53	0.83 ± 0.10
M33 inner	8.72	26	45	0.58 ± 0.09
M31	8.93	62	92	0.67 ± 0.11

There are two primary sets of evolutionary models currently used in the massive star community. The first is the Geneva Evolutionary Models [1] that model the evolution of single stars. The other is the Binary Population and Spectral Synthesis (BPASS) models that focus on binary evolution [153,154].

Besides the obvious difference between the two of modeling single vs. binary stars, the models also have some important differences. In the case of the Geneva models, there are only results for a few metallicities, as is shown in the figure below. This makes comparisons between the observations and the models quite difficult because there are only a few points. However, these models have been created with different initial rotation rates as it plays quite a large effect on the resulting physics. Conversely, the BPASS models have results spanning a wide range of metallicities, but these models do not include rotation. Thus, due to these differences, it is difficult to compare the observations directly to either set of models. However, in time, the models will continue to improve.

In Figure 11, we show the agreement between the WC/WN ratios and the evolutionary models. We have not included NGC 6822 or IC 1613 in this diagram, as they each have too few WRs for meaningful statistics (4 and 1, respectively). We also have not included IC 10, as we feel the current value is, at best, an upper limit. We have included the value for the MW determined as described above, although we suspect that this too is an upper limit. As for the predictions: The solid line is from the older Geneva evolutionary models, the first to include rotation [1]. The green dashed line is an updated version of the predictions from BPASS2 [5], and these 2.2.1 predictions were kindly provided by Eldridge (2019, private communication). The models assume continuous star formation, a Salpeter IMF slope, and an upper mass limit of $300 M_\odot$. The BPASS models also include the effects of binary evolution. Finally, the two ×'s denote results from the latest single-star evolutionary models. The higher metallicity value comes from [155], while the lower metallicity point was computed by Cyril Georgy from preliminary Geneva z = 0.006 models, and used in [77]. There is good agreement between the newer Geneva single-star models and the binary evolution models; this may simply be that the BPASS models do not yet include the effects of rotation. Including rotation can reduce the expected ratio of WC/WN stars; see Figure 10 in [77]. Although the observational data at all metallicities are now in relatively good shape, improvements are still pending in the evolutionary models. Still, we can conclude that the large issue at high metallicity with the oldest models has largely gone away.

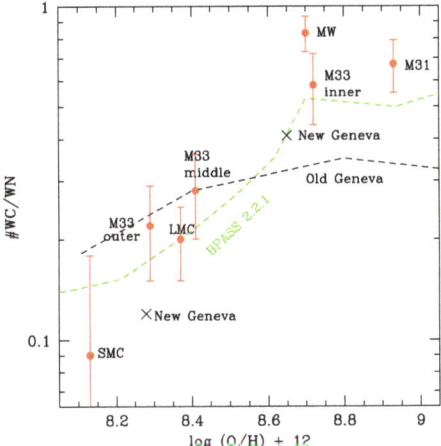

Figure 11. WC/WN ratio vs. metallicity compared to both BPASS2.2.1 and Geneva Evolutionary models. Notice the improved results between the observed WC/WN ratio and the Geneva Evolutionary models, but the lack of models at a variety of metallicities. In addition, notice the good agreement between the BPASS2.2.1 models and both the observed results. The error bars come from $\sqrt{N}$ statistics; see [73,77].

8. Conclusions and the Future of WRs

WRs are the bare stellar cores of massive stars, and the last stage in a massive star's lifetime before they turn into supernovae. Observing a complete set of both the nitrogen and carbon rich WRs within a galaxy allows for important comparisons between the observed WC/WN ratio and that predicted by the evolutionary models. Because the evolution of WRs is highly dependent on the metallicity of the surrounding environment, it is important to do these comparisons across a wide range of galaxies with different metallicities, such as the galaxies in the Local Group.

Finding WRs observationally is done using a combination of interference filters and photometric techniques before the identified candidates are confirmed spectroscopically. This method has been used with great success over the past few decades and led to the discovery of hundreds of WRs in both our galaxy and even those far enough away that we can only observe the integrated light coming from clusters of WRs. While this method has led to the discovery of mostly complete samples of WRs within the Local Group, there is still much progress to be made in more distant galaxies.

The binary fraction of WRs is still highly contested with current observations putting it somewhere between 30–40% for the close binary frequency. However, as the distance between binaries expands, and the effect of binarity on the evolution of WRs in the past is considered, it is difficult to fully understand what role binaries play in the evolution of WRs. Modeling the spectra of the currently known WRs using sophisticated modeling codes such as PoWR and CMFGEN allow us to get a better handle on the physical properties of both the binaries and single stars and compare them across a wide range of metallicities.

As discussed in Section 2, while much progress has been made in the field of WR research, there is still much to be done. With *Gaia*, it is now possible to determine distances to nearby WRs within our own Galaxy leading to better observations of their reddenings and better modeling of their physical properties. We are additionally learning more about the content of other types of massive stars (such as O/B stars, RSGs, etc.) that allow us to compare the ratio of those stars vs. WRs to the evolutionary model predictions placing further constraints on the models. Finally, we are continuing to push the observational boundaries to further and further galaxies in an attempt to observe complete samples of WRs in both the galaxies of the Local Group and beyond!

Author Contributions: K.N. wrote the majority of the paper with P.M. contributing primarily to Sections 2 and 4.

Funding: This research was partially funded by the National Science Foundation, most recently through AST-1612874, as well as through Lowell Observatory.

Acknowledgments: The authors acknowledge all of their dear friends, family, and collaborators who have supported them though many years of Wolf–Rayet research. They additionally thank J.J. Eldridge for her help with better understanding WR binary evolution, Paul Crowther for useful information on Galactic WR surveys, as well as an anonymous referee whose suggestions improved the paper.

Conflicts of Interest: The authors declare no conflict of interest.

References

1. Meynet, G.; Maeder, A. Stellar evolution with rotation. XI. Wolf–Rayet star populations at different metallicities. *Astron. Astrophys.* **2005**, *429*, 581–598. [CrossRef]
2. Conti, P.S. On the relationship between Of and WR stars. *Mem. Soc. R. Sci. Liege* **1975**, *9*, 193–212.
3. Maeder, A.; Conti, P.S. Massive Star Populations in Nearby Galaxies. *Annu. Rev. Astron. Astrophys.* **1994**, *32*, 227–275. [CrossRef]
4. Moffat, A.F.J.; Niemela, V.S.; Marraco, H.G. Wolf–Rayet Stars in the Magellanic Clouds. VI. Spectroscopic Orbits of WC Binaries and Implications for W-R Evolution. *Astrophys. J.* **1990**, *348*, 232. [CrossRef]
5. Eldridge, J.J.; Stanway, E.R.; Xiao, L.; McClelland, L.A.S.; Taylor, G.; Ng, M.; Greis, S.M.L.; Bray, J.C. Binary Population and Spectral Synthesis Version 2.1: Construction, Observational Verification, and New Results. *Publ. Astron. Soc. Aust.* **2017**, *34*, e058. [CrossRef]

6. Leitherer, C.; Schaerer, D.; Goldader, J.D.; Delgado, R.M.G.; Robert, C.; Kune, D.F.; de Mello, D.F.; Devost, D.; Heckman, T.M. Starburst99: Synthesis Models for Galaxies with Active Star Formation. *Astrophys. J. Suppl. Ser.* **1999**, *123*, 3–40. [CrossRef]
7. Levesque, E.M.; Kewley, L.J. The Host Galaxy of GRB 060505: Host ISM Properties. *Astrophys. J. Lett.* **2007**, *667*, L121–L124. [CrossRef]
8. Meynet, G.; Mowlavi, N.; Maeder, A. Massive-star evolution at high metallicity. In *The Metal-Rich Universe*; Israelian, G., Meynet, G., Eds.; Cambridge University Press: Cambridge, UK, 2008; p. 341.
9. Maeder, A.; Lequeux, J.; Azzopardi, M. The numbers of red supergiants and WR stars in galaxies—An extremely sensitive indicator of chemical composition. *Astron. Astrophys.* **1980**, *90*, L17–L20.
10. Massey, P. A UBVR CCD Survey of the Magellanic Clouds. *Astrophys. J. Suppl. Ser.* **2002**, *141*, 81–122. [CrossRef]
11. Russell, S.C.; Dopita, M.A. Abundances of the Heavy Elements in the Magellanic Clouds. II. H II Regions and Supernova Remnants. *Astrophys. J. Suppl. Ser.* **1990**, *74*, 93. [CrossRef]
12. Sanders, N.E.; Caldwell, N.; McDowell, J.; Harding, P. The Metallicity Profile of M31 from Spectroscopy of Hundreds of H II Regions and PNe. *Astrophys. J.* **2012**, *758*, 133. [CrossRef]
13. Beals, C.S. The Wolf–Rayet stars. *Publ. Dom. Astrophys. Obs. Vic.* **1930**, *4*, 271–301.
14. Beals, C.S. On the nature of Wolf–Rayet emission. *Mon. Not. R. Astron. Soc.* **1929**, *90*, 202–212. [CrossRef]
15. Cassinelli, J.P.; Hartmann, L. The subsonic structure of radiatively driven winds of early-type stars. *Astrophys. J.* **1975**, *202*, 718–732. [CrossRef]
16. Conti, P.S.; De Loore, C.W.H. (Eds.) Mass loss and evolution of O-type stars. In Proceedings of the Symposium, Vancouver Island, BC, Canada, 5–9 June 1978; Volume 83.
17. Smith, L.F. A revised spectral classification system and a new catalogue for galactic Wolf–Rayet stars. *Mon. Not. R. Astron. Soc.* **1968**, *138*, 109. [CrossRef]
18. van der Hucht, K.A.; Conti, P.S.; Lundstrom, I.; Stenholm, B. The Sixth Catalogue of Galactic Wolf–Rayet Stars—Their Past and Present. *Space Sci. Rev.* **1981**, *28*, 227–306. [CrossRef]
19. van der Hucht, K.A. The VIIth catalogue of galactic Wolf–Rayet stars. *New Astron. Rev.* **2001**, *45*, 135–232. [CrossRef]
20. Crowther, P.A.; De Marco, O.; Barlow, M.J. Quantitative classification of WC and WO stars. *Mon. Not. R. Astron. Soc.* **1998**, *296*, 367–378. [CrossRef]
21. Conti, P.S.; Niemela, V.S.; Walborn, N.R. A radial velocity study of three WN stars and an O3f star in the Carina Nebula. *Astrophys. J.* **1979**, *228*, 206–219. [CrossRef]
22. Massey, P.L. Spectroscopic Studies of Wolf–Rayet Stars with Absorption Lines. Ph.D. Thesis, University of Colorado, Boulder, CO, USA, 1980.
23. Scheiner, J.; Frost, E.B. *A Treatise on Astronomical Spectroscopy*; Ginn: Boston, MA, USA, 1894.
24. Campbell, W.W. The Wolf–Rayet stars. *Astron.-Astro-Phys.* **1894**, *13*, 448–476.
25. van der Hucht, K.A. New Galactic Wolf–Rayet stars, and candidates. An annex to the VIIth Catalogue of Galactic Wolf–Rayet Stars. *Astron. Astrophys.* **2006**, *458*, 453–459. [CrossRef]
26. Hadfield, L.J.; van Dyk, S.D.; Morris, P.W.; Smith, J.D.; Marston, A.P.; Peterson, D.E. Searching for hidden Wolf–Rayet stars in the Galactic plane—15 new Wolf–Rayet stars. *Mon. Not. R. Astron. Soc.* **2007**, *376*, 248–262. [CrossRef]
27. Mauerhan, J.C.; Van Dyk, S.D.; Morris, P.W. Red Eyes on Wolf–Rayet Stars: 60 New Discoveries via Infrared Color Selection. *Astron. J.* **2011**, *142*, 40. [CrossRef]
28. Mauerhan, J.C.; Van Dyk, S.D.; Morris, P.W. 12 New Galactic Wolf–Rayet Stars Identified via 2MASS + Spitzer/GLIMPSE. *Publ. Astron. Soc. Pac.* **2009**, *121*, 591. [CrossRef]
29. Faherty, J.K.; Shara, M.M.; Zurek, D.; Kanarek, G.; Moffat, A.F.J. Characterizing Wolf–Rayet Stars in the Near- and Mid-infrared. *Astron. J.* **2014**, *147*, 115. [CrossRef]
30. Rosslowe, C.K.; Crowther, P.A. A deep near-infrared spectroscopic survey of the Scutum-Crux arm for Wolf–Rayet stars. *Mon. Not. R. Astron. Soc.* **2018**, *473*, 2853–2870. [CrossRef]
31. Shara, M.M.; Moffat, A.F.J.; Gerke, J.; Zurek, D.; Stanonik, K.; Doyon, R.; Artigau, E.; Drissen, L.; Villar-Sbaffi, A. A Near-Infrared Survey of the Inner Galactic Plane for Wolf–Rayet Stars. I. Methods and First, Results: 41 New WR Stars. *Astron. J.* **2009**, *138*, 402–420. [CrossRef]

32. Shara, M.M.; Faherty, J.K.; Zurek, D.; Moffat, A.F.J.; Gerke, J.; Doyon, R.; Artigau, E.; Drissen, L. A Near-infrared Survey of the Inner Galactic Plane for Wolf–Rayet Stars. II. Going Fainter: 71 More New W-R Stars. *Astron. J.* **2012**, *143*, 149. [CrossRef]
33. Westerlund, B.E.; Rodgers, A.W. Wolf–Rayet stars and planetary nebulae in the Large Magellanic Cloud. *Observatory* **1959**, *79*, 132–134.
34. de Vacuouleurs, G.; Buscombe, W.; Gascoigne, S.C.B. Large Magellanic Cloud. *Aust. J. Sci.* **1954**, *17*, 1.
35. Feast, M.W.; Thackeray, A.D.; Wesselink, A.J. The Magellanic Clouds—Spectroscopy and photometry of bright stars. *Observatory* **1958**, *78*, 156–165.
36. Azzopardi, M.; Breysacher, J. New Wolf–Rayet stars in the Large Magellanic Cloud. *Astron. Astrophys.* **1979**, *75*, 243–246.
37. Breysacher, J. Spectral Classification of Wolf–Rayet Stars in the Large Magellanic Cloud. *Astron. Astrophys. Suppl. Ser.* **1981**, *43*, 203.
38. Neugent, K.F.; Massey, P.; Morrell, N. A Modern Search for Wolf–Rayet Stars in the Magellanic Clouds. IV. A Final Census. *Astrophys. J.* **2018**, *863*, 181. [CrossRef]
39. Breysacher, J.; Azzopardi, M.; Testor, G. The fourth catalogue of Population I Wolf–Rayet stars in the Large Magellanic Cloud. *Astron. Astrophys. Suppl. Ser.* **1999**, *137*, 117–145. [CrossRef]
40. Hunter, D.A.; Shaya, E.J.; Holtzman, J.A.; Light, R.M.; O'Neil, J.; Earl, J.; Lynds, R. The Intermediate Stellar Mass Population in R136 Determined from Hubble Space Telescope Planetary Camera 2 Images. *Astrophys. J.* **1995**, *448*, 179. [CrossRef]
41. Melnick, J. The 30 Doradus nebula. I. Spectral classification of 69 stars in the central cluster. *Astron. Astrophys.* **1985**, *153*, 235–244.
42. Massey, P.; Hunter, D.A. Star Formation in R136: A Cluster of O3 Stars Revealed by Hubble Space Telescope Spectroscopy. *Astrophys. J.* **1998**, *493*, 180–194. [CrossRef]
43. Drissen, L.; Moffat, A.F.J.; Walborn, N.R.; Shara, M.M. The Dense Galactic Starburst NGC 3603. I. HST/FOS Spectroscopy of Individual Stars in the Core and the source of Ionization and Kinetic Energy. *Astron. J.* **1995**, *110*, 2235. [CrossRef]
44. de Koter, A.; Heap, S.R.; Hubeny, I. On the Evolutionary Phase and Mass Loss of the Wolf–Rayet–like Stars in R136a. *Astrophys. J.* **1997**, *477*, 792. [CrossRef]
45. Massey, P. Massive stars in the galaxies of the Local Group. *New Astron. Rev.* **2013**, *57*, 14–27. [CrossRef]
46. Breysacher, J.; Westerlund, B.E. Wolf–Rayet stars in the Small Magellanic Cloud. *Astron. Astrophys.* **1978**, *67*, 261–265.
47. Azzopardi, M.; Breysacher, J. A search for new Wolf–Rayet stars in the Small Magellanic Cloud. *Astron. Astrophys.* **1979**, *75*, 120–126.
48. Morgan, D.H.; Vassiliadis, E.; Dopita, M.A. A new Wolf–Rayet star in the Small Magellanic Cloud. *Mon. Not. R. Astron. Soc.* **1991**, *251*, 51P. [CrossRef]
49. Massey, P.; Duffy, A.S. A Search for Wolf–Rayet Stars in the Small Magellanic Cloud. *Astrophys. J.* **2001**, *550*, 713–723. [CrossRef]
50. Massey, P.; Olsen, K.A.G.; Parker, J.W. The Discovery of a 12th Wolf–Rayet Star in the Small Magellanic Cloud. *Publ. Astron. Soc. Pac.* **2003**, *115*, 1265–1268. [CrossRef]
51. Conti, P.S.; Massey, P.; Garmany, C.D. Spectroscopic Studies of Wolf–Rayet Stars. V. Optical Spectrophotometry of the Emission Lines in Small Magellanic Cloud Stars. *Astrophys. J.* **1989**, *341*, 113. [CrossRef]
52. Wray, J.D.; Corso, G.J. Wolf–Rayet Stars in M33. *Astrophys. J.* **1972**, *172*, 577. [CrossRef]
53. Boksenberg, A.; Willis, A.J.; Searle, L. Observations of three Wolf–Rayet stars in M33. *Mon. Not. R. Astron. Soc.* **1977**, *180*, 15P–19P. [CrossRef]
54. Bohannan, B.; Conti, P.S.; Massey, P. A GRISM search of Messier 33 for emissionline objects. *Astron. J.* **1985**, *90*, 600–605. [CrossRef]
55. Conti, P.S.; Massey, P. Wolf-rayet stars and giant HII regions in M 33: Casual associations or meaningful relationships? *Astrophys. J.* **1981**, *249*, 471–480. [CrossRef]
56. Rosa, M.; Dodorico, S. Wolf–Rayet stars in extragalactic H II regions. II - NGC604 - A giant H II region dominated by many Wolf–Rayet stars. *Astron. Astrophys.* **1982**, *108*, 339–343.
57. Dodorico, S.; Rosa, M.; Wampler, E.J. Search for Wolf–Rayet features in the spectra of giant HII regions. I. Observations in NGC 300, NGC 604, NGC 5457 and He 2-10. *Astron. Astrophys. Suppl. Ser.* **1983**, *53*, 97–108.
58. Massey, P.; Conti, P.S. Wolf-rayet stars in M 33. *Astrophys. J.* **1983**, *273*, 576–589. [CrossRef]

59. Massey, P.; Conti, P.S.; Armandroff, T.E. The Spectra of Extra-Galactic Wolf–Rayet Stars. *Astron. J.* **1987**, *94*, 1538. [CrossRef]
60. Armandroff, T.E.; Massey, P. Wolf–Rayet stars in NGC 6822 and IC 1613. *Astrophys. J.* **1985**, *291*, 685–692. [CrossRef]
61. D'Odorico, S.; Rosa, M. Wolf–Rayet stars in extragalactic HII regions: Discovery of a peculiar WR in IC 1613/#3. *Astron. Astrophys.* **1982**, *105*, 410–412.
62. Davidson, K.; Kinman, T.D. Data on an unusual Wolf–Rayet star in the nearby Galaxy IC 1613. *Publ. Astron. Soc. Pac.* **1982**, *94*, 634–639. [CrossRef]
63. Westerlund, B.E.; Azzopardi, M.; Breysacher, J.; Lequeux, J. Discovery of a Wolf–Rayet star in NGC 6822. *Astron. Astrophys.* **1983**, *123*, 159–161.
64. Armandroff, T.E.; Massey, P. Wolf–Rayet Stars in Local Group Galaxies: Numbers and Spectral Properties. *Astron. J.* **1991**, *102*, 927. [CrossRef]
65. Massey, P.; Armandroff, T.E.; Conti, P.S. IC 10: A "Poor Cousin" Rich in Wolf–Rayet Stars. *Astron. J.* **1992**, *103*, 1159. [CrossRef]
66. Massey, P.; Armandroff, T.E. The Massive Star Content, Reddening, and Distance of the Nearby Irregular Galaxy IC 10. *Astron. J.* **1995**, *109*, 2470. [CrossRef]
67. Rieke, G.H.; Lebofsky, M.J.; Thompson, R.I.; Low, F.J.; Tokunaga, A.T. The nature of the nuclear sources in M82 and NGC 253. *Astrophys. J.* **1980**, *238*, 24–40. [CrossRef]
68. Massey, P.; Holmes, S. Wolf–Rayet Stars in IC 10: Probing the Nearest Starburst. *Astrophys. J.* **2002**, *580*, L35–L38. [CrossRef]
69. Tehrani, K.; Crowther, P.A.; Archer, I. Revealing the nebular properties and Wolf–Rayet population of IC10 with Gemini/GMOS. *Mon. Not. R. Astron. Soc.* **2017**, *472*, 4618–4633. [CrossRef]
70. Moffat, A.F.J.; Shara, M.M. Wolf–Rayet stars in the Local Group galaxies M 31 and NGC 6822. *Astrophys. J.* **1983**, *273*, 544–561. [CrossRef]
71. Moffat, A.F.J.; Shara, M.M. Wolf–Rayet Stars in the Andromeda Galaxy. *Astrophys. J.* **1987**, *320*, 266. [CrossRef]
72. Massey, P.; Armandroff, T.E.; Conti, P.S. Massive stars in M31. *Astron. J.* **1986**, *92*, 1303–1333. [CrossRef]
73. Massey, P.; Johnson, O. Evolved Massive Stars in the Local Group. II. A New Survey for Wolf–Rayet Stars in M33 and Its Implications for Massive Star Evolution: Evidence of the "Conti Scenario" in Action. *Astrophys. J.* **1998**, *505*, 793–827. [CrossRef]
74. Conti, P.S.; Massey, P. Spectroscopic studies of Wolf–Rayet stars. IV—Optical spectrophotometry of the emission lines in galactic and large Magellanic Cloud stars. *Astrophys. J.* **1989**, *337*, 251–271. [CrossRef]
75. Zaritsky, D.; Kennicutt, R.C., Jr.; Huchra, J.P. H II Regions and the Abundance Properties of Spiral Galaxies. *Astrophys. J.* **1994**, *420*, 87. [CrossRef]
76. Magrini, L.; Perinotto, M.; Mampaso, A.; Corradi, R.L.M. Chemical abundances of Planetary Nebulae in M 33. *Astron. Astrophys.* **2004**, *426*, 779–786. [CrossRef]
77. Neugent, K.F.; Massey, P.; Georgy, C. The Wolf–Rayet Content of M31. *Astrophys. J.* **2012**, *759*, 11. [CrossRef]
78. Neugent, K.F.; Massey, P. The Wolf–Rayet Content of M33. *Astrophys. J.* **2011**, *733*, 123. [CrossRef]
79. Massey, P.; Neugent, K.F.; Morrell, N.; Hillier, D.J. A Modern Search for Wolf–Rayet Stars in the Magellanic Clouds: First Results. *Astrophys. J.* **2014**, *788*, 83. [CrossRef]
80. Yuan, F.; Quimby, R.M.; Wheeler, J.C.; Vinkó, J.; Chatzopoulos, E.; Akerlof, C.W.; Kulkarni, S.; Miller, J.M.; McKay, T.A.; Aharonian, F. The Exceptionally Luminous Type Ia Supernova 2007if. *Astrophys. J.* **2010**, *715*, 1338–1343. [CrossRef]
81. Yuan, F.; Akerlof, C.W. Astronomical Image Subtraction by Cross-Convolution. *Astrophys. J.* **2008**, *677*, 808–812. [CrossRef]
82. Becker, A.C.; Wittman, D.M.; Boeshaar, P.C.; Clocchiatti, A.; Dell'Antonio, I.P.; Frail, D.A.; Halpern, J.; Margoniner, V.E.; Norman, D.; Tyson, J.A.; et al. The Deep Lens Survey Transient Search. I. Short Timescale and Astrometric Variability. *Astrophys. J.* **2004**, *611*, 418–433. [CrossRef]
83. Stetson, P.B. DAOPHOT: A Computer Program for Crowded-Field Stellar Photometry. *Publ. Astron. Soc. Pac.* **1987**, *99*, 191. [CrossRef]
84. Stetson, P.B.; Davis, L.E.; Crabtree, D.R. Future development of the DAOPHOT crowded-field photometry package. In *CCDs in Astronomy*; Astronomical Society of the Pacific Conference Series; Jacoby, G.H., Ed.; Astronomical Society of the Pacific: San Francisco, CA, USA, 1990; Volume 8, pp. 289–304.

85. Smith, L.F. The distribution of Wolf–Rayet stars in the Galaxy. *Mon. Not. R. Astron. Soc.* **1968**, *141*, 317. [CrossRef]
86. van den Bergh, S. Stellar Associations in the Andromeda Nebula. *Astrophys. J. Suppl. Ser.* **1964**, *9*, 65. [CrossRef]
87. Shara, M.M.; Mikołajewska, J.; Caldwell, N.; Iłkiewicz, K.; Drozd, K.; Zurek, D. The first transition Wolf–Rayet WN/C star in M31. *Mon. Not. R. Astron. Soc.* **2016**, *455*, 3453–3457. [CrossRef]
88. Massey, P.; Olsen, K.A.G.; Hodge, P.W.; Jacoby, G.H.; McNeill, R.T.; Smith, R.C.; Strong, S.B. A Survey of Local Group Galaxies Currently Forming Stars. II. UBVRI Photometry of Stars in Seven Dwarfs and a Comparison of the Entire Sample. *Astron. J.* **2007**, *133*, 2393–2417. [CrossRef]
89. de Vaucouleurs, G.; Buta, R. The galactic extinction of extragalactic objects. I - The csc B law and the extinction coefficient. *Astron. J.* **1983**, *88*, 939–961. [CrossRef]
90. Neugent, K.F.; Massey, P.; Morrell, N. The Discovery of a Rare WO-type Wolf–Rayet Star in the Large Magellanic Cloud. *Astron. J.* **2012**, *144*, 162. [CrossRef]
91. Massey, P.; Neugent, K.F.; Morrell, N. A Modern Search for Wolf–Rayet Stars in the Magellanic Clouds. II. A Second Year of Discoveries. *Astrophys. J.* **2015**, *807*, 81. [CrossRef]
92. Massey, P.; Neugent, K.F.; Morrell, N. A Modern Search for Wolf–Rayet Stars in the Magellanic Clouds. III. A Third Year of Discoveries. *arXiv* **2017**, arXiv:astro-ph.SR/1701.07815.
93. Neugent, K.F.; Massey, P.; Hillier, D.J.; Morrell, N. The Evolution and Physical Parameters of WN3/O3s: A New Type of Wolf–Rayet Star. *Astrophys. J.* **2017**, *841*, 20. [CrossRef]
94. Bailer-Jones, C.A.L.; Rybizki, J.; Fouesneau, M.; Mantelet, G.; Andrae, R. Estimating Distance from Parallaxes. IV. Distances to 1.33 Billion Stars in Gaia Data Release 2. *Astron. J.* **2018**, *156*, 58. [CrossRef]
95. Rizzi, L.; Bresolin, F.; Kudritzki, R.P.; Gieren, W.; Pietrzyński, G. The Araucaria Project: The Distance to NGC 300 from the Red Giant Branch Tip Using HST ACS Imaging. *Astrophys. J.* **2006**, *638*, 766–771. [CrossRef]
96. Deharveng, L.; Caplan, J.; Lequeux, J.; Azzopardi, M.; Breysacher, J.; Tarenghi, M.; Westerlund, B. H II regions in NGC 300. *Astron. Astrophys. Suppl. Ser.* **1988**, *73*, 407–423.
97. Schild, H.; Testor, G. New Wolf–Rayet stars near the nucleus of NGC 300. *Astron. Astrophys.* **1991**, *243*, 115–117.
98. Schild, H.; Testor, G. Ten little Wolf–Rayet stars in NGC 300. *Astron. Astrophys.* **1992**, *266*, 145–149.
99. Breysacher, J.; Azzopardi, M.; Testor, G.; Muratorio, G. Wolf–Rayet stars detected in five associations of NGC 300. *Astron. Astorphys.* **1997**, *326*, 976–981.
100. Bresolin, F.; Kudritzki, R.P.; Najarro, F.; Gieren, W.; Pietrzyński, G. Discovery and Quantitative Spectral Analysis of an Ofpe/WN9 (WN11) Star in the Sculptor Spiral Galaxy NGC 300. *Astrophys. J.* **2002**, *577*, L107–L110. [CrossRef]
101. Schild, H.; Crowther, P.A.; Abbott, J.B.; Schmutz, W. A large Wolf–Rayet population in NGC 300 uncovered by VLT-FORS2. *Astron. Astrophys.* **2003**, *397*, 859–870. [CrossRef]
102. Crowther, P.A.; Carpano, S.; Hadfield, L.J.; Pollock, A.M.T. On the optical counterpart of NGC 300 X-1 and the global Wolf–Rayet content of NGC 300. *Astron. Astrophys.* **2007**, *469*, L31–L34. [CrossRef]
103. Kennicutt, R.C., Jr.; Lee, J.C.; Funes, J.G.; Sakai, S.; Akiyama, S. An Hα Imaging Survey of Galaxies in the Local 11 Mpc Volume. *Astrophys. J. Suppl. Ser.* **2008**, *178*, 247–279. [CrossRef]
104. Crowther, P.A.; Bibby, J.L. On the massive star content of the nearby dwarf irregular Wolf–Rayet galaxy IC 4662. *Astron. Astrophys.* **2009**, *499*, 455–464. [CrossRef]
105. Bibby, J.L.; Crowther, P.A. A Very Large Telescope imaging and spectroscopic survey of the Wolf–Rayet population in NGC7793. *Mon. Not. R. Astron. Soc.* **2010**, *405*, 2737–2753. [CrossRef]
106. Hadfield, L.J.; Crowther, P.A. A survey of the Wolf–Rayet population of the barred, spiral galaxy NGC 1313. *Mon. Not. R. Astron. Soc.* **2007**, *381*, 418–432. [CrossRef]
107. Hadfield, L.J.; Crowther, P.A.; Schild, H.; Schmutz, W. A spectroscopic search for the non-nuclear Wolf–Rayet population of the metal-rich spiral galaxy M 83. *Astron. Astrophys.* **2005**, *439*, 265–277. [CrossRef]
108. Bibby, J.L.; Crowther, P.A. The Wolf–Rayet population of the nearby barred spiral galaxy NGC 5068 uncovered by the Very Large Telescope and Gemini. *Mon. Not. R. Astron. Soc.* **2012**, *420*, 3091–3107. [CrossRef]
109. Bibby, J.; Crowther, P.; Sandford, E. Location of WR stars in NGC 6744. In Proceedings of the Massive Stars: From Alpha to Omega, Rhodes, Greece, 10–14 June 2013; p. 140.

110. Bibby, J.; Shara, M.; Zurek, D.; Crowther, P.A.; Moffat, A.F.J.; Drissen, L.; Wilde, M. The Distribution of Massive Stars in M101. In *Wolf–Rayet Stars: Proceedings of an International Workshop Held in Potsdam*; Universitätsverlag Potsdam: Potsdam, Germany, 2015; p. 355.
111. Shara, M.M.; Bibby, J.L.; Zurek, D.; Crowther, P.A.; Moffat, A.F.J.; Drissen, L. The Vast Population of Wolf–Rayet and Red Supergiant Stars in M101. I. Motivation and First, Results. *Astron. J.* **2013**, *146*, 162. [CrossRef]
112. Pledger, J.L.; Shara, M.M.; Wilde, M.; Crowther, P.A.; Long, K.S.; Zurek, D.; Moffat, A.F.J. The first optical spectra of Wolf–Rayet stars in M101 revealed with Gemini/GMOS. *Mon. Not. R. Astron. Soc.* **2018**, *473*, 148–164. [CrossRef]
113. Monreal-Ibero, A.; Walsh, J.R.; Iglesias-Páramo, J.; Sandin, C.; Relaño, M.; Pérez-Montero, E.; Vílchez, J. The Wolf–Rayet star population in the dwarf galaxy NGC 625. *Astron. Astrophys.* **2017**, *603*, A130. [CrossRef]
114. Testor, G.; Schild, H. Wolf–Rayet stars beyond 1 Mpc: Why we want to find them and how to do it. *Messenger* **1993**, *72*, 31–34.
115. Crowther, P.A.; Hadfield, L.J. VLT/FORS Surveys of Wolf–Rayet Stars beyond the Local Group: Type Ib/c Supernova Progenitors? *Messenger* **2007**, *129*, 53–57.
116. Bibby, J.L.; Shara, M.M.; Crowther, P.A.; Moffat, A.F.J. Searching for Wolf–Rayet Stars Beyond the Local Group. In *Proceedings of a Scientific Meeting in Honor of Anthony F. J. Moffat*; Astronomical Society of the Pacific Conference Series; Drissen, L., Robert, C., St-Louis, N., Moffat, A.F.J., Eds.; Astronomical Society of the Pacific: San Francisco, CA, USA, 2012; Volume 465, p. 478.
117. Eldridge, J.J.; Fraser, M.; Maund, J.R.; Smartt, S.J. Possible binary progenitors for the Type Ib supernova iPTF13bvn. *Mon. Not. R. Astron. Soc.* **2015**, *446*, 2689–2695. [CrossRef]
118. Fraser, M. The progenitors of core-collapse supernovae. In *The Lives and Death-Throes of Massive Stars*; Eldridge, J.J., Bray, J.C., McClelland, L.A.S., Xiao, L., Eds.; International Astronomical Union: Paris, France, 2017; Volume 329, pp. 32–38.
119. Allen, D.A.; Wright, A.E.; Goss, W.M. The dwarf emission galaxy He 2-10. *Mon. Not. R. Astron. Soc.* **1976**, *177*, 91. [CrossRef]
120. Kunth, D.; Sargent, W.L.W. Observations of Wolf-rayet stars in the emissionline galaxy Tololo 3. *Astron. Astrophys.* **1981**, *101*, L5–L8.
121. López-Sánchez, Á.R.; Esteban, C. Massive star formation in Wolf–Rayet galaxies. III. Analysis of the O and WR populations. *Astron. Astrophys.* **2010**, *516*, A104. [CrossRef]
122. Sokal, K.R.; Johnson, K.E.; Indebetouw, R.; Reines, A.E. An Emerging Wolf–Rayet Massive Star Cluster in NGC 4449. *Astron. J.* **2015**, *149*, 115. [CrossRef]
123. Sokal, K.R.; Johnson, K.E.; Indebetouw, R.; Massey, P. The Prevalence and Impact of Wolf–Rayet Stars in Emerging Massive Star Clusters. *Astrophys. J.* **2016**, *826*, 194. [CrossRef]
124. Garmany, C.D.; Conti, P.S.; Massey, P. Spectroscopic studies of O type stars. IX. Binary frequency. *Astrophys. J.* **1980**, *242*, 1063–1076. [CrossRef]
125. Sana, H.; de Koter, A.; de Mink, S.E.; Dunstall, P.R.; Evans, C.J.; Hénault-Brunet, V.; Maíz Apellániz, J.; Ramírez Agudelo, O.H.; Taylor, W.D.; Walborn, N.R. The VLT-FLAMES Tarantula Survey. VIII. Multiplicity properties of the O-type star population. *Astron. Astophys.* **2013**, *550*, A107. [CrossRef]
126. Gies, D.R. Binaries in Massive Star Formation. In *Massive Star Formation: Observations Confront Theory*; Astronomical Society of the Pacific Conference Series; Beuther, H., Linz, H., Henning, T., Eds.; Astronomical Society of the Pacific: San Francisco, CA, USA, 2008; Volume 387, p. 93.
127. Sana, H.; de Mink, S.E.; de Koter, A.; Langer, N.; Evans, C.J.; Gieles, M.; Gosset, E.; Izzard, R.G.; Le Bouquin, J.B.; Schneider, F.R.N. Binary Interaction Dominates the Evolution of Massive Stars. *Science* **2012**, *337*, 444. [CrossRef]
128. Massey, P.; Conti, P.S.; Niemela, V.S. Spectroscopic studies of Wolf-rayet stars with absorption lines. VII.HD 156327 and HD 192641 and the question of WR duplicity. *Astrophys. J.* **1981**, *246*, 145–152. [CrossRef]
129. Foellmi, C.; Moffat, A.F.J.; Guerrero, M.A. Wolf-Rayet binaries in the Magellanic Clouds and implications for massive-star evolution - I. Small Magellanic Cloud. *Mon. Not. R. Astron. Soc.* **2003**, *338*, 360–388. [CrossRef]
130. Foellmi, C.; Moffat, A.F.J.; Guerrero, M.A. Wolf-Rayet binaries in the Magellanic Clouds and implications for massive-star evolution - II. Large Magellanic Cloud. *Mon. Not. R. Astron. Soc.* **2003**, *338*, 1025–1056. [CrossRef]

131. Neugent, K.F.; Massey, P. The Close Binary Frequency of Wolf–Rayet Stars as a Function of Metallicity in M31 and M33. *Astrophys. J.* **2014**, *789*, 10. [CrossRef]
132. Stevens, I.R.; Blondin, J.M.; Pollock, A.M.T. Colliding Winds from Early-Type Stars in Binary Systems. *Astrophys. J.* **1992**, *386*, 265. [CrossRef]
133. Rauw, G.; Nazé, Y. X-ray emission from interacting wind massive binaries: A review of 15 years of progress. *Adv. Space Res.* **2016**, *58*, 761–781. [CrossRef]
134. Crowther, P.A.; Walborn, N.R. Spectral classification of O2-3.5 If*/WN5-7 stars. *Mon. Not. R. Astron. Soc.* **2011**, *416*, 1311–1323. [CrossRef]
135. Tehrani, K.A.; Crowther, P.A.; Bestenlehner, J.M.; Littlefair, S.P.; Pollock, A.M.T.; Parker, R.J.; Schnurr, O. Weighing Melnick 34: The most massive binary system known. *Mon. Not. R. Astron. Soc.* **2019**, *484*, 2692–2710. [CrossRef]
136. Garofali, K.; Levesque, E.M.; Massey, P.; Williams, B.F. The First, Candidate Colliding-Wind Binary in M33. *arXiv* **2019**, arXiv:1906.03274.
137. Moffat, A.F.J.; Breysachar, J.; Seggewiss, W. The WO 4 + O4 V binary Sk 188 in the SMC. In *Wolf–Rayet Stars: Progenitors of Supernovae?* Observatoire de Paris: Paris, France, 1983; p. 23.
138. Tramper, F.; Gräfener, G.; Hartoog, O.E.; Sana, H.; de Koter, A.; Vink, J.S.; Ellerbroek, L.E.; Langer, N.; Garcia, M.; Kaper, L. On the nature of WO stars: A quantitative analysis of the WO3 star DR1 in IC 1613. *Astron. Astrophys.* **2013**, *559*, A72. [CrossRef]
139. Talent, D.L. A Spectrophotometric Study of HII Regions in Chemically Young Galaxies. Ph.D. Thesis, Rice University, Houston, TX, USA, 1980.
140. Gräfener, G.; Vink, J.S.; de Koter, A.; Langer, N. The Eddington factor as the key to understand the winds of the most massive stars. Evidence for a Γ-dependence of Wolf–Rayet type mass loss. *Astron. Astrophys.* **2011**, *535*, A56. [CrossRef]
141. Gräfener, G.; Koesterke, L.; Hamann, W.R. Line-blanketed model atmospheres for WR stars. *Astron. Astrophys.* **2002**, *387*, 244–257. [CrossRef]
142. Hillier, D.J.; Miller, D.L. The Treatment of Non-LTE Line Blanketing in Spherically Expanding Outflows. *Astrophys. J.* **1998**, *496*, 407–427. [CrossRef]
143. Crowther, P.A. Physical Properties of Wolf–Rayet Stars. *Annu. Rev. Astron. Astrophys.* **2007**, *45*, 177–219. [CrossRef]
144. Hillier, D.J. Spectrum formation in Wolf–Rayet stars. In *Wolf–Rayet Stars: Proceedings of an International Workshop Held in Potsdam*; Hamann, W.R., Sander, A., Todt, H., Eds.; Universitätsverlag Potsdam: Potsdam, Germany, 2015; pp. 65–70.
145. Sander, A.A.C. Recent advances in non-LTE stellar atmosphere models. In *The Lives and Death-Throes of Massive Stars*; Eldridge, J.J., Bray, J.C., McClelland, L.A.S., Xiao, L., Eds.; International Astronomical Union: Paris, France, 2017; Voume 329, pp. 215–222,
146. Sander, A.; Todt, H.; Hainich, R.; Hamann, W.R. The Wolf–Rayet stars in M 31. I. Analysis of the late-type WN stars. *Astron. Astrophys.* **2014**, *563*, A89. [CrossRef]
147. Hainich, R.; Rühling, U.; Todt, H.; Oskinova, L.M.; Liermann, A.; Gräfener, G.; Foellmi, C.; Schnurr, O.; Hamann, W.R. The Wolf–Rayet stars in the Large Magellanic Cloud. A comprehensive analysis of the WN class. *Astron. Astrophys.* **2014**, *565*, A27. [CrossRef]
148. Shenar, T.; Sablowski, D.P.; Hainich, R.; Todt, H.; Moffat, A.F.J.; Oskinova, L.M.; Ramachandran, V.; Sana, H.; Sander, A.A.C.; Schnurr, O. The Wolf–Rayet binaries of the nitrogen sequence in the Large Magellanic Cloud: Spectroscopy, orbital analysis, formation, and evolution. *arXiv* **2019**, arXiv:1905.09296,
149. Hainich, R.; Pasemann, D.; Todt, H.; Shenar, T.; Sander, A.; Hamann, W.R. Wolf–Rayet stars in the Small Magellanic Cloud. I. Analysis of the single WN stars. *Astron. Astrophys.* **2015**, *581*, A21. [CrossRef]
150. Shenar, T.; Hainich, R.; Todt, H.; Sander, A.; Hamann, W.R.; Moffat, A.F.J.; Eldridge, J.J.; Pablo, H.; Oskinova, L.M.; Richardson, N.D. Wolf–Rayet stars in the Small Magellanic Cloud. II. Analysis of the binaries. *Astron. Astrophys.* **2016**, *591*, A22. [CrossRef]
151. Massey, P.; Neugent, K.F.; Hillier, D.J.; Puls, J. A Bake-off between CMFGEN and FASTWIND: Modeling the Physical Properties of SMC and LMC O-type Stars. *Astrophys. J.* **2013**, *768*, 6. [CrossRef]
152. Lamers, H.J.G.L.M.; Levesque, E.M. *Understanding Stellar Evolution*; IOP: Bristol, UK, 2017. [CrossRef]
153. Eldridge, J.J.; Izzard, R.G.; Tout, C.A. The effect of massive binaries on stellar populations and supernova progenitors. *Mon. Not. R. Astron. Soc.* **2008**, *384*, 1109–1118. [CrossRef]

154. Eldridge, J.J.; Stanway, E.R. BPASS predictions for binary black hole mergers. *Mon. Not. R. Astron. Soc.* **2016**, *462*, 3302–3313. [CrossRef]
155. Georgy, C.; Ekström, S.; Meynet, G.; Massey, P.; Levesque, E.M.; Hirschi, R.; Eggenberger, P.; Maeder, A. Grids of stellar models with rotation. II. WR populations and supernovae/GRB progenitors at Z = 0.014. *Astron. Astrophys.* **2012**, *542*, A29. [CrossRef]

© 2019 by the authors. Licensee MDPI, Basel, Switzerland. This article is an open access article distributed under the terms and conditions of the Creative Commons Attribution (CC BY) license (http://creativecommons.org/licenses/by/4.0/).

MDPI
St. Alban-Anlage 66
4052 Basel
Switzerland
Tel. +41 61 683 77 34
Fax +41 61 302 89 18
www.mdpi.com

Galaxies Editorial Office
E-mail: galaxies@mdpi.com
www.mdpi.com/journal/galaxies

www.ingramcontent.com/pod-product-compliance
Lightning Source LLC
LaVergne TN
LVHW070741100526
838202LV00013B/1283